Applying Data Science

Arthur K. Kordon

Applying Data Science

How to Create Value with Artificial
Intelligence

 Springer

Arthur K. Kordon
CEO, Kordon Consulting LLC
Fort Lauderdale, FL, USA

ISBN 978-3-030-36377-2   ISBN 978-3-030-36375-8   (eBook)
https://doi.org/10.1007/978-3-030-36375-8

This Springer imprint is published by the registered company Springer Nature Switzerland AG.
The registered company address is: Gewerbestrasse 11, 6330 Cham, Switzerland

To my Teachers

Preface

Artificial Intelligence is no match for natural stupidity.

Anonymous

Like it or not, artificial intelligence (AI) has invaded our lives like a storm. Some are thrilled by the new opportunities this magical technology may open up. Others are scared for their jobs, expecting the invasion of the robots. Most of us, especially those in the business, are confused by the AI hype and concerned about losing our competitive edge. A package of new buzzwords, such as big data, advanced analytics, and the Internet of Things (IoT), under the rising scientific star of Data Science are contributing to the growing level of confusion.

Often the confusion leads to a mess after incompetent attempts to push this technology without the needed infrastructure and skillset. Unfortunately, the results are disappointing and the soil is poisoned for future applications for a long period of time. Handling the confusion and navigating businesses through this maze of new technologies and buzzwords related to AI is a well-identified need in the business world. Business practitioners and even academics are not very familiar with the capabilities of Data Science as a research discipline, linked to AI. Another issue is that the hype about AI in the media is not supported with adequate popular training in its technical basis. For many practitioners, the alternative of using academic references, with their highly abstract level and emphasis on pure theory, is a big challenge.

There is a growing need for a book that demonstrates the value of using AI-driven technologies within the Data Science framework, explains clearly the main principles of the different approaches with less of a focus on theoretical details, offers a methodology for developing and deploying practical solutions, and suggests a roadmap for how to introduce these technologies into a business.

The alternative is to generate natural stupidity by incompetent application of AI.

Fig. 1 Business landscape
around AI Olympus

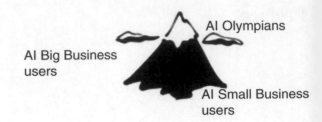

Motivation

Applying Data Science: How to Create Value with Artificial Intelligence is one of the first books on the market that fills this need. There are several factors that contributed to the decision to write such a book.

The first was a desire to emphasize the "giant leap" of AI from a narrow research area to a central business issue. The author is not aware of any other recent academic approach capable of such an amazing breakthrough. However, we need to clarify the big differences in this process for various types of businesses. To start, we will use a visual analogy between the current business landscape related to AI capabilities (Fig. 1) and ancient Greek mythology.[1]

As shown in Fig. 1, the current mountainous landscape of AI-related capabilities is divided into three categories: AI Olympians, AI big business users, and AI small business users. Several leaders in AI, such as Google, Amazon, Facebook, Microsoft, Apple, and NVIDIA, occupy the peak, the AI Olympus (in analogy to Greek mythology, we name them the AI "gods," the Olympians). They are the center of excellence that concentrates most of the world's R&D potential in this area, the generator of new approaches and infrastructures, and a shining example of business success.

Big businesses are the current target for applying AI-driven technologies. They have data and the potential to allocate resources and identify profitable application areas. The book is focused on this business category. Unfortunately, AI-driven technologies will need some time to "trickle down" to small businesses, with a reduced cost and simplified infrastructure to fit their limited capacity and become a new profit generator.

The second factor that contributed to the decision to write this book is the need to challenge the "Data is the new oil" paradigm. The objective is not to question the growing value creation potential of data and of all data-driven methods for analysis and modeling. We challenge the absolute favoritism of these methods at the expense of other well-established approaches. It leads to the dangerous acceptance of context-free data analysis and of attempts of full automation of the whole process

---

[1]In the ancient Greek world, the 12 great gods and goddesses of the Greeks were referred to as the Olympian Gods, or the Twelve Olympians. The name of this powerful group of gods comes from Mount Olympus, where the council of 12 met to discuss matters.

of model generation. "Give us the data and we'll turn it into gold," some vendors promise. The problem is that many believe and pay for this hype.

Unfortunately, obsession with data is built into the foundation of Data Science. One area for improvement is data interpretation and acquisition of knowledge about analyzing the data. Problem definition and all related issues of integrating business knowledge into the whole Data Science process need much attention as well.

The third factor was the need to challenge the narrow machine learning vision of AI that dominates the research community at large and the media. It is true that machine learning, and recently deep learning, has shown tremendous progress. However, these accomplishments are mostly related to the top class of Olympian companies. AI includes several other technologies with great application potential, such as evolutionary computation, swarm intelligence, decision trees, and intelligent agents, to name a few. We owe the business community a broader perspective of AI-driven technologies, with more application options for performance improvement.[2]

The fourth factor that contributed to the decision to write this book is the author's unique experience in developing new AI-driven technologies and applying them in big global corporations, such as Dow Chemical, Georgia Pacific, and Steelcase. As R&D Leader at Dow Chemical's research labs, the author knows very well the "kitchen" for transferring a new emerging technology from a pure academic idea to a successful industrial application. Later in his career, as Advanced Analytics Leader at Dow Chemical and CEO of Kordon Consulting LLC, he has accumulated exceptional experience in introducing and applying AI-driven technologies in manufacturing and business analytics. Sharing this unique professional experience is priceless.

Purpose of the Book

Data Science and AI-driven technologies are relatively new to industry. They are still a fast-growing research area in the category of emerging technologies. On top of that, the business community needs help in distinguishing hype from reality, and guidance to navigate through the maze of fashionable buzzwords, technological haze, and gazillions of tools offering voodoo. Business leaders are interested in understanding the capabilities of AI-driven technologies in order to assess the competitive advantage they may achieve if they apply them in their business. Data scientists and business practitioners need a robust methodology to turn a data-driven business problem into an actionable solution that delivers long-term value.

The purpose of the book is to fill this need, to address these issues, and to give guidelines to a broad business and academic audience on how to successfully

---

[2]Some AI-driven technologies are described in J. Keller, D. Liu, and D. Fogel, *Fundamentals of Computational Intelligence*, Wiley-IEEE Press, 2016.

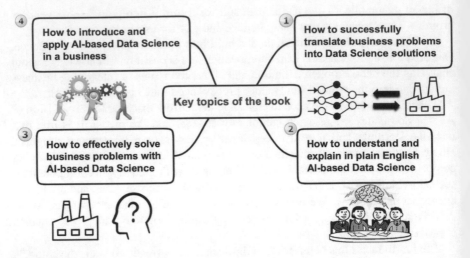

Fig. 2 Main topics of the book

introduce and apply AI-based Data Science. The key topics of the book are shown in the mind map in Fig. 2 and are discussed next:

1. *How to successfully translate business problems into Data Science solutions.* The first main topic of the book emphasizes the leading role of defining, understanding, and communicating business problems for the purpose of developing data-driven solutions. It requires business knowledge acquisition and avoidance of the lost-in-translation trap between model developers and business users. The proposed approach challenges the practical relevance of the dominant "data drives everything" mindset in the Data Science community.

2. *How to understand and explain in plain English AI-based Data Science.* The second key topic of the book focuses on the ambitious task of explaining a selected set of AI-driven technologies with minimal mathematical and technical details, and with an emphasis on their unique application capabilities. The reader is advised on proper technology selection, and its realistic value creation opportunities are illustrated with corresponding use cases.

3. *How to successfully solve business problems with AI-based Data Science.* The third key topic of the book covers the central point of interest—the practical methodology for the development and deployment of robust solutions to business problems. It is based on the generic Data Science method with the addition of the specific requirements of selected AI-based approaches. The key steps of proposed methodology include project definition, domain knowledge acquisition, data preparation and analysis, and model development and deployment. They are discussed in detail, with practical recommendations and real-world examples. In addition, the critical issues of building the needed infrastructure and developing people skills are addressed.

4. *How to introduce and apply AI-based Data Science in a business.* The fourth main topic of the book is devoted to an issue of major interest to the business community at large: making a decision about how relevant these technologies are for the future of their business and how to integrate them into their capabilities. Several potential scenarios for introduction and growth of AI-based Data Science in a business are discussed and a corresponding roadmap is suggested. The book also gives practical advice on how to become an effective data scientist in a business setting.

Who Is This Book for?

The target audience is much broader than the existing scientific community working in AI and the data analytics, big data, and Data Science business communities. The readers who can benefit from this book are presented in the mind map in Fig. 3 and are described below:

- *Data scientists.* This group represents the most appropriate readers of the book. They can broaden their skillset with the new AI-driven technologies, navigate management through the buzzword and hype maze, and use the methodology described to generate effective business solutions. Reading all chapters, except Chap. 16, is recommended.
- *Data analysts.* This group includes the people who use Data Science-generated solutions for gaining insight and improving business decisions. The book will broaden their knowledge about the new opportunities offered by AI-driven technologies and may spawn ideas for business improvement. Reading all chapters, except Chap. 16, is recommended.
- *Industrial researchers.* This group includes scientists in industrial labs who create new products and processes. They will benefit from the book by understanding the impact of AI technologies on industrial research and using the proposed application strategy to broaden and improve their performance. Reading all chapters, except Chap. 16, is recommended.
- *Practitioners in different businesses.* This group consists of the key potential final users of the technology, such as process engineers, supply chain organizers, economic analyzers, and medical doctors. This book will introduce the main AI-driven technologies in an understandable language and will encourage these people to find new applications in their businesses. Reading all chapters, except Chap. 16, is recommended.
- *Managers.* Top-level and R&D managers will benefit from the book by understanding the mechanisms of value creation and the competitive advantages of AI-based Data Science. Middle- and low-level managers will find in the book a practical and nontechnical description of these emerging technologies that can make the organizations they lead more productive. In addition, the book will clarify the current confusing technology landscape and will help them to

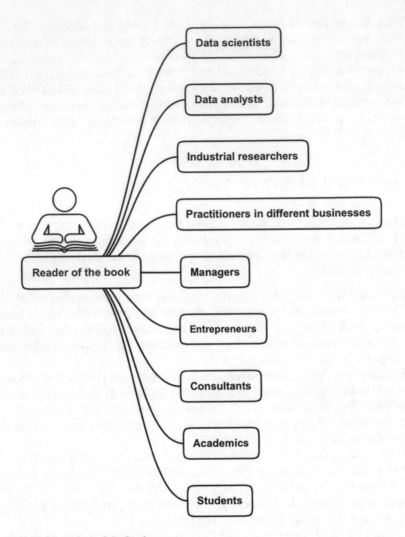

Fig. 3 Potential readers of the book

introduce the needed skillset into their businesses. Reading Chaps. 1, 2, 3, 4, 6, 12, 13, 14, 15, 16, and 17 is recommended.

- *Entrepreneurs.* This class consists of enthusiastic professionals who are looking to start new high-tech businesses and venture capitalists who are searching for the Next Big Thing in technology investment. Recently, AI has been a very attractive bait for venture capital investment. However, the lack of technical competency in the field has become one of the key obstacles to entrepreneurs giving their blessing to startups. The book will give entrepreneurs substantial information about the nature of AI technologies, their potential for value creation, and many use cases. This will be a good basis for adding the missing technical depth in the development of business plans and investment strategy analysis. Reading Chaps. 1, 2, 3, 4, 6, 12, 13, 14, 15, 16, and 17 is recommended.

- *Consultants.* This diverse group includes professionals with various technical backgrounds and skillsets. They will benefit significantly from using the book as a reliable source for the whole process of translating data-driven business problems into effective solutions by using advanced technologies such as AI-based methods. The book is written by one of the top consultants in the field and covers the key topics in a style appropriate for this large group of potential readers. Reading all chapters is recommended.
- *Academics.* This group includes the broad class of academics who are not familiar with the research in and technical details of the field and the small class of academics who are developing AI and moving it ahead. The first group will benefit from the book by using it as an introduction to the field and thereby understanding the specific requirements for successful practical applications, defined directly by industrial experts.

  The second group will benefit from the book by gaining better awareness of the economic impact of AI, and understanding the industrial needs and all the details for successful practical application. Reading all chapters is recommended.
- *Students.* Undergraduate and graduate students in technical, economic, medical, and even social disciplines can benefit from the book by understanding the advantages of AI-driven methods and their potential for implementation in their specific field. In addition, the book will help students to gain knowledge about the practical aspects of industrial research and the issues faced in real-world applications. For those students, who are taking regular Data Science classes, the book will broaden their knowledge of AI-driven methods and will prepare them for the nasty reality of applying Data Science in the real world. Reading all chapters is recommended.

How This Book Is Structured

The structure of the book, with its organization into parts and chapters, is shown in Fig. 4.

Part I of the book focuses on the practical methodology of how to translate a data-driven business problem into an effective solution by using the powerful AI technologies according to the Data Science paradigm. Chapter 1 clarifies the current technological landscape, defines AI-based Data Science, and identifies its key competitive advantages and issues. Chapter 2 emphasizes the importance of identifying and defining appropriate business problems for AI-driven technologies. Chapter 3 gives an overview of the recommended set of AI technologies. Chapter 4 focuses on the importance of integrating and selecting different

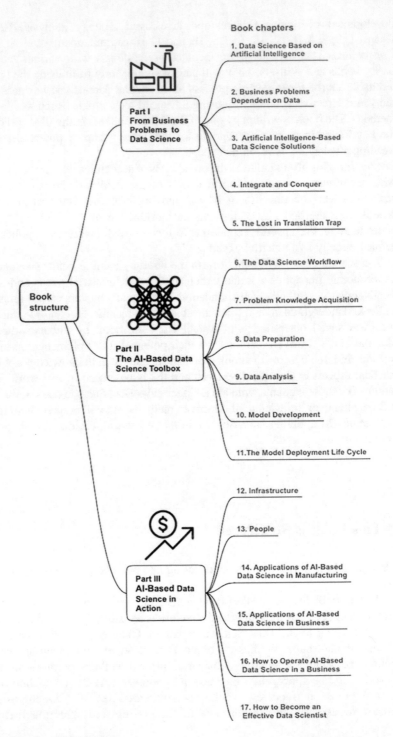

Fig. 4 Structure of the book

AI-based methods, and Chap. 5 focuses on the significance of clear communication between tech-savvy data scientists and business-savvy users.

Part II defines the Data Science workflow and covers its key steps. Chapter 6 gives an overview of the Data Science work process, and Chap. 7 describes the various methods to capture and integrate existing domain knowledge. Chapter 8 includes the necessary data preparation steps, and Chap. 9 covers the data analysis sequence. Chapter 10 presents the various modeling methods, and Chap. 11 is focused on the important model deployment life cycle, including performance tracking and model maintenance.

Part III of the book demonstrates AI-based Data Science in action. Chapter 12 gives recommendations for the needed hardware, software, and database infrastructure, while Chap. 13 discusses the required skillset and team composition for effective AI-based Data Science applications. Chapter 14 gives examples of such applications in manufacturing, while business applications are given in Chap. 15. Chapter 16 gives recommendations for how to introduce, apply, and leverage AI-based Data Science in a business, and Chap. 17 includes practical advice on how to become an effective data scientist.

What This Book Is NOT About

- *Detailed theoretical description of AI approaches*. The book does not include a deep academic presentation of the different AI methods. The broad target audience requires the descriptions to have a minimal technical and mathematical burden. The focus of the book is on the application issues of artificial intelligence, and all the methods are described and analyzed at a level of detail that will help with their broad practical implementation.
- *Introduction of new AI methods*. The book does not propose new AI methods or new algorithms for the known approaches. The novelty in the book is on the application side of AI in the context of the Data Science methodology.
- *A software manual for AI-based Data Science-related software products or programming languages*. This is not an instruction manual for a particular software product. The author's purpose is to define a generic methodology for AI-based Data Science applications, independent of any specific programming language or software.
- *Introduction of databases*. Issues related to modeling of data, database design, data warehousing, data manipulation, data harmonization, and data collection are not covered in detail in the book. The book is focused on the analysis and model development side of Data Science.
- *A book on statistics*. This is not a "standard" Data Science book where statistics has the lion's share of the methods for model development that are covered. Neither is the book a detailed reference on statistical approaches. The focus is on AI-based methods in Data Science.

Features of the Book

The key features that differentiate this book from the other titles on AI and Data Science are shown in Fig. 5 and defined below as follows:

1. *Emphasis on AI methods in Data Science.* The book demonstrates the benefits of applying a selected set of AI technologies, such as machine learning, deep learning, evolutionary computation, swarm intelligence, decision trees, and intelligent agents, to solve data-driven business problems by using the Data Science method.
2. *The Troika of Applied Data Science.* One of the main messages in the book is that focusing only on the technology and ignoring the other aspects of real-world applications is a recipe for failure. The winning application strategy is based on three key components—scientific methods, infrastructure, and people. We call these the Troika of Applied Data Science, shown in Fig. 6.

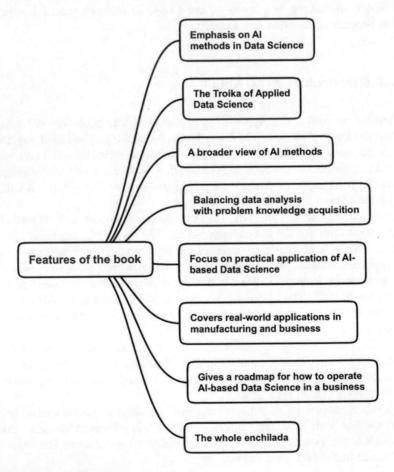

Fig. 5 Main features of the book

Fig. 6 The Troika of
applied data science

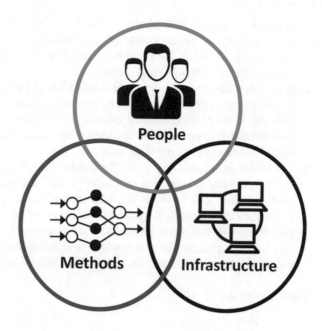

The first component (people) represents the most important factor—the people
involved in the whole implementation cycle, such as data scientists, managers,
programmers, and the different types of final users. The second component
(methods), where most of the current attention is focused, includes the theoretical
basis of AI and statistics. The third component (infrastructure) represents the
necessary infrastructure for implementing a developed business solution. It
includes the needed hardware, software, and all organizational work processes
for development, deployment, and support. Promoting this broader view of Data
Science and, especially, clarifying the critical role of the human component in the
success of practical applications is the leading philosophy in the book.

3. *A broader view of AI methods.* Contrary to the dominant vision of AI as a
   machine/deep learning kingdom, a broader selection of various AI-driven technologies has been used in the book. This should give the business community
   more opportunities to solve a wide range of complex business problems and
   improve productivity.
4. *Balancing data analysis with problem knowledge acquisition.* Another key
   feature of the book is the need to counteract data fetishism. One of the key
   messages in the book is that problem knowledge acquisition plays a leading role
   in solving data-driven problems. We believe that context-free data rarely brings
   insight and value.
5. *Focusing on practical application of AI-based Data Science.* Application issues
   are central in the discussion of all topics in the book. They define the technical

depth of the AI methods described, the details of each step for developing Data Science solutions, and estimation of the potential value. The main objective of the approach described in the book is to help businesses to generate solutions with sustainable long-term value generation.

6. *A roadmap for how to operate AI-based Data Science in a business.* The book gives some answers to one of the key questions in the business community—how to build AI-based Data Science capabilities if needed. It includes guidelines on how to introduce and develop an appropriate infrastructure and skillset. Several organizational options are discussed as well.

7. *The whole enchilada.* The book offers an integrated approach and covers most of the key issues in applying AI-based Data Science over a wide range of manufacturing and business applications. It is the perfect handbook for an industrial data scientist. It gives a broad view of the methods, implementation processes, and application areas. The book is about a high-level view of this field, i.e., we hope that the interested reader will focus on the forest. It is not a book about the leaves. There are too many books about those leaves.

Fort Lauderdale, FL, USA   Arthur K. Kordon

Acknowledgments

The author would like to acknowledge the contributions of the following former colleagues from the Dow Chemical Company as pioneers in introducing AI-based technologies into industry: Guido Smits, Alex Kalos, and Mark Kotanchek. He also would like to recognize the leadership of Tim Rey in founding the Advanced Analytics group in the company and opening the door of this emerging technology to the author.

Above all, the author would like to emphasize the enormous influence of his mentor academician Mincho Hadjiski in building the author's philosophy as a scientist with a strong focus on real-world applications. His creative ideas and constructive suggestions, including for this book, have been an important compass in the author's mindset.

Special thanks go to the Bulgarian cartoonist Stelian Sarev for his kind gesture of giving as a gift the humorous caricatures used in Chap. 17 and to Gregory Piatetsky-Shapiro (kdnuggets.com) for his kind permission to use the caricature in Fig. 17.1.

The author would like to thank many people whose constructive comments improved substantially the structure of the book and the final manuscript, including Michael Affenzeller, Carlos Aizpurua, Muffasir Badshah, Karel Cerny, Leo Chiang, Victor Christov, Sam Coyne, Michael Dessauer, Miroslav Enev, Steven Frazier, Petia Georgieva, Steve Gustafson, Lubomir Hadjiski, Joseph Johnson, Janusz Kacprzyk, Theresa Kotanchek, Bill Langdon, Randy Pell, Daniel Ramirez, Nestor Rychtyckyj, Tariq Samad, Subrata Sen, George Stanchev, Zdravko Stefanov, and Zhongying Xiao. He is especially grateful for the full support and enthusiasm of Ronan Nugent, which was critical for the success of the whole project.

Fort Lauderdale, FL, USA

Arthur K. Kordon

February 2020

Contents

Part I
From Business Problems to Data Science

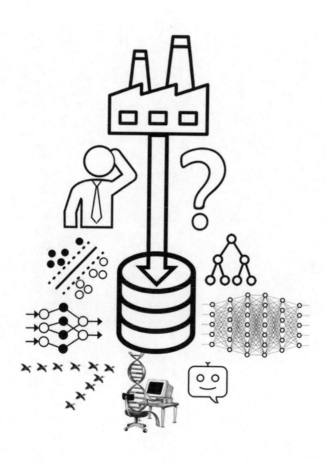

Chapter 1
Data Science Based on Artificial Intelligence

There are three great events in history. One, the creation of the universe. Two, the appearance of life. The third one, which I think is equal in importance, is the appearance of Artificial Intelligence.

Edward Fredkin

Artificial intelligence (AI) represents a paradigm shift that is driving at the same time scientific progress and the evolution of industry. Of special interest to industry is the fast transition of AI from a narrow research area in a limited number of academic labs to a key topic in the business world. This process is combined with an invasion of other new paradigms, such as analytics, big data, and the Internet of Things (IoT). On top of that, a new discipline, named Data Science, has gradually occupied a central place as the leading branch of knowledge that covers all relevant methods and work processes for translating data into actionable business solutions. Learning its basis and key capabilities is becoming a new academic and business challenge. Another critical factor is the appearance of data scientists, who have become the drivers of this transformation.

Unfortunately, this technological avalanche has caught most businesses unprepared. The confusion begins with a bombardment of catchy buzzwords, combined with an invasion of gazillions of vendors offering AI voodoo, and growing anxiety about the future of the business if the new promised opportunities are missed.

The first objective of this chapter is to reduce the confusion by clarifying the key buzzwords and summarizing the big opportunities for the proper use of this package of technologies. The second is to define Data Science based on AI, introducing its key methods and special features. The third is to describe the competitive advantages and challenges of applying AI-based Data Science.

© Springer Nature Switzerland AG 2020
A. K. Kordon, *Applying Data Science*,
https://doi.org/10.1007/978-3-030-36375-8_1

1.1   Big Data, Big Mess, Big Opportunity

The exponential growth of data is a constant challenge in the business world. If properly handled, it opens up new opportunities for competitive advantage and future growth. Of special importance is the current boom in several paradigms related to data, such as big data, analytics, and the Internet of Things (IoT), with the leading role played by AI. Understanding the nature of these technologies and their value creation capabilities is the first step in exploring their great potential.

1.1.1   From Hype to Competitive Advantage

Very often the first introduction to AI and related technologies is driven by hype. On the one hand, this hype contributes to the diffusion of both information and misinformation alike as the technology evolves from the lab to the mainstream. On the other hand, the hype seeds enthusiasm for something new and exciting rising on the technology horizon. Ignorance about the advertised approaches, however, blurs the real assessment required to differentiate hype from reality. As a result, a lot of businesses are hesitant to make decisions about embracing the technology. They need to be convinced with more arguments about the value creation opportunities of AI-driven technologies and their real applicability.

The Hype
Usually when a new technology starts getting noticed, subsequent hype in the industry is inevitable. The hype is partly created by the vendor and consulting communities looking for new business, partly created by professionals in industry wanting to keep up with the latest trends, and partly created by companies with an ambition to be seen as visionary and to become early champions.

Recently, AI has experienced a "Big Bang" as a prophecy generator. Some examples follow: an AI-Nostradamus claims that "Of all the trends, artificial intelligence is the one that will really impact everything." Another one continues in the same line "You can think of AI like we thought of electricity 100 years ago. When we wanted to improve a process or a procedure, we added electricity. When we wanted to improve the way we made butter, we added electric motors to churns. As a single trend that is going to be overriding, that's pretty much it." A third one concludes "Artificial intelligence is going to have a massive impact on organizations globally."

One medicine for dealing with hype is to consider the famous Amara's law:

> We tend to overestimate the effect of a technology in the short run and underestimate the effect in the long run.

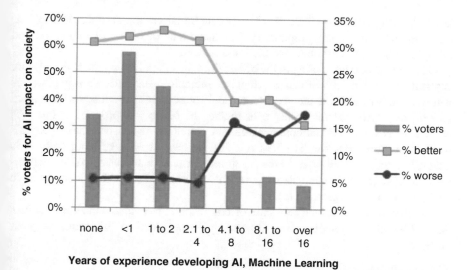

Fig. 1.1 Response to the question "Will AI and automation change society for better or worse?" against years of experience in applying AI

An interesting result in accordance with Amara's law is a recent survey based on 588 votes by the popular website KDNuggets.[1] While about 60% of KDnuggets readers think AI and automation will improve society, the optimism drops significantly among those with four or more years of experience in developing AI systems (see Fig. 1.1). Obviously, the reality correction after gaining experience with AI cools the initial inflated enthusiasm.

Another pill against the AI hype is some lessons from the past. At the dawn of AI in 1957, the economist Herbert Simon predicted that computers would beat humans at chess within 10 years. (It took 40.) In 1967 one of the founding fathers of AI, Marvin Minsky, said "Within a generation the problem of creating Artificial Intelligence will be substantially solved." Simon and Minsky were both intellectual giants, but they were bad AI prophets. Thus it is understandable that dramatic claims about future breakthroughs meet with a certain amount of skepticism.

The Mess

When the hype gets out of control, the facts get lost in the haze and the real promise of AI regresses into a high-tech version of snake oil. As a result, companies—including those most likely to benefit from AI-driven solutions—are starting to doubt the claims of AI gurus and vendors promising the Ultimate Intelligence Machine. On the other hand, a side effect of the hype is a growing anxiety about potential lost opportunities. The business-related media amplify these concerns with the image of a ubiquitous AI presence in the business of the future. The key recipient of this message is top management. In order to look active and visionary, often the

---

[1]https://www.kdnuggets.com/2017/07/optimism-ai-impact-experience.html

leadership triggers a mess by having large-scale initiatives to introduce AI into their organizations without any preparation. The usual lack of knowledge about AI and related technologies contributes to the mess as well.

One myth that is circulating in the public domain and poisoning the soil for future applications is that AI will cause extensive unemployment and that organizations will need fewer people in the years to come. From current experience with applied AI systems, such systems hardly ever replace an entire job, process, or business model. Most often they complement human activities, which can make people's work even more effective. The best rule for the new division of labor with AI is rarely, if ever, "give all tasks to the machine." Instead, a successful AI implementation of a process will automate some repetitive tasks while it will be more valuable for humans to do the creative tasks.

The first step in reducing this state of confusion is to evaluate the opportunities that AI-based technologies offer, especially those close to the nature of your business.

The Opportunity

Beyond the hype, AI-driven technologies have shown tremendous success in many applications in manufacturing and business. The long list of such applications in manufacturing includes process monitoring and optimization, preventive maintenance, and energy cost reduction. The list of business applications includes price forecasting, customer churn prevention, fraud detection, and human resources optimization. The key application areas will be illustrated with several examples in Chap. 14 for manufacturing and in Chap. 15 for business applications. The most important competitive advantages of AI-driven technologies, such as delivering solutions with "objective intelligence," dealing with uncertainty and complexity, generating novelty, and delivering low-cost modeling and optimization, are discussed later in this chapter. Readers are also recommended to search for the most recent appropriate publicly available AI-driven use cases that can make their business more efficient. Use cases found in this way are the best starting point in the long journey of introducing, exploring, and applying the benefits of AI-driven technologies. The key opportunity hypothesis can be defined as follows: once AI-based systems surpass human performance at a given task, they are much likelier to spread quickly. The first challenge, however, is finding this opportunity in your business.

A key question in estimating the opportunities is the perception of the high cost of these systems. On the hardware side, the concern is that big data and complex AI algorithms will require significant investment. Fortunately, the necessary algorithms and hardware for modern AI can be bought or rented as needed. Google, Amazon, Microsoft, and other companies are making powerful AI-related technology infrastructure available via the cloud. It is assumed that the severe competition among these rivals will cause the prices to drop over time.

Of bigger concern is the cost of gaining knowledge of and learning these new capabilities. The dominant perception is that "AI is rocket science," with very high requirements for math, programming tricks, and bunch of complex algorithms. This is indeed the costliest component of introducing AI-driven technologies. However, AI is not rocket science, and has been implemented in many different places without special requirements for mass-scale training in the technical details. The availability of data scientists, the key drivers of these technologies, is increasing. The training opportunities are exploding, especially online. Training in the era of AI will probably be a continuous process that will gradually include the majority of the labor force. Those who refuse to reskill will surely fail to fit into the new business world order and hence must be ready to miss the boat.

1.1.2   Key Buzzwords Explained

A wide chasm exists between those who build AI-driven technologies and those who use them. Communicating AI using technical jargon will not only confuse potential customers, but also scare them away. The other extreme, of the car-dealer-type pitch, creates a bullshit version of AI that alienates knowledgeable potential users. The best strategy is to explain the new technologies after tailoring one's conversations to the level of the customers by referencing concepts they already understand, such as brain, network, and tree, and by making AI approachable with friendly examples that clearly demonstrate important use cases.

We would like to follow this strategy in this book, beginning with a condensed explanation of the key buzzwords related to AI-driven technologies.

AI

The classical, and still valid, definition of AI, according to *The Handbook of Artificial Intelligence,* is as follows:

Artificial Intelligence is the part of computer science concerned with designing intelligent computer systems, that is, systems that exhibit the characteristics we associate with intelligence in human behavior—understanding language, learning reasoning, solving problems, and so on.[2]

The key message is that AI is designed to simulate human thinking. In order to accomplish this goal, an AI system has to interact with the environment, memorize, learn, and reason. A generic structure of an AI system is shown in Fig. 1.2. The following is a short description of the key functional modules:

- *Communication Module.* This includes several methods that mimic human interactions with the outside world through various means—written and spoken language, vision, and signs. One of the key methods is natural language processing (NLP), which identifies the syntax and semantics of written language (a clear example is the Google translator). Spoken language is handled by speech recognition algorithms (Amazon's Alexa is a clear example). Recently, AI

---

[2]A. Barr and E. Feigenbaum, *The Handbook of Artificial Intelligence*, Morgan Kaufmann, 1981.

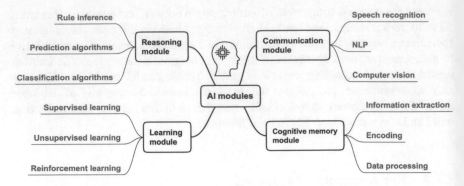

Fig. 1.2 Key modules of an AI system

capabilities for image processing have grown tremendously, especially in the area of face recognition and self-driving cars.

- *Cognitive memory module.* Similar to the human brain, this module supports the functionality to keep the necessary information for learning and reasoning by the AI system. It includes not only the data but also the parameters of the developed models and the rules of the defined cognitive models.
- *Learning module.* This includes a broad range of algorithms that allow an AI system to learn from available data, knowledge, and a changing environment. The learning can be guided by a teacher (supervised), who prepares training examples in advance and validates the results. The machine can even discover unknown patterns in the data (unsupervised learning) and learn new behavior in a hard way by responding to reward/punishment actions (reinforcement learning). Machine learning capabilities have been available since the late 1980s but recently a novel approach, deep learning, has significantly improved this critical capability of AI systems.
- *Reasoning module.* This is the least developed module. In the early days of AI, this module included knowledge bases with rules defined by domain experts. This type of AI system is called an expert system and was popular in the late 1980s. However, the reasoning was static and based on subjective knowledge. It is expected that the growing learning capabilities of AI and the new potential of cognitive computing will significantly increase the ways in which machines reason and make decisions.

To summarize, AI is an attempt to make computers as smart, as or even smarter than human beings. It is about giving computers human-like behaviors, thought processes, and reasoning abilities. As a result of this, smart computers enhance human intelligence and increase the potential for value creation.

It is not a surprise that such a complex research area has several different types. One key division into two types, (1) general and (2) narrow AI, is based on the level of generalization. Artificial general intelligence (AGI)—also called strong artificial

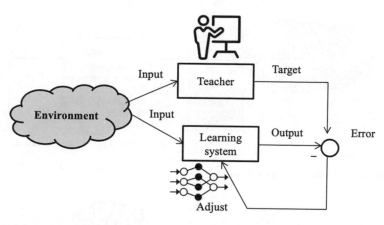

Fig. 1.3 Supervised machine learning

intelligence—is the intelligence of a machine that can successfully perform any intellectual task that a human being can. Narrow AI—also called weak AI—is designed to perform a narrow task (e.g., only facial recognition, only NLP, or only fault detection). Unfortunately, AGI is still in the front line of research and not ready for industry. The focus in the book is on the narrow AI that has demonstrated enormous capabilities to solve complex problems in a variety of businesses.

Machine Learning

The most widespread AI-driven technology is machine learning. According to Wikipedia, "Machine learning is the scientific study of algorithms and statistical models that computer systems use in order to perform a specific task effectively without using explicit instructions, relying on patterns and inference instead."[3] The biggest benefit of this technology is that a machine learning algorithm identifies patterns and relationships in data that are used to make predictions about data it has not seen before. It includes several different methods, such as neural networks, support vector machines, random forests, and K-means. Machine learning algorithms are classified based on the desired outcome of the algorithm. The most frequently used algorithm types include supervised learning, unsupervised learning, and reinforcement learning:

- *Supervised learning.* In this, the algorithm generates a function or a classifier that maps inputs to desired outputs. The key assumption is the existence of a "teacher" who provides knowledge about the environment by delivering input–target training samples or labels (Fig. 1.3). The parameters of the learning system are adjusted by reference to the error between the target and the actual response (the output). Supervised learning is the key method in the most popular machine learning approach, neural networks.

---

[3] Accessed on June 22, 2019.

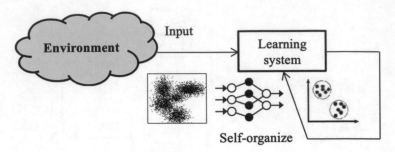

Fig. 1.4 Unsupervised machine learning

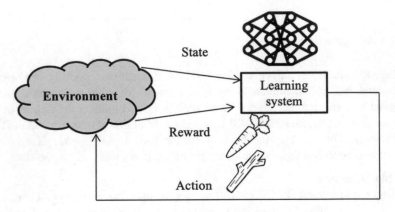

Fig. 1.5 Reinforcement machine learning

Unsupervised learning. The algorithm generates patterns from a set of inputs, since target examples are not available or do not exist at all. Unsupervised learning does not need a teacher and requires the learner to find patterns based on self-organization (Fig. 1.4). The learning system is self-adjusted by the structure discovered in the input data. A typical case of unsupervised learning is that of self-organized maps obtained using neural networks.

- *Reinforcement learning.* The concepts underlying reinforcement learning come from animal behavior studies. One of the most commonly used examples is that of the newborn baby gazelle. Although it is born without any understanding or model of how to use its legs, within minutes it is standing and within 20 min it is running. This learning has come from rapidly interacting with its environment, learning which muscle responses are successful, and being rewarded by survival. In the same way the reinforcement learning algorithm is based on the idea of learning by interacting with an environment and adapting one's behavior to maximize an *objective function* specific to this environment (Fig. 1.5). The learning mechanism is based on the trial-and-error of actions and evaluating the reward. Every action has some impact on the environment, and the environment provides a carrot-and-stick type of feedback that guides the learning algorithm.

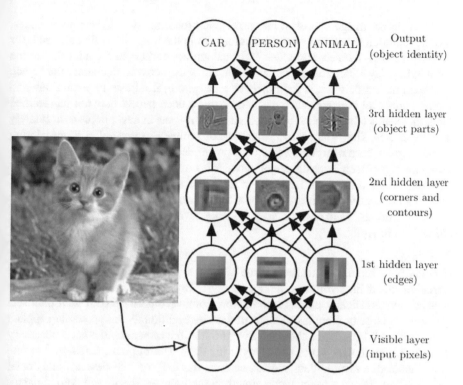

Fig. 1.6 A deep learning neural network structure for image recognition

The aim is to find the *optimal behavior*: the one whose actions maximize the long-term reinforcement. Reinforcement learning is often used in intelligent agents and has recently been used in deep learning.

Deep Learning

Deep learning is the most sophisticated machine learning approach. The key improvement is that it automates feature selection—the process of learning more abstract representations of the input data (feature extraction) through a sequence of layers This is one of the most time-consuming and critical tasks in developing high-quality models with standard machine learning techniques. Deep learning overcomes this challenge by automatically extracting the most suitable features. In essence, deep learning algorithms learn how to learn.

Usually, a deep learning algorithm is based on a deep neural network. In contrast to the standard neural network with one or two layers, it has many layers of artificial neurons. When a neuron fires, it sends signals to connected neurons in the next layer. During deep learning, connections in the network are strengthened or weakened as needed to improve the performance of the system. An example of a deep learning neural network for image recognition is shown in Fig. 1.6 (adapted from Fig. 1.2 in http://www.deeplearningbook.org).

The original image is broken into pixels, and the first layer detects pixel values. The next hidden layers capture different features in the image. The first hidden layer identifies edges, the second hidden layer recognizes corners and contours, and the third hidden layer identifies object parts that allow the network to classify the object. If a deep neural network is fed pixels of a cat photo, it adjusts its parameters and learns high-level concepts such as "cat." After a deep neural network has learned from thousands of sample cat photos, it can identify cats in new photos as accurately as people can. The "giant leap" from special samples to general concepts during learning gives deep neural networks their power. However, for complex images, the structure of the deep learning neural network becomes very complicated, with more than one hundred layers and millions of tuning parameters. Training such structures takes a lot of time and requires additional hardware, such as graphic processing units (GPUs). Another weakness of deep learning is that generated features are very difficult to interpret.

Big Data

To qualify as big data, data must come into the system at high velocity, with large variation, and at high volumes. Wikipedia says: "Big data is a field that treats ways to analyze, systematically extract information from, or otherwise deal with data sets that are too large or complex to be dealt with by traditional data-processing application software." [4] Big is not about the absolute size, but rather about what is necessary to collect, harmonize, store, and analyze the data. The big data paradigm became fashionable due to new data sources such as the IoT, mobile devices, and social media. They created a tremendous growth in the volume, speed, and type of data.

An accepted definition of big data includes the three Vs—Volume, Velocity, and Variety (see Fig. 1.7 for a visualization) [5]:

- *Volume.* The amount of data that organizations need for making decisions has grown tremendously. The first criterion for categorizing data as big is when the size becomes part of the problem. The critical size is business-specific and depends on technological progress. The key business issue is the cost of data collection. The absolute size could be in the range of one petabyte. However, the growing popularity of cloud services has increased significantly the relevant data size and has eased the cost burden.
- *Velocity.* Another feature of recent data is that it is being generated at a much faster rate than data in the past. The growing penetration of the IoT requires data streams in real time.
- *Variety.* Data comes in all types of formats—from structured, numeric data in traditional databases to unstructured text documents, emails, pictures, video, and audio.

---

[4] Accessed on June 23, 2019.

[5] Adapted from a post by Michael Walker on 28 November 2012, http://www.datascienceassn.org/blogs/michaelwalker

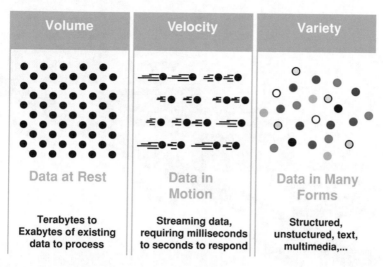

Fig. 1.7 The 3 Vs of big data—Volume, Velocity, and Variety

IoT

The basic idea behind the IoT is that everything consumers and businesses do is now leaving a digital trace—which can be turned into insight and has a potential to create value.

According to Wikipedia, "The Internet of things (IoT) is the extension of Internet connectivity into physical devices and everyday objects. Embedded with electronics, Internet connectivity, and other forms of hardware (such as sensors), these devices can communicate and interact with others over the Internet, and they can be remotely monitored and controlled." [6] Each device or object is uniquely identifiable through its embedded computing system but is able to inter-operate within the existing Internet infrastructure. Experts estimate that the IoT will consist of about 30 billion objects by 2020.

Data being analyzed in near real time provides a lot of valuable IoT use cases. One example is intelligent health-monitoring equipment that can enable faster action than that of doctors or nurses monitoring the same signals manually. Another example is smart homes equipped with smart sensors that can simultaneously increase safety by reducing risks such as fire and flooding, and bring down operational costs by switching heating and air-conditioning on and off at the right times to exploit off-peak rates. There are great expectations that body sensors will track people's activity levels and help change their behavior to improve well-being, while medical sensors can support overall health, for example by monitoring blood sugar levels and dispensing insulin when necessary.

---

[6]Accessed on June 23, 2019.

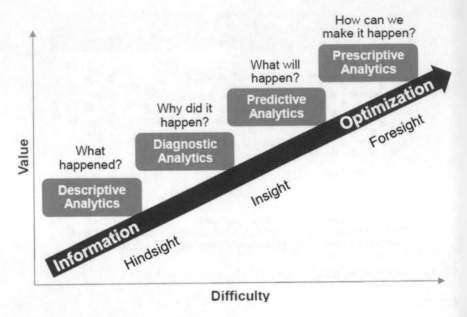

Fig. 1.8 Key types of analytics according to Gartner

The ongoing advance of AI is causing a natural integration with the IoT. The current trend is toward replacing almost all "dumb" devices with smart machines. Hence the need for AI.

Data Analytics

According to Techopedia, "Data analytics refers to qualitative and quantitative techniques and processes used to enhance productivity and business gain. Data is extracted and categorized to identify and analyze behavioral data and patterns, and techniques vary according to organizational requirements."[7]

This generic buzzword has many specific meanings, the most popular of which have been systematized by Gartner depending on their difficulty and value creation potential as descriptive, diagnostic, predictive, and prescriptive analytics (see Fig. 1.8)[8]:

- *Descriptive analytics: What happened?* This type of analytics, which is the most used, delivers solutions that help a business to understand better what is going on based on historical data. It provides insight based on patterns and relationships found between the key metrics and measures within the business. Utilizing effective visualization tools enhances the message of descriptive analytics. An

---

[7] Accessed on June 23, 2019.

[8] Adapted from https://www.zdnet.com/article/googling-prescriptive-analytics-youtube-recommen dations-and-the-analytics-continuum//

example of descriptive analytics is analyzing the key factors and patterns in increasing the energy cost of a business.

- *Diagnostic analytics: Why did it happen?* The purpose of diagnostic analytics is to empower an analyst to drill down and isolate the root cause of a problem. An example of diagnostics analytics is identifying the key factors in and potential sources of equipment failure from available process data.
- *Predictive analytics: What is likely to happen?* Predictive analytics is used to make predictions about unknown future events. Predictive models typically utilize a variety of variables related to a problem to make a forecast. It uses many techniques, such as statistical algorithms, data mining, and machine learning, to analyze available historic data and make predictions about the future. The forecasts are based on discovering trends and patterns from the past that could be valid in the future, as well as identifying the key drivers that influence what we are trying to predict. A typical example of this type of analytics is forecasting of raw materials prices based on selected economic drivers and historical patterns.
- *Prescriptive analytics: What do we need to do?* The highest level of complexity and potential value creation is developing a prescriptive model. This utilizes the insight gained from what has happened, why it has happened, and a variety of "what-might-happen" analysis to help the user determine the course of action to take. Often the recommended actions are derived by sophisticated optimization algorithms that guarantee the best set of decisions for solving a defined problem considering the existing constraints of real-world problems. An example of prescriptive analytics is to use predicted prices in the purchasing of raw materials and minimizing the cost.

Data Mining

Data mining is the process of processing data sets to identify trends, and patterns and discover relationships, to solve business problems or generate new opportunities through the analysis of the data. This discipline is about designing algorithms to extract insights from rather large and potentially unstructured data (for example, text mining). Techniques include pattern recognition, feature selection, clustering, and supervised classification.

Business Intelligence

Business intelligence (BI) is focused on dashboard creation, selection of key performance indicators (KPIs), producing and scheduling data reports based on statistical summaries.

How are the Buzzwords Related?

An attempt to link the buzzwords discussed above in a systematic way is shown in Fig. 1.9.

The buzzwords are classified into three categories: Technologies, Applications, and Data. AI is the broadest area in the Technologies. Machine learning is only one of several technologies that are part of AI. Deep learning is part of the machine learning set of algorithms. Analytics is the broadest buzzword in the Applications area. It includes data mining and business intelligence as special steps in the analytics process. Big data is the broadest buzzword in the Data category, and it contains the IoT as one of the key generators of large-volume data in real time.

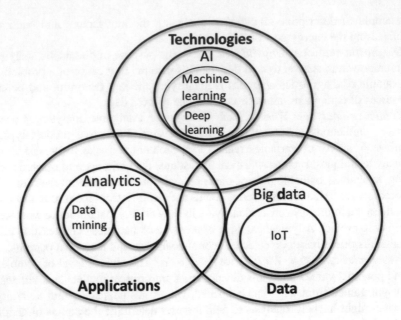

Fig. 1.9 Relationships between key buzzwords

1.1.3 Why Now?

Several factors are contributing to the recent growing interest in AI-driven approaches. The unique combination of a data and a social media invasion with a tremendous infrastructure growth has opened up new value creation opportunities. A lot of businesses are looking at AI-driven technologies as a key component of their future competitiveness. These factors are discussed briefly below.

Data Invasion

The volume of data generated by digital platforms, wireless sensors, social media, and billions of mobile phones continues to double every 3 years. It is estimated that the world creates about 2.2 billion gigabytes every day. This data avalanche, however, creates new opportunities for extracting value from unknown patterns and relationships derived from data analysis and modeling. AI-driven approaches are the most powerful tools for translating big data into more insights and more complex relationships. Most of these technologies, especially machine learning, are data-hungry and need a lot of data to accomplish the learning process. The growth of data is having a very positive effect on the development of AI-driven approaches. It is pushing the research toward novel methods, such as deep learning, that can deliver insight and solutions from large amounts of diverse data such as images, video, and audio.

According to Barry Smyth, a professor of computer science at University College Dublin "Data is to AI what food is to humans."[9] In the same way as different types of food have diverse effects on human body, different qualities of data influence significantly the final result. Just as junk food deteriorates human health, low-quality data leads to models with poor performance.

Social Media Invasion

Social media is more integral to our lives than ever before, and nobody expects it to stop growing anytime soon. Facebook, YouTube, Instagram, and Twitter make it easy for anybody to participate in online communication and online networking, and these platforms have led to a massive global proliferation across demographics. It is expected that the number of users accessing social media will soon surpass 2.5 billion, with 91% of social media users accessing platforms from mobile devices.

Extracting insight and delivering solutions for business use from the available data in social media is becoming one of the key objectives of applied Data Science. AI-driven methods offer the best portfolio of algorithms to fulfill this challenging task.

Infrastructure Growth

Another important factor contributing to the current explosion of AI-driven technologies is the continuous growth of the capabilities of the computing infrastructure. Even as Moore's law[10] is nearing its physical limits, other innovations are fueling this continued progress. Computational power has gotten a boost from an unlikely source: video game consoles. The need for computational power to power even more sophisticated video games has spurred the development of graphics processing units. GPUs have enabled image processing in neural networks that is 10–30 times as fast as what conventional CPUs can do.

The growth of cloud-based platforms has given virtually every company the tools and storage capacity to deploy Data Science solutions. Cloud-based storage and analytics services enable even small firms to store their data and process it on distributed servers. Companies can purchase as much disk space as they need, greatly simplifying their data architecture and IT requirements and lowering capital investment. As computation capacity and data storage alike have largely been outsourced, many tools have become accessible, and data can now be more easily combined across sources.

The growth is even more substantial on the software front. The trend of offering free open source software and free versions of popular packages (such as RapidMiner) allows AI-based methods to be introduced at a minimal software cost. In parallel, the established vendors, such as SAS, IBM, and Microsoft have updated their key products with improved AI-driven methods that allow development and deployment of complex analytical systems at an enterprise-wide level.

---

[9]http://www.ucd.ie/research/people/computerscience/professorbarrysmyth/

[10]Moore's law states that the number of transistors included in an integrated circuit doubles approximately every 2 years.

Value Creation Opportunities of AI-Driven Technologies

The value creation opportunities of AI-driven technologies are based on three unique features unlike those of traditional automation solutions. The first feature is their ability to automate complex physical-world tasks that require adaptability and agility. Whereas traditional automation technology is task-specific, the second distinct feature of AI-generated solutions is their ability to solve problems across industries and job titles. The third and most powerful feature of AI-driven technologies is self-learning, enabled by repeatability at scale.

The self-learning aspect of AI is a fundamental change. Whereas traditional automation capital degrades over time, intelligent automation assets constantly improve.

A significant part of the value obtained from AI-driven technologies will come not from replacing existing labor and capital, but in enabling them to be used much more effectively. For example, AI can enable humans to focus on the parts of their role that add the most value. Also, AI augments labor by complementing human capabilities, offering employees new tools to enhance their natural intelligence.

Another opportunity for value creation from applying AI-driven technologies is their ability to propel innovations. A good example is driverless vehicles. Using a combination of global positioning systems, cameras, computer vision, and machine learning algorithms, driverless cars can enable a machine to sense its surroundings and act accordingly. AI-driven technologies are critical in this process but they allow traditional companies to build new partnerships to stay relevant. As innovation begets innovation, the potential impact of driverless vehicles on economies could eventually extend well beyond the automotive industry. For example, the insurance industry could create new revenue streams from the masses of data that self-driving vehicles generate and the new advanced AI algorithms it uses.

AI-Driven Technologies as a Key Component of Business Competitiveness

Probably the most important shift is that AI-driven technologies, starting from a pure research discipline have become an area of strategic business interest. A recent survey showed that 59% of organizations are still gathering information to build an AI strategy, while 40% are piloting or adopting AI technologies. Tech giants, including Baidu and Google, spent between $20 billion and $30 billion on AI in 2016, with 90% of this spent on R&D and deployment and 10% on AI acquisitions.[11]

The report of that survey cites many examples of internal development, including Amazon's investments in robotics and speech recognition, and Salesforce's investment in intelligent agents and machine learning. BMW, Tesla, and Toyota lead auto manufacturers in their investments in robotics and machine learning for use in driverless cars. Toyota is planning to invest $1 billion in establishing a new research institute devoted to AI for robotics and driverless vehicles.

---

[11]McKinsey Global Institute, *Artificial Intelligence: The Next Digital Frontier?* 2017.

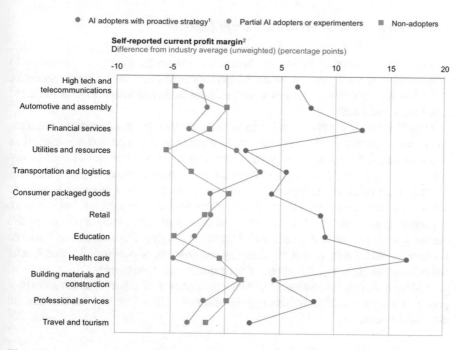

Fig. 1.10 AI early adopters with proactive strategies have higher profit margins

The competitive advantage of early adopters of AI-driven technologies has already been demonstrated for several sectors in the economy. The results are shown in Fig. 1.10, where the profit margins of companies that apply AI-driven technologies and non-adopters in different sectors of the economy are compared.

As is shown, healthcare, financial services, and professional services are seeing the greatest increase in their profit margins as a result of AI adoption. McKinsey found that companies that benefit from senior management support for AI initiatives, have invested in infrastructure to support its scale, and have clear business goals achieve a 3–15 percentage point higher profit margin than non-AI adopters.

1.2   What Is AI-Based Data Science?

The growing influence of data and AI-driven technologies on science and industry is having an impact on the new rising-star discipline of Data Science. In order to reflect this new reality, a broader version of this discipline, named AI-based Data Science, is proposed in the book. The definition and specific features of AI-based Data Science are discussed below.

1.2.1   Definition of AI-Based Data Science

Defining such a broad discipline as Data Science is not an easy task. According to Wikipedia: "Data science is a multi-disciplinary field that uses scientific methods, processes, algorithms and systems to extract knowledge and insights from structured and unstructured data."[12]

Data Science is a very complex field that incorporates mathematics, statistics, computer science and programming, math modeling, database technologies, data visualization, data analytics, and so on. From a business point of view, Data Science is a three-legged stool that combines business acumen, data wrangling, and analytics to create high value. Focusing on the hard science skills such as machine learning is not enough. It is one thing to know how to play with fancy algorithms, but it is more important to understand what insights these mathematical models reveal about the business, and what actions to take based on those insights. Experience and business knowledge play a role, as well as curiosity and passion. Sometimes the best results come from nonexperts just because of their desire and persistence.

AI-based Data Science uses AI-driven approaches in addition to statistics to turn data into insight and actions. The specific features of this type of Data Science are discussed below.

1.2.2   Features of AI-Based Data Science

AI-Focused

The main difference between "standard" Data Science and its AI-based sister is the leading role of AI-driven methods. They add some critical new capabilities to turn data into insight and actions. The key methods of this sort are discussed briefly below:

- *Different knowledge extraction mechanisms.* AI-driven methods offer several options to extract knowledge from available data, such as by using machine learning, simulated evolution, and swarm intelligence. Machine learning algorithms allow a computer to discover patterns and relationships in historical data that could be used for decision-making with unseen data in the future. Simulated evolution automatically generates knowledge from data based on math models fighting with each other. The winners are the best predictors or classifiers derived from the available data. The algorithms of the third option for knowledge extraction from data, swarm intelligence, are based on simulating social interactions among swarms of biological species, for example ants and bees.
- *Decision generation.* Several AI-driven methods, such as decision trees and intelligent agents, allow data to be transformed into automatically generated

---

[12] Accessed on June 23, 2019.

decisions based on proper quantitative criteria. These decisions can be either executed automatically or used by humans in their decision-making process.

- *Partially automated scientific process.* Several AI-driven approaches, such as machine learning, deep learning, and evolutionary computation, generate their solutions by automatically executing the key steps of the scientific process. In machine learning, the hypothesis is defined either by the labeled data in the case of supervised learning or by automatically discovered patterns (clusters) in the case of unsupervised learning. During the machine learning process, the hypothesis is tested on validation data and a decision is made about its correctness based on model performance.

Broad Set of Technologies

These unique capabilities are based on the new AI-driven technologies. Since AI is an active area of research, the list is continuously growing. Some selected methods will be discussed briefly below, and in more detail in Chap. 3:

- *Machine learning.* The classical machine learning algorithms are based on artificial neural networks, inspired by the capabilities of the brain to process information. A neural network consists of a number of nodes, called neurons, which are a simple mathematical model of the real biological neurons in the brain. The neurons are connected by links, and each link has a numerical weight associated with it. The learned patterns in biological neurons are memorized by the strength of their synaptic links. In a similar way, the learned knowledge in an artificial neural network can be represented by the numerical weights of the mathematical links. In the same way as biological neurons learn new patterns by readjusting the synapse strengths based on positive or negative experience, artificial neural networks learn by readjustment of the numerical weights based on a defined fitness function.
- *Deep learning.* This new technology is also based on neural networks but has very high complexity (sometimes with more than 100 layers and millions of parameters). Deep learning algorithms are used to detect objects in images, analyze sound waves to convert spoken speech to text, or process natural human language into a structured format for analysis. This technology is still in its early days for business use but it is expected to play a key role in the major application area of driverless vehicles.
- *Evolutionary computation.* This automatically generates solutions of a given problem with defined fitness by simulating natural evolution in a computer. Some of the generated solutions have entirely new features, i.e., the technology is capable of creating novelty. In simulated evolution, it is assumed that a fitness function is defined in advance. The process begins with the creation in the computer of a random population of artificial individuals, such as mathematical expressions, binary strings, symbols, or structures. In each phase of simulated evolution, a new population is created by genetic computer operations, such as mutation, crossover, and copying. As in natural evolution, only the best and the brightest survive and are selected for the next phase. Due to the random nature of simulated evolution, it is repeated several times before the final solutions are selected. Very often the constant fight for high fitness during simulated evolution delivers solutions beyond the existing knowledge about the problem explored.

- *Swarm intelligence.* This explores the advantages of the collective behavior of an artificial flock of computer entities by mimicking the social interactions of animal and human societies. A clear example is the performance of a flock of birds. Of special interest is the behavior of ants, termites, and bees. This approach is a new type of dynamic learning, based on continuous social interchange between the individuals. As a result, swarm intelligence delivers new ways to optimize and classify complex systems in real time. This capability of AI-based Data Science is of special importance for industrial applications in the area of scheduling and control in dynamic environments.
- *Decision trees.* These represent rules, which can be understood by humans and used as knowledge extracted from data. Decision tree output is very easy to interpret even for people from a nonanalytical background. It does not require any statistical knowledge to read and interpret the output. Its graphical representation is very intuitive, and users can easily relate their assumptions.
- *Intelligent agents.* Intelligent agents are artificial entities that have several intelligent features, such as being autonomous, responding appropriately to changes in their environment, persistently pursuing goals, and being flexible, robust, and social by interacting with other agents. Of special importance is the interactive capability of the intelligent agents, since it mimics types of human interaction, such as negotiation, coordination, cooperation, and teamwork. A popular version of intelligent agents, chatbots, are used in machine–human communication.

Usually modern AI is identified by the research and business communities at large with machine learning, and recently has been identified with deep learning. In fact, the whole package of technologies, not only machine learning, is contributing to the current success of AI. One of the objectives of this book is to describe them, demonstrate their applicability, and encourage data scientists to use them.

Broad Applicability
Due to the enhanced capabilities delivered by the broad set of new technologies, AI-based Data Science is highly applicable to many fields, including social media, medicine, security, healthcare, the social sciences, biological sciences, engineering, defense, business, economics, finance, marketing, and many more. Examples of specific applications will be given in Chap. 14 for manufacturing and Chap. 15 for businesses.

Advanced Skillset Required
The new opportunities for improvement that AI-based Data Science offers require additional knowledge about the key AI technologies. The needed skillset includes basic knowledge about the principles of the methods, training in appropriate software platforms of their implementation, and awareness for their application potential.

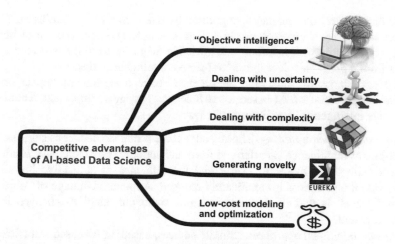

Fig. 1.11 Key competitive advantages of AI-based Data Science

1.3   Competitive Advantages of AI-Based Data Science

Data Science is becoming a critical factor to maintain competitiveness in the increasingly data-rich business environment. Much like the application of simple statistics, organizations that embrace Data Science will be rewarded, while those that do not will be challenged to keep pace. As more complex, disparate data sets become available, the chasm between these groups will only continue to widen. It is believed that the new, powerful technologies of AI-based Data Science will increase further the competitive advantage of the first group.

The competitive landscape for these technologies includes the following well-established approaches: first-principles modeling, statistics, classical optimization, and heuristics. They are referred to as competitors in this book. The key competitive advantages of AI-based Data Science are shown in Fig. 1.11 and discussed in detail below.

1.3.1   Creating "Objective Intelligence"

The most important feature that boosts AI-based Data Science ahead of the competition is the "objective" nature of the "smart" solutions delivered. "Objective intelligence" is similar to first-principles and statistical models in that those are also "objective," since they are based on the laws of nature and laws of numbers. However, "objective intelligence" is differentiated by its capability to automatically extract solutions from data through machine learning, simulated evolution, or emergent phenomena. In contrast, "subjective intelligence"is based on human assessment only. It is not derived from "objective" sources, such as the laws of nature or empirical dependencies, supported by data. "Subjective intelligence" can be very dangerous if the expertise in a field is scarce. The combination of limited numbers of experts with insufficient knowledge may transform the expected application from a problem solver into a "subjective intelligence" disaster.

The advantages of "objective intelligence" have a significant impact on the application potential of AI-based Data Science. The most important features of "objective intelligence" are discussed below:

- *Consistent decision-making.* The key advantage is that the decisions suggested by "objective intelligence" are derived from and supported by data. As a result, the rules defined are closer to reality and the influence of subjective biases and individual preferences is significantly reduced. Another advantage of "objective intelligence" is that its decisions are not static but adapt to changes in the environment.
- *Nonstop intelligent operation.* "Smart" devices, based on AI-based Data Science, operate continuously and reliably for long periods of time in a wide range of process conditions. As we all know, human intelligence cannot endure a 24/7 mode of intensive intellectual activity. Even the collective intelligence of rotating shifts, typical in manufacturing, has significant fluctuations due to wide differences in the operators' expertise and their attention to the process at a given moment. In contrast, "objective intelligence" continuously refreshes itself by learning from data and knowledge streams. This is one of the key differences between AI-based Data Science and the competition. The competitive solutions can also operate nonstop, but they cannot continuously, without human interference, maintain, update, and enhance their own intelligence.
- *Handling high dimensionality and hidden patterns.* AI-based Data Science can infer solutions from multidimensional spaces with thousands of factors (variables) and millions of records. This feature is beyond the capabilities of human intelligence. Another advantage of "objective intelligence" is its ability to capture unknown complex patterns from available data. It is extremely difficult for human intelligence to detect patterns with many variables and on different time scales.
- *Continuous self-improvement by learning.* Several learning approaches, such as neural networks, statistical learning theory, and reinforcement learning, are the engines of almost perpetual progress in the capabilities of "objective intelligence." The competitive statistical methods lack this unique feature.
- *No politics.* One can look at "objective intelligence" as an honest and loyal "employee" who works tirelessly to fulfill her/his duties while continuously improving her/his qualifications. Political maneuvering, growing pretensions, and flip-flopping, so typical in the behavior of human intelligence, is unknown. It is not a surprise that this feature sounds very appealing to management.

1.3.2   Dealing with Uncertainty

A key strength of AI-based Data Science is in handling technical uncertainty. The economic benefits of this competitive advantage are substantial. Reduced technical uncertainty leads to tighter control around process quality, faster new product design, and less frequent incidents. All of these benefits explicitly translate technical advantages into value.

One of the advantages of statistics is that uncertainty is built into its foundations. Of special practical importance are statistical estimates of the uncertainty of model predictions, represented by their confidence limits. The different ways in which AI-based Data Science handles uncertainty are discussed below:

- *Minimum a priori assumptions.* Fundamental modeling deals with uncertainty only within strictly defined a priori assumptions, dictated by the validity regions of the laws of nature; statistics handles uncertainty by calculating confidence limits within the ranges of available data; and heuristics explicitly builds the boundaries of validity of the rules. All of these options significantly narrow down the assumption space of the potential solutions and make it very sensitive to changing operating conditions. As a result, their performance lacks robustness and leads to gradually evaporating credibility and imminent death of the application outside the assumption space. In contrast, AI-based Data Science has a very open assumption space and can operate with almost any starting data or pieces of knowledge. The methods that allow AI-based Data Science to operate with minimum a priori information are highlighted below.
- *Reducing uncertainty through learning.* One of the possible ways to deal with unknown operating conditions is through continuous learning. By using several machine learning methods, AI-based Data Science can handle and gradually reduce wide uncertainty. This allows adaptive behavior and low cost.
- *Reducing uncertainty through simulated evolution.* Another approach to fighting unknown conditions is by evolutionary computation. This technology is one of the rare cases when modeling can begin with no a priori assumptions at all. Uncertainty is gradually reduced by the evolving population of potential solutions, and the fittest winners in this process are the final result of this fight with the unknown.
- *Handling uncertainty through self-organization.* In self-organizing systems, such as intelligent agents, new patterns occur spontaneously by interactions, which are internal to the system. As in simulated evolution, this approach operates with no a priori assumptions. Uncertainty is reduced by the new emerging solutions.

1.3.3   Dealing with Complexity

Big data and the Internet of Things have pushed the complexity of real-world applications to levels that were unimaginable even a couple of years ago. A short list includes the following changes: (1) the number of interactive components has risen by several orders of magnitude; (2) the dynamic environment requires solutions that are capable of both continuous adaptation and abrupt transformations; and (3) the nature of interactions is depending more and more on time-critical relationships between the components. Another factor that has to be considered in dealing with the increased complexity of practical applications is the required simple dialog with the final user. The growing complexity of the problem and the generated solution must be transparent to the user.

The competitive approaches face significant problems in dealing with complexity. First-principles models have relatively low dimensionality; even statistics has difficulties in dealing with thousands of variables and millions of records; heuristics is very limited in representing large numbers of rules; and classical optimization has computational and convergence problems with complex search spaces of many variables.

The different ways in which AI-based Data Science handles complexity better than the competition are discussed below:

- *Reducing dimensionality through learning.* AI-based Data Science can cluster the data by learning automatically how it is related. This condensed form of information significantly reduces the number of entities representing the system.
- *Reducing complexity through simulated evolution.* Evolutionary computation delivers distilled solutions with low complexity (especially when a complexity measure is included in the fitness function). One side effect of simulated evolution is that the unimportant variables are gradually removed from the final solutions, which leads to automatic variable selection and dimensionality reduction.
- *Handling complex optimization problems.* Evolutionary computation and swarm intelligence may converge and find optimal solutions in noisier and more complex search spaces than the classical approaches can handle.

1.3.4   Generating Novelty

Probably the most valuable competitive advantage of AI-based Data Science is its unique capability to automatically create innovative solutions. In the classical method, before shouting "Eureka," the inventor goes through a broad hypothesis search, and trials of many combinations of different factors. Since the number of hypotheses and factors is close to infinity, the expert also needs "help" from nonscientific forces such as luck, inspiration, a "divine" spark, or even a bathtub or a falling apple. As a result, classical discovery of novelty is an unpredictable process.

AI-based Data Science can increase the chances of success and reduce the overall effort in innovation discovery. Since generating intellectual property is one of the key components of economic competitive advantage, this unique strength of AI-based Data Science may have an enormous economic impact.

The three main ways of generating novelty by AI-based Data Science are discussed next:

- *Capturing emergent phenomena from complex behavior.* Self-organized complex adaptive systems mimic the novelty discovery process by means of their property of *emergence.* This property is a result of coupled interactions between the parts of a system. As a result of these complex interactions, new, unknown patterns emerge. The features of these novel patterns are not inherited or directly derived from any of the parts. Since the emergent phenomena are unpredictable discoveries, they require to be captured, interpreted, and defined by an expert with high imagination.
- *Extracting new structures by simulated evolution.* One specific method in evolutionary computation, genetic programming, can generate almost any types of new structure based on a small number of given building blocks.
- *Finding new relationships.* The most widespread use of AI-based Data Science, however, is in capturing unknown relationships between variables. Of special importance are the complex dependencies derived, which are difficult to reproduce using classical statistics. The development time for finding these relationships is significantly shorter than for building first-principles or statistical models. In the case of simulated evolution, even these dependencies are derived automatically, and the role of the expert is reduced to selection of the most appropriate solutions based on performance and interpretability.

1.3.5 Low-Cost Modeling and Optimization

Finally, what really matters for practical applications is that all the technical competitive advantages of AI-based Data Science discussed above lead to costs of modeling and optimization that are lower than the competition. The key ways in which AI-based Data Science accomplishes those of important advantage are given below:

- *High-quality empirical models.* The models derived by AI-based Data Science, especially by symbolic regression via genetic programming, have optimal accuracy and complexity. On the one hand, they represent adequately the complex dependencies among the influential process variables and deliver accurate predictions. On the other hand, their relatively low complexity allows robust performance in the presence of minor process changes, when most competitive approaches collapse. In general, empirical models have minimal development cost. In addition, high-quality symbolic regression models, with their improved robustness, significantly reduce the deployment and maintenance cost.
- *Optimization for a broad range of operating conditions.* Two popular AI-based Data Science technologies, evolutionary computation and swarm intelligence,

broaden the capabilities of classical optimization in conditions with complex surfaces and high dimensionality. As a result, AI-based Data Science gives more technical opportunities to operate with minimal cost in new, previously not optimized areas. Of special importance are the dynamic optimization options of swarm intelligence, where the process could track continuously, in real time, the economic optimum.

• *Low total cost of ownership of modeling and optimization.* All of the advantages discussed above contribute to an overall reduction of the total cost of ownership of the combined modeling and optimization efforts driven by AI-based Data Science. Some competing technologies may have smaller components in the cost. For example, the development and deployment cost of statistics is much lower. However, considering all the components, especially the growing share of maintenance costs in the cost of modeling and optimization, AI-based Data Science is a clear winner. The more complex the problem, the bigger the advantages of using this emerging technology. All known technical competitors have very limited capabilities to handle imprecision, uncertainty, and complexity and to generate novelty. As a result, they operate inadequately in new operating conditions, reducing profit and pushing maintenance costs through the roof.

Still, the biggest issue in estimating the total cost of ownership is the high introductory cost of the technology. Since AI-based Data Science is virtually unknown in industry at large, significant marketing and training efforts are needed. One of the purposes of this book is to suggest solutions that will reduce this cost.

1.4   Key Challenges in Applying AI-Based Data Science

In order to give an objective assessment of AI-based Data Science we need also to identify the potential issues of technical and nontechnical nature.

1.4.1   Technical Issues in Applying AI-Based Data Science

The important technical issues that may reduce the efficiency of AI-based Data Science applications are shown in Fig. 1.12 and discussed next.

Data Quality

It is not a surprise that data quality becomes a critical factor for the success of AI-based Data Science. Firstly, data availability must be checked very carefully. It is possible that the historical records may be too short to capture seasonal effects or trends. Secondly, the ranges of the most important factors in the data have to be as broad as possible to represent nonlinear behavior. Data-driven models developed on narrow data ranges have low robustness and require frequent readjustment. Thirdly, the frequency of data collection must be adequate for the nature of the modeling. For

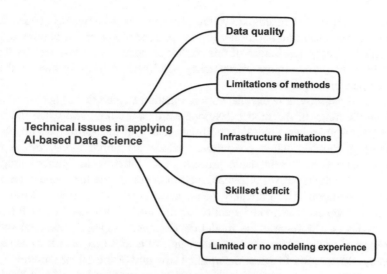

Fig. 1.12 Key technical issues in applying AI-based Data Science

example, dynamic modeling requires more frequent collection and data sampling. Steady-state models, on the other hand, assume a low data collection frequency that filters out dynamic effects. Fourthly, the noise level has to be within acceptable limits to avoid the classical Garbage-In-Garbage-Out (GIGO) effect. In cases where some of these requirements are not met, it is recommended to create an adequate data collection infrastructure and to begin the application only after collecting the right data. Making a compromise on data quality is one of the most frequent mistakes in applying AI-based Data Science.

Limitations of AI-Based Methods

Each AI-based Data Science method has its own limitations, which will be discussed in detail in Chap. 4. The major issues that these weaknesses create are listed below. Some of these limitations can be compensated by combination with other AI-based Data Science approaches:

- *Black-box models.* Many users view neural networks (the dominant machine learning technology) as magic pieces of software that represent unknown patterns or relationships in the data. The difficult interpretation of the magic, however, creates a problem. The purely mathematical description of even simple neural networks is not easy to understand. A black box links the input parameters to the outputs and does not give any insight into the nature of the relationships. As a result, black boxes are not well accepted by the majority of users, especially in manufacturing, having in mind the big responsibility of controlling plants.
- *Tacit knowledge generation.* In many cases deep neural networks cannot be interpreted at all. Their structure may have tens of millions of connections, each of which contributes a small amount to the developed model. In this case we have a reverse version of the famous Polanyi's paradox: machines know more than

they can tell us.[13] Unfortunately, the generated tacit knowledge coded in the layers of deep learning neural net cannot be understood, and it is risky to reuse it. This limits the application of this method in some industries, such as finance and insurance, that require transparency and interpretability of models that are developed and applied.

- *Poor extrapolation.* The excellent approximation capabilities of neural networks within the range of the model development data are not valid when the model operates in unknown process conditions. It is true that empirical models also cannot guarantee reliable predictions outside the initial model development range, defined by the available process data. However, the various empirical modeling methods deliver different levels of degrading performance in unknown process conditions. Unfortunately, neural networks are very sensitive to unknown process changes. The model quality significantly deteriorates even for minor deviations (<10% outside the model development range). A potential solution for improving the extrapolation performance of neural networks is to use evolutionary computation to select the optimal structure of the neural network.
- *Maintenance nightmare.* The combination of poor extrapolation and black box models significantly complicates the maintenance of neural networks. The majority of industrial processes experience changes in their operating conditions of more than 10% during a typical business cycle. As a result, the performance of deployed neural network models degrades, and triggers frequent model retuning, and even complete redesign. Since the maintenance and support of neural networks requires special training, this inevitably increases maintenance cost.
- *Computationally intensive.* Deep learning and evolutionary computation require substantial number-crunching power. Fortunately, the continuous growth of computational power according to Moore's law and the new hardware options offered by GPUs and specialized neural net chips are gradually resolving this issue. Improved algorithms are making additional gains in productivity and the companies driving the AI field are continuously developing new, more powerful tools. The third way to reduce computational time is to shrink down the dimensionality of the data by effective variable/feature selection.
- *Time-consuming solution generation.* An inevitable effect of computationally intensive AI-driven methods, such as deep learning and evolutionary computation, is slow model generation. Depending on the dimensionality and the nature of the application, this may take hours, or even days.

Infrastructure Limitations

The success of the application of AI-based Data Science depends on the integration capabilities of the existing hardware, software, and work process infrastructure. Usually most offline applications do not require significant changes in the existing infrastructure. A special case is users' addiction to Excel and their requirement to interact only within its environment. For real-time implementation of AI-based Data

---

[13]Polanyi's paradox states that "we know more than we can tell, i.e., many of the tasks we perform rely on tacit, intuitive knowledge that is difficult to codify and automate."

Science in the area of the IoT, however, a careful analysis of the software limitations and the maintenance infrastructure is a must.

Skillset Deficit
The business and academic world have understood that the key obstacle to materializing the potential of Data Science is the big deficit of data scientists. It is estimated that the number of graduates from Data Science programs could increase by a robust 7% per year. However, it is projected that even greater (12%) annual growth in demand, which would lead to a shortfall of some 250,000 data scientists. It has to be considered that in many courses this discipline is taught using examples far away from business reality, without messy data where nothing is obvious and one gap in data follows another. Another limitation of courses is a lack of business knowledge, so critical for the success of data scientists. The options for training of AI-based Data Scientists are very limited, since the AI-driven technologies covered in most courses are restricted to machine learning, with a focus on neural networks and decision trees.

Another missing skillset is that of the business translator, who serves as the link between data scientists and business users. In addition to being data-savvy, business translators need to have deep organizational knowledge and industrial or functional expertise. It may be possible to outsource analytics activities, but business translator roles require proprietary knowledge and should be more deeply embedded in the organization. It is estimated that there could be a demand for approximately 2–4 million business translators in the United States alone over the next decade.[14]

Limited or no Modeling Experience
The success and the speed of implementing AI-based Data Science depend also on the previous record of modeling applications. Even lessons from applying simple statistical models are of help, since a modeling culture has been introduced. As a result, users have some experience in using and maintaining models, as well as assessment of the value created. On the other hand, one of the issues of limited modeling experience is the risk of unrealistic expectations, which paves the way for an application fiasco. Usually the lack of a modeling culture is combined with limited infrastructure for implementation and support, which additionally raises the total cost of ownership due to the necessary investment in infrastructure and training.

1.4.2   Nontechnical Issues in Applying AI-Based Data Science

The key nontechnical issues that may lead to unsuccessful AI-based Data Science applications are shown in Fig. 1.13 and discussed below.

---

[14]McKinsey Global Institute, *The Age of Analytics: Competing in a Data-Driven World*, 2016.

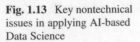

Fig. 1.13 Key nontechnical
issues in applying AI-based
Data Science

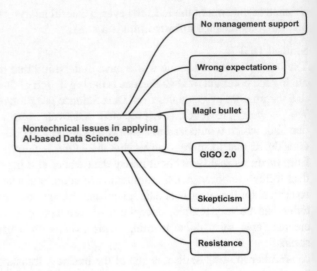

No Management Support

As with any emerging technology, AI-based Data Science needs initial management
blessing before it can begin to deliver sustainable value. Of critical importance is
also the consistency of support for a period of at least 3 years. Unfortunately, this
requirement may be unrealistic in businesses with frequent restructuring and management changes. The best strategy to address this issue is to find application areas
with fast demonstration of value creation and to promote AI-based Data Science with
effective marketing.

Wrong Expectations

Probably the most difficult issue in applied AI-based Data Science is how to help the
final user in defining the proper expectations of the technology. Very often the
dangerous combination of lack of knowledge, technology hype, and negative recep-
tion from some research and business communities creates incorrect anticipation
about the real capabilities of AI-based Data Science. The two extremes of wrong
expectations, either by exaggeration or by underestimation of the capabilities of
AI-based Data Science, cause almost equal damage to the promotion of the technology in industry.

Magic Bullet

The expectation of technical magic is based on the unique capabilities of applied
AI-based Data Science to handle uncertainty and complexity and to generate novelty. The impressive features of the broad diversity of methods, such as machine
learning, deep learning, evolutionary computation, and swarm intelligence, contribute to such a Harry Potter-like image even when most of the users do not understand
the principles behind them. Another factor adding to the silver bullet perception of
AI-based Data Science is the technology hype from the vendors, the media, and
some high-ranking managers.

As a result, potential users approach AI-based Data Science as the last hope to resolve very complex and difficult problems. Often, they begin looking at the technology after several failed attempts of using other methods. In some cases, however, the problems are ill-defined and not supported by data and expertise. In order to avoid the magic bullet trap, it is strongly recommended to identify the requirements, communicate the limitations of the appropriate methods, and define realistic expectations in the very early phase of potential AI-based Data Science applications.

GIGO 2.0

The worst-case scenario of the magic bullet image is the GIGO 2.0 effect. In contrast to the classical meaning of GIGO 1.0 (Garbage-In-Garbage-Out), which represents ignorant expectations of a potential solution in the case of bad data, GIGO 2.0 embodies the next level of arrogant hope, defined as Garbage-In-Gold-Out. In essence, this is the false belief that low-quality data can be compensated for with sophisticated data analysis. Unfortunately, AI-based Data Science, with its diverse capabilities to analyze data, is one of the top-ranking technologies that create GIGO 2.0 arrogant expectations. It is observed that the bigger the disarray of the data, the higher the hope that exotic, unknown technologies will be able to clean up the mess. Usually top management who are unaware of the nasty reality of the mess initiate this behavior.

It is strongly recommended to protect potential AI-based Data Science applications from the negative consequences of the GIGO 2.0 effect. The best winning strategy is to define the requirements and the expectations in advance and to communicate clearly to the user the limitations of the methods. Better to reject an impossible implementation due to low-quality data than to poison the soil for many feasible AI-based Data Science applications in the future.

Skepticism

In contrast to the magic bullet optimistic euphoria, disbelief and lack of trust in the capabilities of applied AI-based Data Science is the other extreme of wrong expectation, but in this case on the negative side. Skepticism is usually the initial response of the final users of the technology on the business side. Several factors contribute to this behavior, such as lack of awareness of the technical capabilities and application potential of AI-based Data Science, lessons from other overhyped technology fiascos in the past, and caution about ambitious high-tech initiatives pushed by management.

Skepticism is a normal attitude if risk is not rewarded. Introducing emerging technologies, such as AI-based Data Science requires a risk-taking culture from all participants in this difficult process. The recommended strategy for success and reducing skepticism is to offer incentives to the developers and the users of the technology.

Resistance

The most difficult form of wrong expectations of AI-based Data Science is that technical or political biases leads people to actively oppose the technology. In most cases the driving forces of resistance are the parts of the business that feel threatened by the new, powerful capabilities of the technology. In order to prevent a future organizational fight, it is very important to communicate a firm commitment not to reduce the labor force as a result of AI-based Data Science applications.

The other part of the resistance movement against AI-based Data Science includes the professional statisticians. Most of them do not accept the statistical validity of some of the AI-based Data Science methods, especially neural-network-based models. The key argument of these fighters for the purity of the statistical theory is the lack of a statistically sound confidence metric for the nonlinear empirical solutions derived by AI-based Data Science.

The third part of the resistance camp against AI-based Data Science includes the active members of the Anything But Model (ABM) movement, who energetically oppose any attempt to introduce new technologies.

1.5   Common Mistakes

1.5.1   Believing the Hype

The biggest mistake that can be made at the beginning of applying AI-based Data Science is to trust blindly in the AI voodoo. There is no doubt that this paradigm has grown significantly in recent years and has begun aggressively to invade industry. However, the progress is driven mostly by the AI Olympians. One of the reasons for this concentration on the development of AI-driven technologies is that they have increased the productivity of these high-tech giants and have opened up new markets (for example, for driverless vehicles). The hype and exuberant enthusiasm about AI is mostly based on the success of this paradigm in the AI Olympian companies.

The rest of the business world, however, has to look more carefully at the hype. The first step toward a reality correction is understanding the limitations of the key AI-driven technologies (details are given in Chap. 4). The next step is gaining awareness about their maturity. For example, deep learning is still a technology in progress and not entirely ready for industry. The third step in replacing the hype with realistic expectations is to assess the requirements for nonstandard infrastructure and additional training.

Only after filling the gaps in one's knowledge about AI by these three steps can one have a more accurate technical view of this fashionable field.

1.5.2   Neglecting to Estimate the Demand for AI-Based Data Science

Another typical mistake is introducing AI-based Data Science without assessing the demand for its capabilities in a specific business. The fact that this package of technologies is successful at the AI Olympian companies does not automatically translate into productivity growth in your business. A comprehensive analysis of the potential need for AI-driven applications needs to be done in parallel with understanding the technical benefits of this paradigm. A good starting point is to look for use cases of similar AI-based Data Science business implementations. These can give more specific knowledge about the necessary AI-driven methods and infrastructure to start with.

1.5.3   Mass-Scale Introduction of AI-Based Data Science in a Business without Required Skillset Availability

When top management believes the AI hype and passionately embraces the paradigm, it often triggers a program of fast implementation of the technology across the organization. Unfortunately, this approach has a very low chance of success, due to the complex nature of AI-based technologies and the higher requirements for training. These technologies are not appropriate for accelerated mass-scale application. A combination of data scientists and business translators is needed to accomplish this task. Without gradually building an AI-based Data Science group with sufficient capacity in terms of of data scientists and business translators and the corresponding technical infrastructure, attempts to impose this technology from above by force are doomed to failure.

Recommendations for how to effectively introduce AI-based Data Science in a business are given in Chap. 16.

1.5.4   Introducing Data Science Bureaucracy

An inefficient way to introduce AI-based Data Science is by creating a bureaucratic structure and filling it with technically incompetent managers. A satirical scenario of this approach is shown in Fig. 1.14.[15]

---

[15]This idea is inspired by a Tom Fishburne cartoon at www.marketoonist.com

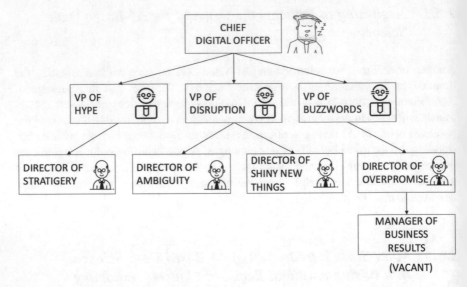

Fig. 1.14 The ideal bureaucratic structure of an AI-based Data Science business

1.6 Suggested Reading

- B. Baesens, *Analytics in a Big Data World*, Wiley, 2014.
- M. Berry and G. Linoff, *Data Mining Techniques*, third edition, Wiley, 2013.
- D. Hardoon and G. Shmueli, *Getting Started with Business Analytics*, CRC Press, 2013.
- M. Negnevitsky, *Artificial Intelligence: A Guide to Intelligent Systems*, third edition, Pearson Education Canada, 2014.
- F. Provost and T. Fawcett, *Data Science for Business*, O'Reilly Media, 2013.
- S. Russell and P. Norvig, *Artificial Intelligence: a Modern Approach*, third edition, Pearson, 2009.
- A. Wodecki, *Artificial Intelligence in Value Creation: Improving Competitive Advantage*, Palgrave Macmillan, 2019.

1.7 Questions

Question 1
Give examples of hype related to Data Science, AI, big data, analytics, and the IoT.

Question 2
Find use cases that can benefit your business.

Driverless cars	GPU	Machine Learning	Pattern	Data Mining
Paradigm	Watson	Data Science	NLP	Value
Analytics	Competitive Advantage	AI BINGO (free square)	The Cloud	Python
Cognitive Computing	Deep Learning	Robots	Framework	AI
Vision	Big Data	IoT	Data Scientist	Chatbots

Fig. 1.15 Example of AI bullshit bingo card

Question 3
What is the difference between descriptive and prescriptive analytics?

Question 4
Discuss the expected challenges in applying AI-based Data Science.

Question 5
AI Bingo. Bullshit Bingo is a very popular game in the corporate world. Usually one has a card with selected buzzwords related to a selected topic. The rule is to click on or mark each block when you see or hear the corresponding word or phrase. When you get five blocks horizontally, vertically, or diagonally, you stand up and shout "BULLSHIT!!!" An example of a card related to AI is given in Fig 1.15.

Create your own version of this card, following the steps in http://www.bullshitbingo.net/byo/.

Chapter 2
Business Problems Dependent on Data

We cannot solve our problems with the same level of thinking that created them.

Albert Einstein

The first step to successfully applying AI-based Data Science is understanding the specific business needs that will benefit the most from these technologies. The process includes an intensive knowledge exchange between data scientists and business experts. On the one hand, the data scientists educate their business partners about the capabilities of the appropriate AI-driven technologies and gradually gain awareness about the current business problems. On the other hand, the business experts share their concerns and the current technical issues they need to fix, and begin to understand the real potential of the new AI magic. It is expected that the final result of this dialog is the selection of a relevant business problem (or problems) that can be resolved effectively with these technologies.

Solving business problems is the best way to introduce and apply AI-based Data Science in a business. The focus of this chapter is on discussing the key issues in finding, identifying, defining, and solving business problems. Special attention will be given to business problems that depend heavily on data and require more sophisticated methods than statistics, i.e., AI-based Data Science.

The first objective of the chapter is to emphasize the leading role of business problems. The second objective of the chapter is to strongly recommend a clear separation between the problem and solution spaces during all steps of the project, while the third objective is to give an overview of typical business problems related to data that can be solved by AI methods. The fourth objective of the chapter is to suggest ideas for how to find data-driven business problems appropriate for AI. The last, fifth objective is to discuss the critical and often neglected issues of clear problem definition and value assessment.

© Springer Nature Switzerland AG 2020
A. K. Kordon, *Applying Data Science*,
https://doi.org/10.1007/978-3-030-36375-8_2

2.1   The Leading Role of Business Problems

Solving business problems more efficiently than can be done by established methods is the best way to introduce a new technology in practice. Finding the appropriate problem, however, is not a trivial task and requires a combination of business and Data Science knowledge. Of special importance is the growing role of data. Recently, it has become a critical factor in problem solving and decision-making. However, the attempt to give the leading place to data, mostly driven by vendors, does not entirely agree with business practice and may lead to wrong decisions.

2.1.1   "Data Is the New Oil" Hype

At the center of the obsession with data is the mantra "Data is the new oil." It has been embraced by the business community due to the hidden message of future wealth, supported by vendors' promises to "transform data into gold." The media liked it, too, and have spread aggressively this simplified vision of the next economic miracle. As a result, an inflated image of data as the ultimate decisive factor in shaping our future has been created.

This famous phrase was announced in 2006 by Clive Humbly, a UK mathematician and the architect of Tesco's Clubcard. However, the slogan has been taken out of context and the key emphasis of the message missed. He said the following:

Data is the new oil. It's valuable, but if unrefined it cannot be used. It has to be changed into gas, plastic, chemicals, etc. to create a valuable entity that drives profitable activity; so must data be broken down, analyzed for it to have value.[1]

In support of the full statement, we suggest an analogy with the internal combustion engine, which was the critical factor in transforming oil into gold (Fig. 2.1).

The physical process that executes this transformation is the typical cycle for most of the internal combustion engines in cars that use gasoline as a fuel. It consists of four major steps: intake, compression, power and exhaust. The process that transforms raw data into value also includes four major steps: data preprocessing, extracting insight from data analysis, building predictive models, and using them for improved business decisions that create the final value. This process is strongly recommended and used in the book.

There is no doubt that data is at the heart of the disruptions occurring across the economy. It has become a critical corporate asset, and business leaders want to know what the information they hold is worth. We'll focus on two key issues that influence the value of data and have to be considered: information content dilution and key data fallacies.

---

[1]https://www.analyticbridge.datasciencecentral.com/profiles/blogs/data-science-simplified-principles-and-process

Fig. 2.1 Analogy between transformations of oil and data into value

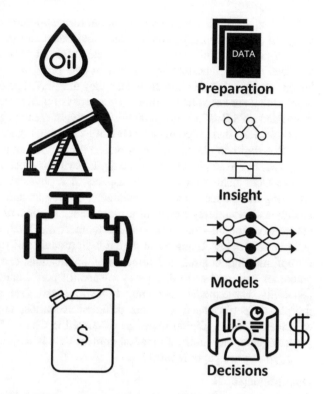

Information Content Dilution

On the one hand, the growing avalanche of big data (in the sense of Volume, Velocity, and Variety) gives more opportunities to find new patterns and dependencies and create value. On the other hand, however, the information content of some data can be significantly diluted. Much of this newly available data is in the form of clicks, images, text, or signals of various sorts, and is very different from structured numeric data that can be easily placed in rows and columns. Finding high information content from this ocean of unstructured data requires sophisticated algorithms and, very often, new data storage technologies. For example, extracting relevant information about a product price from comments on Twitter and Facebook requires a tremendous amount of sentiment analysis of a very large volume of data. Even if it can bring complementary insight to model development or decision-making, the high cost of its extraction may not justify its use.

Another form of information content dilution is "dark data." According to Gartner, dark data is "the information assets organizations collect, process, and store during regular business activities, but generally fail to use for other purposes."[2] Though the categories of dark data may vary across companies, the following categories of unstructured data are usually considered dark data: old versions of

---

[2]https://www.gartner.com/it-glossary/dark-data

documents, emails, log files, and data about former employees. Most dark data is not analyzed (90% according to the International Data Corporation). One of the reasons that explain this phenomenon is that the informational content depends on the problem defined (another argument that the problem should lead!). When the businesses collected this historical data, they may have had other criteria or problems in mind that are not valid anymore. There are not many business incentives to open this "Pandora's Box" of old content which might create legal issues as well.

Fortunately, there is counter process of generating high-information-content data. Recently, digitizing important information flows relevant to the business has been a key trend. For example, GE creates a digital twin for most of its assets. These digital twin models include all necessary aspects of a physical asset or a larger system, including mechanical, electrical, chemical, economic, and statistical aspects. These models also accurately represent a plant under significant conditions of variations related to operation—fuel mix, ambient temperature, load, weather forecast models, and market pricing. Using these digital twin models and state-of-the-art techniques for optimization, control, and forecasting, applications can more accurately predict outcomes with respect to different measures of performance, reliability, and maintainability. These models, in conjunction with sensor data, give the ability to predict the plant's performance, evaluate different scenarios, understand trade-offs, and enhance efficiency (digital twins are discussed in Chap. 14). Another example of a process generating high-information-content data is the current trend in healthcare for digitizing old paper records.

Data Fallacies

The other issue that can reduce the growing impact of data and lead to wrong conclusions is data fallacies. The key such fallacies are shown in Fig. 2.2 and discussed below:

- *Data dredging*. This is the practice of repeatedly testing new hypotheses against the same set of data, failing to acknowledge that most correlations are spurious and the result of chance. Tests for statistical significance only work if a hypothesis has been defined upfront. This is also sometimes known as data fishing or data snooping.[3]
- *Data cherry-picking*. The practice of selecting results that fit your hypothesis and excluding those that do not. This is the worst and most harmful example of being dishonest with data. People often only highlight data that backs their case, rather than the entire body of results.
- *Survivorship bias*. This is the effect of drawing conclusions from an incomplete data set, because that data has "survived" some selection criterion. When analyzing data, it is important to ask a question about missing data. Sometimes the full picture is obscured because the available data is only possible because it has survived a selection of some sort.

---

[3]More detailed information about data fallacies can be found at https://data-literacy.geckoboard. com

Fig. 2.2 Key types of data
fallacy

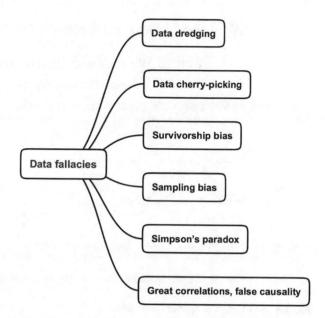

- *Sampling bias.* Drawing conclusions from a set of data that is not representative of the population of interest. This is a classic problem in marketing surveys where potential customers taking part in a poll are not representative of the total population, due to either self-selection or bias by the analysts.
- *Simpson's paradox.* A phenomenon in which a trend appears in different groups of data but disappears or reverses when the groups are combined.
- *Great correlations, false causality.* This data fallacy is the false assumption that when two events occur together, one must have caused the other. Correlation does not imply causation. Often, correlations between two things tempt us to believe that one caused the other. However, it is often a coincidence or there is a third factor causing both of the effects that we observe. An example of a spurious correlation between worldwide noncommercial space launches and sociology doctorates awarded is shown in Fig. 2.3. It shows not only an impressive correlation coefficient of 0.79 but very similar patterns of both variables over the whole data range. There is no way to claim, however, that noncommercial space launches cause sociology doctorates to be awarded or vice versa.[4]

---

[4]This example and many others are collected in https://www.tylervigen.com/spurious-correlations

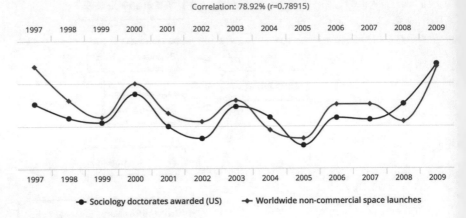

Fig. 2.3 An example of spurious correlation

2.1.2 Problems-First Approach

The exponentially growing volume of data gives businesses new opportunities to deliver value. These possibilities, however, are business-specific. They depend strongly on the type of business in general and the business problems in particular. That is why the driving force for deciding to introduce Data Science in general or to apply a specific modeling technique in particular is the value creation potential of solving business problems. For example, if a business evaluates that more accurate price-forecasting models can be developed by use of bigger and more information-ally rich economic time series data, a decision can be made to introduce Data Science capabilities and explore this opportunity.

Why Problems Should Lead Data
The arguments why business problems should be the leading factor in making strategic or tactical decisions about applying Data Science in general are discussed briefly below:

- *The problem defines the value hypothesis.* The decisive factor in decision-making about applying any new technology in a business is the value creation hypothesis. It is assumed that the return on the investment in the technology will justify the cost. This rule is valid for any level of decision, strategic or tactical, for any specific problem of fixing technical issues. Business experts define the value hypothesis. An example of such a hypothesis is that an estimate of the steam consumption of a manufacturing plant in real time can allow the energy cost to be reduced by 5%. Data is used to validate the assumption, but the hypothesis is not defined by the data itself but by the domain experts. The chances of implementation of new technologies in a business without a credible value hypothesis are very low.

- *The problem defines the technical hypothesis.* The second decisive factor in decision-making about applying a new technology in a business is the technical hypothesis. This includes assumptions about potential technical factors and mechanisms that can contribute to the solution of the problem. Usually several competing hypotheses are defined by technical experts and evaluated during project development. In fact, defining the technical objectives drives the first decision about a potential project. Continuing with the value hypothesis example, the business may want a continuous estimate of its steam consumption based on the available sensors. The key technical hypothesis is that linear or nonlinear relationships between measurements from those sensors and steam consumption can be found and used for the estimate. This is the first step toward the final business goal of minimizing energy cost. The definition of the value hypothesis follows if key decision-makers accept the relevance of the technical hypothesis.
- *The role of data depends on the business problem.* All data-related issues, such as data sources, the data collection structure, and data quality assessment, are driven by the project definition. In the previous example, only data related to steam consumption may have been collected and analyzed.
- *Data-first strategy is an academic exercise.* The current trend to reverse the roles, giving the leadership to data, is mostly driven by academics and advertised by vendors. One possible explanation is that there are many diverse data sets publicly available that can be used to explore almost any research idea (with www.kaggle.com as the leading site). However, the academics forget that most of these data sets were created due to a business or research problem. Some of these data sets have been used as a benchmark in a competition to solve a business problem.

Problem and Solution Spaces

A common mistake that inexperienced data scientists make is to start and navigate the dialog with their business partners in the direction of AI technologies without understanding the problem. We strongly recommend avoiding this habit. The first step in dealing with this issue is to make a clear distinction between the questions related to the business problem (we called this the problem space) and the corresponding questions related to the technical options for solving the problem (we call this the solution space). An example of a visualization of this separation is shown in Fig. 2.4.

The problem space in Fig. 2.4 includes examples of specific business problems, such as raw materials forecasting, steam consumption reduction, and machine breakdown reduction. By default, this is the territory of business experts, who communicate using business-specific language.

The solution space in Fig. 2.4 includes examples of relevant technologies, such as evolutionary computation, neural networks (NN), deep learning, and decision trees, which can be used in solving the problem. By default, this is the territory of data scientists, who communicate using the technical jargon of the corresponding technologies.

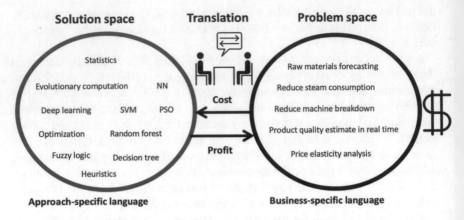

Fig. 2.4 A possible visualization of problem and solution spaces

One of the biggest challenges in applying AI-based Data Science is to achieve a smooth dialog between these two communities. Different techniques to avoid this lost-in-translation-trap and improve communication are discussed in Chap. 5. The McKinsey Global Institute has also identified the importance of this dialog. They suggest a new role of business translator in the future AI-driven economy.[5]

The other challenge is not to follow the "problem first" approach, and instead to impose a solution based on gut feeling and technology preferences. This is the approach driven by a category of very aggressive and arrogant data scientists who have a "universal" solution for any problem and do not spend time listening to their business partners. Often the final result, however, is an inefficient solution and disappointment. The suggestion is to avoid jumping to the solution space before extracting available knowledge about the problem. The right strategy is to focus on the problem space and deliver, as detailed a problem definition as possible before discussing the options in the solution space.

2.2 Typical Business Problems Related to AI-Based Data Science

The growing popularity of AI is mostly based on mass-scale applications from the Olympian companies, such as personal assistants like Siri, Cortana, or Alexa. AI algorithms are also behind the scene in almost any activity in the most popular social media tools. For example, during any Google search, an analytic engine determines which search results to show and which ads to display. Another example is related to posting something on Facebook, when an AI algorithm is automatically run to

[5]McKinsey Global Institute, *The Age of Analytics: Competing in a Data-Driven World*, 2016.

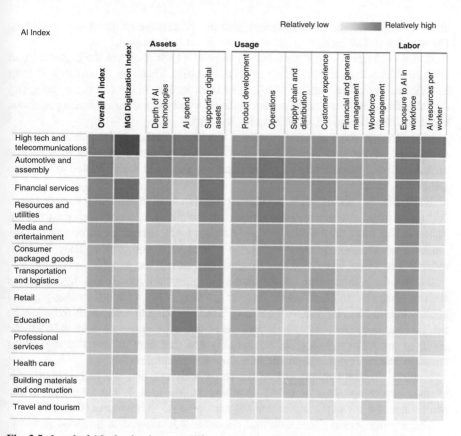

Fig. 2.5 Level of AI adoption in some different sectors of the economy in the US and Europe

determine if it must be rejected due to promotion, spam, or porn. A similar approach is used in Gmail each time an email is sent, where an analytic algorithm decides whether or not to put it in the spam box. It is expected that the Olympians will continue to deliver new mass-scale applications in the future. They'll spread AI built into their products at low cost to billions of customers.

Our focus, however, is on the AI application landscape in the rest of the business world, especially the big business. We'll start with an interesting study by the McKinsey Global Institute about the level of AI adoption in key sectors of the US and European economy.[6] The results representing the state of the art in 2017 are summarized in Fig. 2.5 and discussed below.

The level of adoption of AI technologies is assessed by an overall AI index. It includes 16 different metrics related to how intensively AI has been applied in terms of assets, usage, and labor. For example, the AI adoption level in terms of assets is estimated by the average number of AI technologies adopted, the average AI spend

---

[6]McKinsey Global Institute, *Artificial Intelligence: The Next Digital Frontier?* 2017.

as a share of total annual investment, and the percentage of cloud services used. Details of the other metrics of the study are given in the report.

It is not a surprise that the leading sectors in AI adoption are the high-tech and telecommunications, and financial services industries. These are industries with long histories of digital investment. They have been leaders in developing or adopting digital tools, both for their core product offerings and for optimizing their operations. For them, investing in AI is the next logical step.

The automotive and assembly sector is also highly ranked in the overall AI index. It was one of the first sectors that implemented advanced robotics at scale for manufacturing, and recently has also been exploring AI technologies to develop self-driving cars.

In the middle ranking are some less digitized industries, including resources and utilities, personal and professional services, and building materials and construction. These sectors have been slow to employ digital tools, except for some parts of the professional services industry and large construction companies. They are also industries in which innovation and productivity growth has lagged.

Toward the bottom of the AI adoption ranking are traditionally less digital sectors such as education and healthcare. Despite growing publicity about cutting-edge AI applications in these industries, the reality is that the real impact appears to be low so far. In healthcare, for example, practitioners and administrators acknowledge the potential for AI to reduce costs but quickly add that they believe that regulatory concerns and customer acceptance will slow down adoption.

Another very important result of the same study is a diagram that estimates the average percentage change in AI spending in the next 3 years relative to the percentage of firms in the sector that have adopted AI technologies (Fig. 2.6).

The leading sectors (high-tech and telecommunications, and financial services) will continue to dominate and grow their AI investment, with a value of around 12%. The AI investment growth in transportation and logistics sector is expected to be 8% while the corresponding growth in the travel, entertainment, and healthcare sectors is around 7%. Most of the remaining sectors of the economy are expected to increase their spending on AI-related technologies by between 3 and 6%. A clear laggard is the construction sector, with anemic growth of around 1%.

A more specific overview of the typical problems relevant to AI-based Data Science in manufacturing and business is presented below.

2.2.1 Typical Problems in Manufacturing

Overall, manufacturing industries have lagged behind the leaders in applying AI-based technologies, with exception of the car industry. The key manufacturing problems that can benefit from AI-based Data Science are shown in the mind map in Fig. 2.7 and discussed briefly below. More detailed examples are given in Chap. 14.

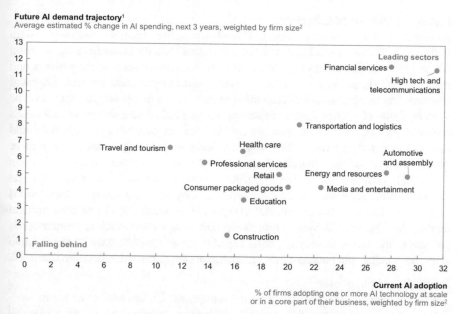

Fig. 2.6 Estimated AI demand in next 3 years by economic sector

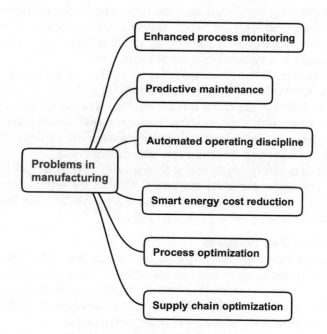

Fig. 2.7 Key problems in manufacturing related to AI-based Data Science

Enhanced Process Monitoring

A central topic in process monitoring is how successfully the manufacturing system is observed. The observability of a process is defined as a data flow from the process, which gives sufficient information to estimate the current state of the process and make correct decisions about future actions, including process control. Classical observability was based mostly on hardware sensors and first-principles models. The modern form of observability, however, includes also data-driven model-based estimates of critical process parameters, as well as process trend analysis and newly discovered patterns. In this way, the decision-making process is much more comprehensive and has a predictive component linked to process trends.

Several AI-driven approaches contribute to enhanced process observability. The key mass-scale application is in inferential sensors, with thousands of applications based on neural networks, and, recently, on genetic programming. These are empirical models that link measurements from cheap hardware sensors, such as temperatures, pressures, and flows, to very expensive quality measurements, such as those emissions, melt index, and biomass. Inferential sensors are discussed in detail in Chap. 14.

Predictive Maintenance

Predictive maintenance predicts potential equipment failures and gives recommendations that include corrective actions, equipment replacement, or even planned failure. The objective of predictive maintenance is to recognize system faults in advance by detecting significant differences in certain parameters or patterns obtained from changed behavior. Sensor networks and statistical/machine learning enable operators to recognize if there is any deviation from the normal operating regime that might be a symptom of a future breakdown.

The decision in predictive maintenance is based on time in use, events, and sensor readings (the system is called a health monitor). In this way, a failure point is identified long before to the actual breakdown, thus delaying the entire process of maintenance, or avoiding unplanned maintenance or replacement costs. Predictive maintenance helps in planned and scheduled maintenance activities. It increases productivity by minimizing downtime and maintenance costs.

According to ta survey conducted by Schneider Electric in the United States on predictive maintenance, industries can save 8–12% of their total revenue, due to a reduction of the chance of equipment breakdown by 70–75%. An example of a process health-monitoring system is given in Chap. 14.

Automated Operating Discipline

An operating discipline integrates all participants in a manufacturing unit into a well-organized work process, which includes managers, engineers, and operators. It can be viewed as the collective knowledge base of a specific manufacturing process. The problem is that it is static and the only link to the dynamic nature of the process is through process operators and engineers in the control room.

AI allows us to develop an adaptive operating discipline that responds adequately to the changing process environment in real time. This is based on the available process knowledge with additional capabilities, such as multivariate monitoring, early fault detection combined with intelligent alarm diagnostics, and optimal performance metrics. An example of an improved operating discipline system using neural networks, fuzzy systems, knowledge-based systems, and evolutionary computation is shown in Chap. 14.

Smart Energy Cost Reduction

The electric utilities sector has great potential to embrace AI in the coming years. At every step of the value chain, from power generation to end consumers, opportunities for machine learning, robotics, and automation of decision-making exist that could help electric utilities better predict supply and demand, balance the grid in real time, reduce downtime, maximize yield, and improve end-users' experience. One of the most useful cases for AI in electricity is for demand and supply prediction. An accurate forecast of the load in the power grid can affect many stakeholders: power generators determine which power sources should be allocated over the next 24 h based on short-term load forecast; transmission grids assign resources based on power transmission requirement; and electricity retailers calculate energy prices based on estimated demand.

Electric utilities are starting to explore the use of AI technologies to produce more accurate short-term load forecasts. DeepMind, the AI startup bought by Google in 2014, is currently working with National Grid to predict supply and demand peaks in the United Kingdom by using weather-related variables and smart meters as exogenous inputs, hoping to cut national energy usage by 10% and maximize the use of renewable power despite its intermittency.[7]

Process Optimization

Optimization is one of the few technologies that directly translate results into value. Needless to say, many optimization techniques have been tried and used in industry. Most of the classical optimization approaches originated from mathematically derived techniques, such as linear programming, direct search, and gradient-based methods. The higher the complexity of a problem, the stronger the limitations of classical optimization methods. In the case of complex high-dimensional landscapes with noisy data, these methods have difficulties in finding a global optimum. Fortunately, AI offers several approaches that can deliver solutions beyond the limitations of classical optimization.

In new product design, evolutionary computation methods, such as genetic algorithms and evolutionary strategies, have been successfully used by companies such as Boeing, BMW, and Rolls Royce. An interesting application area is optimal formulation design, where genetic algorithms, and recently, particle swarm optimization (PSO) have been implemented for optimal color matching at The Dow Chemical Company.

In process optimization, genetic algorithms, evolutionary strategies, ant colony optimization, and particle swarm optimizers deliver unique capabilities for finding a steady state and a dynamic optimum.

---

[7]McKinsey Global Institute, *Artificial Intelligence: The Next Digital Frontier?* 2017.

Supply Chain Optimization

A company's supply chain comprises geographically dispersed facilities where raw materials, intermediate products, or finished products are acquired, transformed, stored, or sold, and transportation links that connect facilities and allow a flow of products. A supply chain network includes vendors, manufacturing plants, distribution centers, and markets. AI may contribute to improved efficiency and reduced cost of supply chain operations by more accurate demand estimation and optimal scheduling.

Demand forecasting is based on quantitative methods for predicting the future demand for products sold by a company. Such predictions are critical for modeling and optimizing product distribution. They are necessary for total-cost minimization of product prices, product sourcing locations, and supply chain expenses. The standard approach to demand forecasting is to use statistical time series analysis, which is limited to linear methods.

AI has the capability to broaden the modeling options for forecasting with nonlinear time series models. One technique with impressive features for representing nonlinear temporal dependencies is recurrent neural networks. It has been successfully applied in voice recognition, process control, and robotics.

In general, optimal scheduling in industry has two key parts—optimal distribution and production planning. Optimal distribution scheduling includes transportation factors, such as vehicle loading and routing, channel selection, and carrier selection. The inventory factors include safety stock requirements, replenishment quantities, and replenishment time. Optimal production scheduling requires actions such as sequencing of orders on a machine, timing of major and minor changeovers, and management of the work-in-progress inventory. The existing optimization models are mostly based on analytical linear programming and mixed integer programming techniques, which have limitations when the problem has high dimensionality and the optimal search landscape is noisy and has multiple optima. Several AI methods, especially some from evolutionary computation and swarm intelligence, are capable of delivering reliable solutions for complex real-world problems.

2.2.2   Typical Problems in Business

The leading sectors in applying AI technologies began solving their business problems 10–15 years ago. Some of the application areas have reached a very high level of sophistication and have become standard solutions. The key business problems that can benefit from AI-based Data Science are shown in the mind map in Fig. 2.8 and discussed briefly below. More detailed examples are given in Chap. 15.

Customer Churn

One of the first and most popular business problems relevant to AI-based Data Science is customer churn analysis. This helps the business to identify lost

Fig. 2.8 Key problems in
business related to AI-based
Data Science

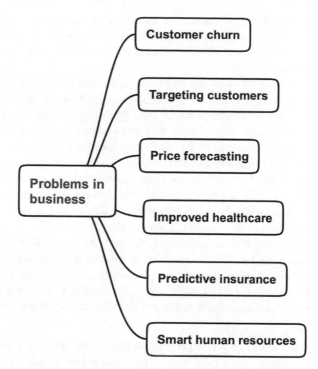

customers, focus on higher-value customers, determine what actions typically pre-
cede a lost customer or sale, and better understand what factors influence customer
retention. The traditional solution is to predict high-propensity churners and address
their needs via a concierge service or marketing campaigns, or by applying special
dispensations. These approaches can vary from industry to industry. They can even
vary from a particular consumer cluster to another within one industry (for example,
telecommunications).

The common factor is that businesses need to minimize these special customer
retention efforts. Thus, a natural methodology would be to score every customer
with a probability of churn and address only the top ones. The top customers might
be the most profitable ones. For example, in more sophisticated scenarios a profit
function is employed during the selection of candidates for special dispensation.
However, these considerations are only a part of the complete strategy for dealing
with churn. Businesses also have to consider risk (and associated risk tolerance), the
level and cost of the intervention, and plausible customer segmentation.

AI technologies, such as decision trees, random forest, and support vector
machines, can complement the statistical tools of logistic regression and survival
analysis for the final identification of the suspected customers.

Understanding and Targeting Customers

This is a much broader business problem than customer churn and the objective is to
use AI-based Data Science to better understand customers and their behaviors and

preferences. Companies are keen to expand their traditional data sets with social media data, and browser logs, as well as text analytics and sensor data, to get a more complete picture of their customers. The big objective, in many cases, is to create predictive models.

One of the most essential ways for businesses today to understand their customers is by categorizing them into cohorts or segments. Segments may vary depending on location, interests, browser and operating system, engagement with a brand or product, and more. Target audiences are no longer an age or gender group alone. Target audiences can be extremely specific and combine multiple criteria to offer customization and personalization. The trend today in AI in digital marketing is all about the "segment of one" and how products and services are marketed to an individual or to a smaller group with more specific interests and goals in mind.

A growing trend in contemporary digital marketing is using chatbots or virtual assistants in this process. They are based on two AI technologies—natural language processing and semantic analysis. Semantic analysis is how machines identify the basic, logical formal meanings of sentences. Although this process is incomplete without considering context, this is still a big step for machine learning. Semantic analysis is a part of AI in digital marketing that is already used in spell checker, social media analysis, sentiment analysis, fact extraction, summarization, and more.

Price Forecasting

Price forecasting is a generic problem with growing popularity in every business. Two factors are contributing to the solution of this problem: (a) many different economic time series are becoming available, and (b) software capabilities for multivariate forecasting are becominging popular. Optimum pricing considers available and predicted inventory, production costs, prices from competitors, and profit margins. Price elasticity models are often used to determine how high prices can be increased before reaching strong resistance. Modern systems offer prices on demand, in real time, for instance when booking a flight or a hotel room. Another business issue is user-dependent pricing which further optimizes pricing and offers different prices based on user segment. In addition to statistical methods for building time series models, some AI-based methods, such as recurrent neural networks, support vector machines, and genetic programming, are contributing to creating accurate nonlinear predictive models. An example of using genetic programming for prediction of raw materials prices is shown in Chap. 15.

Improving Healthcare and Public Health

Healthcare is a promising market for AI. There is enormous potential in the ability of AI to draw inferences from and recognize patterns in large volumes of patient histories, medical images, epidemiological statistics, and other data. AI has the potential to help doctors improve their diagnoses, forecast the spread of diseases, and customize treatments. AI combined with healthcare digitization can allow providers to monitor or diagnose patients remotely, as well as transform the way we treat the chronic diseases that account for a large share of healthcare budgets. Several AI technologies appear to be suitable for use in medical practices. Machine learning is starting to be applied in payments and claims management, but its further application in healthcare may arrive at scale soon. This technology is suited to analyzing the data in millions of medical histories to forecast health risks at the population level. This could be an early win for AI because it brings the potential for large savings and would not require the regulatory scrutiny that would be expected when trying to anticipate individual health risks.

Unfortunately, using AI to diagnose illnesses may not happen so fast. While machine learning is able to use data to make a diagnosis, completely automated diagnosis is not likely to happen quickly, partly because of questions about whether patients will accept it, and partly because of the technical difficulty of integrating data from multiple sources and complying with strong regulatory requirements.[8]

Predictive Insurance

AI-based Data Science can make forecasts and projections more accurate and precise. This can help businesses with a wide array of problems: predicting which customers will prove to be the most valuable, preventing employee attrition, scheduling maintenance in time to prevent breakdowns, and detecting fraud. These capabilities have valuable applications for insurers. Over time, insurance companies have sought ways to collect additional data that they believe could have predictive power. For example, behavioral data from sensors embedded in cars can provide insurers with much more direct information about an individual's driving patterns and risk than the regular data on educational attainments, age, and make of car. Some insurers incorporate credit scores because of empirical evidence that people who pay their bills on time are also better drivers.

Companies offering property insurance are beginning to add sensors to water pipes or in the kitchen to identify predictors of pipe leaks or breaks, flooding, or fire, and to warn the homeowners. Similarly, sophisticated AI-based methods that use behavioral data can transform life insurance models. Insurers can use these models to better price and customize coverage options–or even to warn customers to take preventive actions.

One UK insurance company that used vehicle sensors reported that better driving habits resulted in a 30% reduction in the number of claims, while another reported a 53% drop in risky driving behavior. Extrapolating that type of result across all auto insurers yields a potential impact on the order of $40 billion. Summing the benefits across other insurance industries and lives saved would indicate economic benefits in the hundreds of billions of dollars globally.[9]

Smart Human Resources

A relatively new business problem that can benefit from AI-based Data Science is increasing the efficiency of human resources. From a passive mode of keeping track of employees' performance, the new smart human resources departments are looking at attrition patterns, prediction models for performance and retention, models for employee absence and grievances, and analysis of many other forms of employee

---

[8]McKinsey Global Institute, *Artificial Intelligence: The Next Digital Frontier?* 2017.
[9]McKinsey Global Institute, *The Age of Analytics: Competing in a Data-Driven World*, 2016.

productivity. The companies concerned are starting to correlate their internal performance data against data available from external social networks and can now learn things about their employees they never before thought possible.

One vendor, for example, has a tool that reads comments from biannual engagement surveys and automatically recommends direct behavioral changes to managers to help improve the engagement and productivity of their team. Another company has built a machine learning algorithm that identifies the behavior of their best sales people to help understand how to train others to perform at a higher rate. Many professional services firms are looking at the communication patterns and travel schedules of the highest-performing consultants to figure out what others can learn.

2.3 How to Find Data-Driven Business Problems

This section is focused on one of the most important practical tasks in the process of applying AI-based Data Science—how to identify and select appropriate business problems. Due to the different levels of digitalization and AI technology introduction in the businesses, it is difficult to suggest a detailed set of steps that a data scientist can follow. Instead, a generic methodology is suggested with two major phases—(1) understand the business needs, and (2) match known AI use cases to identified business needs. It is assumed that the data scientist has an assigned role in the business and is aware of the key application areas of AI-based Data Science.

2.3.1 Understand Business Needs

Understanding specific business needs is one of the first challenges a data scientist will face. This requires access to a broad network of business experts, managers, IT specialists, and business clients. In addition, good communication skills are a plus. The following sequence for effective business needs identification is recommended: start with understanding the management vision and potential support for AI-based Data Science, interview key stakeholders in that vision, define appropriate business problems with the help of experts, and prioritize them. This is discussed briefly below.

Understand Management Vision
The critical step in applying AI-based Data Science across a business is getting clear support from top management. Ideally, a top manager takes the lead and forms a steering committee with key experts and executives. The mandate for introducing the technology assumes strategic support for several years and corresponding organizational decisions. The most important factor is the vision and commitment of the top manager in charge.

One of the important documents that navigates the AI application efforts is the statement of direction. It explicitly defines the appropriate uses of AI, data analytics, and Data Science in the business (for example, in process monitoring and control, data mining, and new product development). Another important section of the document is the scope of applicability (for example, a list of businesses, software environments, data warehouses, and control systems). The statement of direction also includes: the appropriate actions that should be taken by the different stakeholders; the next steps to be taken to further leverage the technology in a specific period of time; the key milestones with specific dates, and the owners, stakeholders, and contacts.

Contacting top management and understanding its vision and related documents is the first step in the process of finding and identifying appropriate business problems. Otherwise, the efforts may be inefficient, unsystematic, and unprofessional.

Interview Key Stakeholders

The next step after understanding the management vision is interviewing the key stakeholders who can define business problems. The list includes, but is not limited to, managers, subject matter experts (SMEs), IT experts, and financial experts.

It is very important to structure the interview process in advance and one potential approach that can help is to develop a mind map.[10]

An example of a mind map for structuring the key topics in identifying current business needs is shown in Fig. 2.9.

The stakeholders have been asked to focus on the top three current issues in the business and to evaluate the needs on different criteria with weights of 1, 3, and 5. The criteria and the weighting factors were discussed in advance. In this case the ranking is based on business importance, the complexity of the problem, the commitment of the business based on the resources it can allocate, the expected time for the delivered solution, and the estimated value. Usually a brainstorming session with all stakeholders triggers the discussion and generates a list of potential business problem candidates. The final identification and prioritization require several iterations of individual meetings until the final decision is made.

Identify Business Issues

An example of identified business needs in a specific manufacturing unit after stakeholder brainstorming is shown in the first block of the mind map in Fig. 2.9. The first issue, predictive maintenance to prevent machine breakdowns, is a long-term problem in the unit, which has not been solved with additional sensors. Machine breakdowns cause unit shutdowns with significant production loss. The hypothesis is that by using sophisticated multivariate and AI-based methods, a machine health monitor can be developed that can warn in advance about potential machine breakdowns.

---

[10]The mind maps in the book are based on the product ConceptDraw Mindap (http://www. conceptdraw.com/products/mind-map-software). The reader can use any available mind map tools.

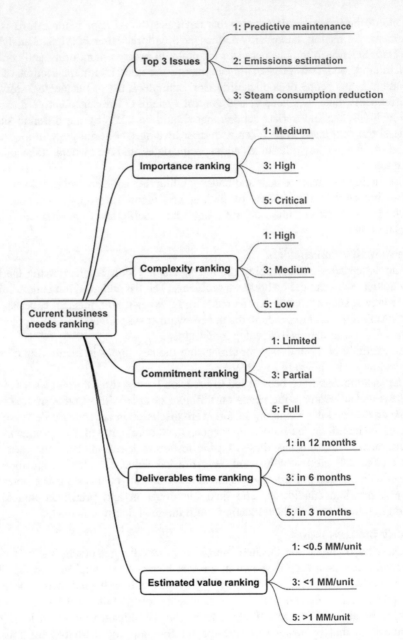

Fig. 2.9 A mind map for identifying business needs

The second identified problem is related to the need for a real-time estimate of the emissions from the unit. Currently, the emissions are measured every 8 h and cannot be used as a responsive factor in the process control system. The advantage of using an emissions estimate in real time is that in this case the unit can operate at an

Table 2.1 An example of a ranking of business problems

Problem	Importance	Complexity	Commitment	Time	Value	Total score
Machine breakdowns	5	1	3	1	5	15
Emissions estimation	3	3	5	3	3	17
Steam reduction	1	5	1	5	1	13

increased rate because the value of the emissions will be kept below the limit all the time using the estimated emissions value.

The third business issue is driven by the current management initiative to reduce the cost of water used in general and of steam consumption in particular. With few exceptions, the stakeholders interviewed agree on the selection of the key business problems of the unit. They are in consensus that the first topic—predictive maintenance to prevent machine breakdowns—is the key issue to fix.

Prioritize Business Issues

In this step, however, stakeholders have changed the focus, after prioritizing the problems on the specific criteria. Their estimates are shown in Table 2.1.

There is full agreement on the importance of the identified problems, with machine breakdowns ranked as critically important and steam consumption as being of medium importance. In the complexity ranking, however, machine breakdowns are an order of magnitude more complex than the other two options. Due to the high complexity, the machine breakdowns problem requires the full commitment of several SMEs in the unit whom the management is not willing to commit full time, and the ranking is for limited commitment. Another issue related to the complexity of this project is that it may take a long time for development, with an initial estimate of around 12 months. The ranking of estimated value gives top priority on the machine breakdowns, the middle ranking to the emissions estimate project, and the lowest expected value to the steam consumption problem.

Contrary to the initial consensus on the importance of the machine breakdowns problem, the detailed prioritization is in favor of another identified issue—development of an emissions estimator. The stakeholders prefer to begin with an important business problem of medium complexity that can be solved within 6 months, and are willing to allocate the necessary resources in the unit.

It is estimated, that, if successful, the project will generate a consistent value in the future due to reduced noncompliance cases and an increased rate of operation.

2.3.2 Match Business Needs with Known Artificial Intelligence-Based Use Cases

The next logical step after defining and ranking the current business problems is to identify the potential solutions based on AI technologies and to communicate the results to all stakeholders for the final decision. This is the responsibility of the data scientist, with the support of corresponding experts in IT and databases services. The following steps are recommended.

Check Competitors' Experience in AI

A strategic step for gaining important information is to analyze competitors' experience in applying AI-based Data Science. Data about invested infrastructure (hardware and software), people skills, and solved problems is critical for comparing your selected strategy in this area with the competitors' strategies. The best sources of information about competitors' experience in AI-based Data Science are the vendors, conference attendance, and publications.

Identify Known Use Cases of Selected Business Problems

The first step in this process is to find the broad category of use cases into which the defined business problem can be categorized. In our example, all top three identified problems are related to manufacturing and can be identified as typical problems related to AI-based Data Science (see Fig. 2.7). The machine breakdown problem is in the category of predictive maintenance, the emissions estimate is in the class of enhanced process monitoring, and the steam reduction is a type of smart energy cost reduction use case.

The next level of identification is based on more specific use cases and technologies within the broad area. In the case of the selected problem of emissions estimation, this is the category of inferential sensors. It includes developing data-driven models that estimate the emissions based on relevant process variables. This is one of the use cases of AI-based Data Science with a high level of maturity and thousands of applications in different manufacturing processes, i.e., the level of risk is relatively low.

A potential solution can be based on linear models developed by use of classical or multivariate statistics. Several AI-based technologies, however, offer the development of nonlinear relationships between the emissions and the process variables. These relationships are usually more accurate than the linear alternatives. The list of technologies includes neural networks, support vector machines, genetic programming, and model ensembles, to name a few. All of these technologies are offered by several vendors and can be developed and deployed with relatively low development, training, and deployment costs.

Evaluate Potential Implementation Issues

A well-known issue with inferential sensors is their low robustness after big process changes. One way to resolve this problem is to collect data with the broadest possible operating range. This includes operating regimes with very low rates when the unit will lose money. Management support is needed for approval of this data collection.

Data availability and quality is another generic source of concern. In this specific case, a long historical record of hourly averaged process data is available in a process historian. The measured emissions have acceptable quality.

Final Recommendation for Starting a Project

The final verdict on whether to start a new project or to move on to other opportunities is made by a consensus among interested stakeholders in a brainstorming session. The decision is based on the problem ranking, suggestions of potential solutions, and evaluation of prospective implementation issues. In our example, the stakeholders decided to start a project for the development of an inferential sensor for emissions at the selected manufacturing unit.

2.4   The Slippery Terrain of Problem Definition

Often projects, especially those related to AI-based Data Science, begin with a large uncertainty due to a lack of knowledge about the new technologies, insufficiently detailed business understanding, and unclear ideas about data availability and quality. As a result, the first important document at the start of the project, the project definition or charter, is ambiguous, with missing quantitative measures, and is easily misinterpreted. It is strongly recommended to focus on reducing uncertainty by clarifying the defined objectives and deliverables, and quantifying as much as possible the performance metrics. Avoiding this step will open the door to future confusion, false expectations, and possible disappointment.

2.4.1   Structure of Problem Definition

A good problem statement or charter has five basic elements:

- Clear project objectives;
- A well-defined scope and boundaries;
- A well-selected stakeholders;
- An initial set of deliverables;
- An initial time/cost estimate.

Clear Project Objectives   The key section of the document that includes a condensed description of the key goals of the project, supported with quantitative information. Qualitative objectives, such as "significant reduction of steam consumption," should be avoided and replaced with quantitative estimates like "a 4% steam consumption reduction on an annual basis." It is recommended that the objectives do not include technical jargon and are clearly understood by all stakeholders.

Well-defined Scope and Boundaries   An equally important section of the document includes the defined limits of the projects in terms of specific manufacturing units or machines, departments, data sources, and ranges.

Well-selected Stakeholders   This includes all stakeholders needed to make the project a reality, such as the project sponsor, key users, key SMEs, data SMEs, and data scientists.

Initial set of Deliverables Even in this early phase, it is recommended to manage expectations with an initial list of potential deliverables, such as data analysis reports, developed and potentially deployed models, and responsibilities for model maintenance and support.

Initial Time/Cost Estimate This includes a very rough estimate of the expected time for development and deployment based on previous experience and experts' opinion. It is needed for estimating the labor cost of the project, which is the basis for planning project funding.

2.4.2   Example of Problem Definition

We'll illustrate this important document with an example of a problem definition for the emissions estimator selected as the business issue to be solved. One of the challenges in defining project objectives is finding a performance metric that is measurable, can be tracked, and is appropriate for claiming success. An example of an appropriate quantitative objective that satisfies these conditions is to develop an inferential sensor for emissions estimation that passes the regulatory tests,[11] and increases the operating rate by 2% while reducing the annual noncompliance cases by 30%.

This definition includes three measurable metrics: (1) the average modeling error of the inferential sensor, which has to be below 15% in order to pass the annual regulatory test; (2) the operating rate, which is directly measurable and proportional to the economic profit; and (3) the number of noncompliance tickets that may add additional cost.

Since the operating rate is proportional to the emissions level, the last two criteria are conflicting, i.e., increasing the rate leads to more emissions and potential noncompliance cases, and vice versa. In the absence of online emissions estimates, the limits on the production rate were made very conservative in order to avoid the risk of violating the environmental constraints, and the plant is losing money. The economic driving force behind the project is that an inferential sensor will allow simultaneously more aggressive control with higher rates and a reduced risk of noncompliance.

Defining the project scope also needs to follow the quantitative metrics as much as possible. Usually this includes the geographical boundaries of the business, the limits on data and available equipment, and work process requirements. For our example, the project scope includes boundaries such as the specific manufacturing unit, the availability of data from a data collection campaign with a range of variation in the production rate, approved by management, control limits defined by the

---

[11] In this case the regulatory tests require an average modeling error <15%.

existing control system, and a requirement for implementation of the project to be done in the existing process information system in the plant.

An important part of this step is also to define the deliverables of the project in terms of products and impact. In the case of the emissions inferential sensor project, the expected deliverable product is an empirical model that has to be integrated within the existing process information system in the plant. The financial impact of the inferential sensor applied can be calculated from the increased profit from higher rates and the reduced number of noncompliance cases.

The project charter includes the necessary stakeholders as well. The list includes the plant manager as the project sponsor, three SMEs from the process control group, eight operators as final users who will need training if the model deployed in operation, one data SME who will be responsible for data collection, two SMEs from IT services responsible for model deployment in the process monitoring system, and two data scientists who will execute the project according to the work process discussed in Part II of this book.

Usually the problem statement document has several iterations after the initial version. The first reality correction happens after the data collection, which triggers major revision of most of the initial assumptions.

2.5   Value Creation Hypothesis

From a business perspective, the key question about the defined project and the criterion for its success is the value it is expected to create. AI-based Data Science generates value when it impacts the profit-making processes. This can only happen by deploying a model in the production environment and changing internal processes and systems to make actionable results accessible to decision-makers. Understanding the potential value of a project requires assessing the factors that influence project cost as well as project return. Unfortunately, this process is not well developed; the corresponding financial estimates are not done in a systematic way and depend on the experience of financial SMEs.[12] Let's not forget, however, that the objective of the starting value estimate is to give a very generic evaluation of the level of return on investment. It is still a hypothesis that needs to be proved during project development and deployment.

In order to begin the value assessment at this early phase of project development, two key issues need to be clarified: defining the sources of potential value creation, and defining the profit metric that will give the measure of financial success.

---

[12]The recent book by W. Verbeke, B. Baesens, and C. Bravo, *Profit Driven Business Analytics*, Wiley, 2017, offers a systematic approach to value assessment of data-driven modeling.

2.5.1   Sources of Value Creation

The first step in exploring the value creation potential of the defined project is to analyze and find the potential sources of value. The second step is to identify the related data sources, to collect the necessary related economic or technical data.

Identification of Value Sources
Usually the value creation sources are well identified by the economics of the business. They are in two broad categories—process/productivity improvement and loss reduction. The specific factors depend on the defined problem. In the example of the emissions estimation problem, there are two key sources of value creation. The first source is reduction of annual noncompliance cases due to emissions higher than the limit. The second source is increasing unit productivity due to safe operation at a high production rate based on estimated emissions in real time. It is expected that the second source will deliver the lion's share of the value.

Value-Related Data Sources
This step includes identifying the specific sources of data that we need to estimate the value based on the defined metrics. Using again the emissions estimation example, for the first source of value creation, reduction of noncompliance cases, we need financial data for the penalties paid for the environmental violation tickets and the values of measured emissions from the process historian. For the second source of value creation, increasing unit productivity, we need economic data for the product price and the measured values of the product, production rate, and emissions from the process historian.

2.5.2   Metrics for Value Creation

The most effective way to demonstrate value from solving a business problem is by developing specific value metrics and tracking their performance.

Definition Value Metrics
Evaluating the performance of predictive models is an important step in any data-driven application. In our example of the emission estimation problem, the value metrics include two components: an environmental penalty and a productivity gain. The environmental penalty includes the monetary value of the penalties and the percentage of time when the measured emissions are above the warning limit (15% below the critical threshold at which the unit is out of compliance). The productivity gain metric includes the profit from the additional product that is generated by operating at higher rates.

Benchmarking of Value Creation
One of the first steps in starting a new project related to AI-based Data Science is to initiate documenting a value creation benchmark before implementing the solution

to the problem. It is recommended that the benchmark should be based on sufficient historical data, especially if the business problem depends on seasonal effects. In the emissions estimation problem, the benchmark includes 2 years of average hourly data for the corresponding process variable in the value metric.

Tracking of Metrics
The proposed metric that has been used for benchmarking of historical data will be implemented after the models are deployed and used. In the emissions estimation problem, the metric is coded as a separate tag in the process information system and will be tracked on a minute-by-minute basis in real time. However, the real comparison of value creation with the benchmark will be done on a quarterly and annual basis.

Definition of Success
It is expected that the definition of success will be quantified in the project objectives. It usually includes both value and technical components. In the case of the emission estimation problem, success of the value creation is defined as:

- Reduction of annual noncompliance cases 30%;
- Increasing the average operating rate by 2% on an annual basis.

The technical success is defined as the average modeling error of the developed models being below 15% in order to pass the annual regulatory test required by the environmental agency. Accomplishing the technical objective is critical, because models with lower accuracy cannot be used and the value hypothesis collapses.

2.6 Common Mistakes

2.6.1 Jumping to Solutions without Defining Business Problems

This is one of the most frequent mistakes, usually driven by inexperienced data scientists. The mindset that AI-based methods are powerful enough to solve any business problem and there is no need to spend much time of understanding the problem is wrong and dangerous. One of the reasons for this mistake is the lost-in-translation trap between data scientists and business experts that reduces communication efficiency. Suggestions for how to avoid this trap are given in Chap. 5.

2.6.2   Neglecting the Importance of a Detailed Realistic Problem Definition

Another common mistake is preparing a very generic and ambiguous problem definition. The missing quantitative details pave the way for a slippery terrain for the whole project development process. As a result, a lot of confusion occurs and negative energy is wasted in clarifying the missing numbers. Time spent on a detailed problem definition prevents a lot of future wasting of time on filling in the missing details. One potential cause of an incomplete and fuzzy problem definition is a lack of support by the business experts. In cases of top management push for a mass-scale implementation of new technologies without a specific justification, that is their natural response.

2.6.3   Ignoring Definition of Value Creation Metrics for the Problem

Focusing on the technical nature of the business problem and neglecting the value creation part is another common mistake. Even the most respected technical accomplishment of introducing new technology should be questioned if it doesn't deliver consistent value. Clarifying the sources of value creation and the performance metrics for solving business problems is a must in the project definition document. The best way to guarantee this important section of the problem definition is to include financial SMEs in the team from the beginning.

2.6.4   Believing in a Data First, Problems Second Approach

The worst mistake is purely ideological, based on a religious belief in the total supremacy of data. It introduces the wrong expectation that the key to defining and solving business problems is not business knowledge but data. A popular slogan in this ideology is "Data will tell us everything we don't know."

Unfortunately, many factors, such as the media, vendors, and many academics, contribute to the growing popularity of this mindset. Our philosophy is based on industrial reality, where business decisions defined by corresponding experts play the leading role in solving business problems. Data plays a critical part in solving these problems, but it is not the force that drives the business. This philosophy is the basis of the methodology proposed in this book.

2.7   Suggested Reading

- McKinsey Global Institute, *The Age of Analytics: Competing in a Data-Driven World*, 2016.
- McKinsey Global Institute, *Artificial Intelligence: The next Digital Frontier?*, 2017.
- W. Verbeke, B. Baesens, and C. Bravo, *Profit Driven Business Analytics*, Wiley, 2017.
- S. Williams and N. Williams, *The Profit Impact of Business Intelligence*, Elsevier, 2007.

2.8   Questions

Question 1
Give arguments why a real-world Data Science project should start with a business problem.

Question 2
What typical business problem is close to your business issues?

Question 3
Prepare a plan for brainstorming session to define a business problem?

Question 4
What is missing in the following problem definition?
    The objective of the project is to reduce significantly the power consumption by better control of process temperature and humidity.

Question 5
When should value creation benchmarking of a new project be done?

Chapter 3
Artificial Intelligence-Based Data Science Solutions

The biggest difficulty with mankind today is that our knowledge has increased so much faster than our wisdom.

Frank Whitmore

After clarifying the problem space as much as possible, it is time to focus our attention on the solution space. Fortunately, due to the growing impact of AI, the options in the solution space have increased significantly. On the one hand, this creates new opportunities for new application areas. On the other hand, however, it requires additional knowledge and training in the new methods. Since AI is a field of intensive research, some of these approaches, such as deep learning and cognitive computing, are still in the exploratory phase and not entirely ready for prime time. Other AI-based approaches, such as neural networks, support vector machines, decision trees, and evolutionary computation, have reached a level of maturity that has opened the door to industrial applications.

The focus of this chapter is on giving a broad overview of the potential solutions that AI-based Data Science offers to businesses. It is beyond the traditional set of machine learning and deep learning methods that currently dominates the references and the media. The first objective of the chapter is to clarify the typical Data Science solutions that businesses can use for delivering value, such as prediction, forecasting, classification, clustering, optimization and association. The second objective is to explain the principles of the following key AI technologies: neural networks, deep neural networks, support vector machines, decision trees, evolutionary computation, swarm intelligence, and intelligent agents (chatbots). The third objective is to give the reader an awareness of advanced AI-based solutions, such as natural language processing, video/image processing, sentiment analysis, and the fundamental area of artificial general intelligence. These very complex technologies are in a continuous mode of development by the AI Olympians and are not discussed in detail in the book.

© Springer Nature Switzerland AG 2020
A. K. Kordon, *Applying Data Science*,
https://doi.org/10.1007/978-3-030-36375-8_3

Fig. 3.1 Typical Data
Science solutions

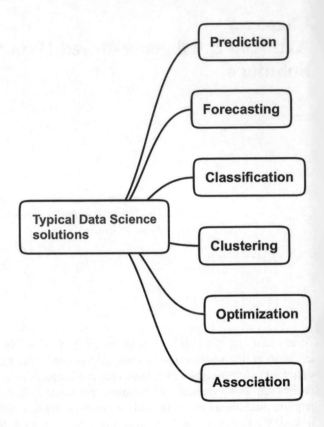

3.1   Typical Solutions Related to Data Science

The first question that needs an answer after defining and understanding the business
problem is to identify the types of solutions expected. The key types are shown in
Fig. 3.1 and discussed next.

3.1.1   Prediction

What Is Prediction?
According to Wikipedia, "predictive modeling uses statistics to predict outcomes."
Most often the event one wants to predict is in the future, but predictive modeling
can be applied to any type of unknown events, regardless of when they occurred. We
would like to broaden the definition to any modeling technique that derives relationships with acceptable predictive accuracy between what we are trying to predict (the
response, target, or dependent variable) and independent or explanatory variables
that describe the properties which we want to use as the basis for making inferences.

Deliverables of Prediction

The ideal deliverable is a numeric prediction of the target variable with low and high confidence limits based on a statistical probability level (very often 95% probability.) For example, the predicted value may be 150 with 95% probability of being between 143 and 158. For many applications, the availability of confidence limits around the prediction is a big plus and gives additional flexibility in the corresponding decision-making based on the prediction.

The prediction quality is measured by the accuracy of prediction with the most popular metric, called the coefficient of determination or r^2. In regression, r^2 is a statistical measure of how well the regression line approximates the real data points. An r^2 of 1 indicates that the regression line perfectly fits the data. The prediction quality depends on the problem, but in most business cases, $r^2 > 0.9$ is a measure of a high-quality model.

Predictive models are divided into three key groups, based on their transparency and potential interpretability: (1) black boxes, (2) gray boxes, and (3) white boxes. Black-box models are opaque to the user. Several AI-based methods, such as neural networks, deep learning, support vector machines, and random forest are in this category. The other extreme is white-box models, where all necessary information is available and the models are interpretable. A typical case is first-principles models. Gray-box models combine a partial theoretical structure with data to complete the model. Thus, almost all equation-based empirical models are gray-box models. Typical examples are statistical models, decision trees, and symbolic regression models generated by genetic programming (described later in this chapter).

Prediction Methods

A predictive mathematical model is used to generate the key deliverable—the prediction from numeric data. The majority of predictive models are based on statistical linear regression where the relationships are easily interpretable. Unfortunately, several other methods for generating predictive models, such as neural networks, support vector machines, and random forest deliver black boxes that are not interpretable. It is possible to use these methods in applications where interpretability of the derived models is not required. When interpretability is a requirement (as is the case with some financial and healthcare applications), two other methods, which generate gray-box models—decision trees and symbolic regression by genetic programming—can be used.

One issue has to be made very clear in all projects that use predictions based on data-driven models. All these models are based on correlations between the target variable and the explanatory variables. Since correlation does not imply causation, these models do not claim causality. Mathematical models that are based on causality are developed by using the first-principles laws of nature.

3.1.2   Forecasting

What Is Forecasting?
In general, forecasting can be grouped into two categories: qualitative and quantitative. Qualitative forecasts are applied when there is no data available and prediction is based only on expert judgment. Quantitative forecasts are based on time series modeling. This kind of model uses historical data and is especially efficient in forecasting some events that occur over periods of time, for example prices, sales figures, and volume of production.

A time series is a sequence that is taken successively at equal steps of time. With the use of time series, it becomes possible to imagine what will happen in the future as future events depend upon the current situation and historical patterns. It is useful to divide a time series into historical and test (holdout) periods. The model is built to make predictions on the basis of historical data and then this model is applied to the test set of observations to validate its performance. It is assumed that the model will perform in a similar way during the forecasting period. An example of division of a time series into historical, test, and forecasting periods is shown in Fig. 3.2.

Deliverables of Forecasting
The key deliverable in forecasting is the so-called Forecasting Troika: the combination of a point forecast with its confidence limits at a specific forecasting step. Examples of such a forecasting troika is given in Table 3.1, where each row

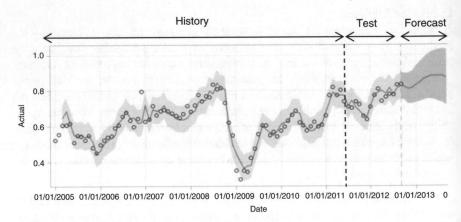

Fig. 3.2 Typical development sequence of forecasting model

Table 3.1 The Forecasting Troika

Date	Forecasts	Lower 95%	Upper 95%
3/1/2011	$0.81	$0.79	$0.87
4/1/2011	$0.80	$0.74	$0.85
5/1/2011	$0.84	$0.78	$0.90
6/1/2011	$0.86	$0.80	$0.92

shows the forecast and the confidence limits for monthly periods at gradually increasing times ahead (the first row is a month ahead forecast, the second row is for number of 2 months ahead, etc.) With increasing forecasting steps, the cone of uncertainty (the difference between the high and low confidence limits) grows, which is clearly seen in Fig. 3.2.

The most popular performance metric for forecasting models is the mean absolute percentage error (MAPE), which measures the absolute error as a percentage of the forecast. If this error is less than 10%, the forecast is accepted as good.

Forecasting Methods

A specific requirement for applying forecasting methods is to understand the underlying pattern of data ordered at a particular time. This pattern is composed of different components that collectively yield the set of observations of a time series. One of the first tasks in developing forecasting solutions is to decompose the time series into these components, the key ones of which are the trend, seasonality/cyclicality, and irregular/random components. An example of such a decomposition is given in Fig. 3.3, and details are given in Chap. 8.

The trend is a long pattern present in the time series. It produces positive, negative, linear, or nonlinear effects on the time series. It represents the variations of low frequency, and if the time series does not contain any increasing or decreasing pattern, then the time series is defined to be stationary in the mean.

The seasonal pattern represents regular fluctuations in the time series. These short-term movements occur owing to seasonal factors and the customs of people. In this case, the data shows regular and predictable changes that occur at regular intervals in the calendar. The seasonal component always contains fixed and known periods due to climate, social habits and practices, the behavior of institutions, etc.

Most of the forecasting methods used, such as exponential smoothing, regression, and auto regression integrated moving average (ARIMA), are statistically based. For example, in exponential smoothing, a forecast is given as a weighted moving average of recent time series observations. In regression, a forecast is given by a linear function of one or more explanatory variables. The most used ARIMA methods, presented by Box and Jenkins in the 1970s, give a forecast as a linear function of past observations (or differences between past observations) and the error values of the time series itself.[1] Recently, several AI-based technologies, such as neural networks, genetic programming, and deep learning recurrent networks, have begun to be used as alternative nonlinear methods for forecasting.[2]

---

[1]G.E.P. Box and G.M. Jenkins, *Time Series Analysis: Forecasting and Control*, Holden Day, 1976.
[2]Some nonlinear methods for forecasting are described in T. Rey, A. Kordon, and C. Wells, *Applying Data Mining for Forecasting Using SAS*, SAS Press, 2012.

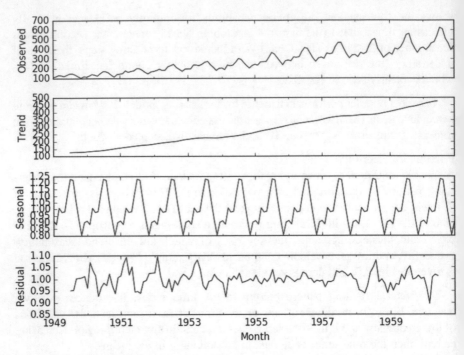

Fig. 3.3 Decomposition of time series into key components

3.1.3 Classification

What Is Classification?

Webster defines classification as "a systematic arrangement in groups or categories according to established criteria." In contrast to regression models, where the target variable is always a number, classification models have qualitative targets, also called categories. In a large number of classification problems, the targets are designed to be binary, i.e., their values can only be 0 or 1. These types of classifier are called binary classifiers. When classifiers need to classify the input into many buckets, they are called multiclass classifiers.

A popular example of a classification problem is when a bank's loan approval department wants to identify potential loan defaulters. It is expected that the classification algorithm, based on the data fed to it, will classify the loan applicants into binary buckets: (1) potential defaulters and (2) potential non-defaulters. The target variable is a binary potential default flag. If the model predicts the potential default flag to be 1, this means that the applicant is likely to default. If the model predicts it to be 0, this means that the applicant is likely not to default.

Predicted

		Yes	No
	Yes	True Positive (TP) = 12	False Negative (FN) = 8
Actual			
	No	False Positive (FP) = 23	True Negative (TN) = 57

Fig. 3.4   The confusion matrix for the bank loan approval example

Deliverables of Classification

The key deliverable is the target classification flag (in our example, the potential default flag). For example, the selected classifier might classify 35 applicants as defaulters and 65 applicants as non-defaulters.

The performance metric for classifiers, however, is not trivial and needs more explanation. First, four raw metrics, namely True Positive (TP), False Positive (FP), True Negative (TN), and False Negative (FN), are defined. Then these four raw metrics are used in a matrix called the confusion matrix (see the example in Fig. 3.4).

From these four raw metrics in the confusion matrix, the final evaluation metrics for a classifier are derived and used for performance assessment. The metrics are briefly defined below.

The accuracy measures how often the classifier is correct in classifying both true positives and true negative. Mathematically, it is defined as

$$Accuracy = (True\ Positive + True\ Negative)/Total\ Predictions$$

The sensitivity or recall measures how many times the classifier got the true positives correct. It is calculated as:

$$Recall = True\ Positive/(True\ Positive + False\ Negative).$$

The specificity measures how many times the classifier got the true negatives correct. Mathematically, it is defined as

$$Specificity = (True\ Negative)/(True\ Negative + False\ Positive).$$

The precision measures, of the total predicted to be positive, how many were actually positive. It is calculated as

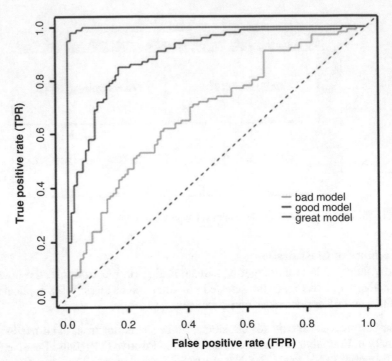

Fig. 3.5 An example of classifier comparison using the ROC (www.stats.stackexchange.com)

$$\text{Precision} = (\text{True Positive})/(\text{True Positive} + \text{False Positive}).$$

The proper selection of evaluation metrics depends on the business problem. It has to be considered, however, that one metric is insufficient for evaluation of classifier performance.

Classification Methods

There are a lot of algorithms that can be used to develop classification models. Some algorithms, such as logistic regression, are good linear classifiers. Others, such as neural networks, random forest, and support vector machines, are good nonlinear classifiers. Usually it is recommended to explore several classification methods and to compare them using the receiver operating characteristics (ROC). This is a two-dimensional plot with the false positive rate (FPR), or one minus the specificity on the x-axis and true positive rate (TPR) or sensitivity on the y-axis. (Fig. 3.5).

In the ROC plot, there is a line that measures how a random classifier will predict the TPR and FPR. This is a straight line, as the classifier has an equal probability of predicting 0 or 1. If a classifier has a better performance, then it should ideally have a higher proportion of TPR as compared with FPR. This will push the curve toward the upper left corner. An example of a comparison of three classifiers with different performance in the ROC is shown in Fig. 3.5.

3.1.4   Clustering

What Is Clustering?

Clustering is the task of dividing data points into a number of groups such that the data points in a group are more similar to other data points in the same group than to those in other groups. The key objective is to segregate groups with similar qualities and assign them to clusters.

Classification and clustering sound similar, but they are two different solutions in applied Data Science. Classification is used in supervised learning (where we have a dependent variable and labeled data), while clustering is used in unsupervised learning, where we don't have any knowledge about the dependent variable and the data is not labeled.

Deliverables of Clustering

In clustering, data instances are grouped together using the concept of "maximizing the intraclass similarity and minimizing the interclass similarity." This translates to the clustering algorithm identifying and grouping instances which are very similar, as opposed to ungrouped instances which are much less similar to one another. Since the task of clustering is subjective, the means that can be used for achieving this goal are plentiful. Every methodology follows a different set of rules for defining the similarity metric between data points. In fact, there are more than 100 clustering algorithms known, but two are applied in most business problems—hierarchical and K-means clustering.

Clustering Methods

Hierarchical clustering, as the name suggests, is an algorithm that builds a hierarchy of clusters. This algorithm starts with each of the data points assigned to a cluster of its own. Then two nearest clusters are merged into the same cluster. In the end, this algorithm terminates when there is only a single cluster left.

The results of hierarchical clustering can be shown using a dendrogram, and an example is shown in Fig. 3.6. This dendrogram can be interpreted as follows: At the bottom, we start with 25 data points, each assigned to separate clusters. The two closest clusters are then merged until we have just one cluster at the top. The height in the dendrogram at which two clusters are merged represents the distance between the two clusters in the data space.

The goal of the other clustering method, K-means clustering, is to find groups in the data, with the number of groups represented by a variable K defined by the user. The algorithm works iteratively to assign each data point to one of K groups based on the features that are provided. Data points are clustered based on feature similarity. An example of K-means clustering with three clusters is shown in Fig. 3.7.

It starts with selecting the number of clusters we want, for example three. In this case the K-means algorithm starts the process at random centers in the data, and then tries to attach the nearest points to these centers. An advantage of this algorithm is that the centroids of the K clusters can be used to label new data.

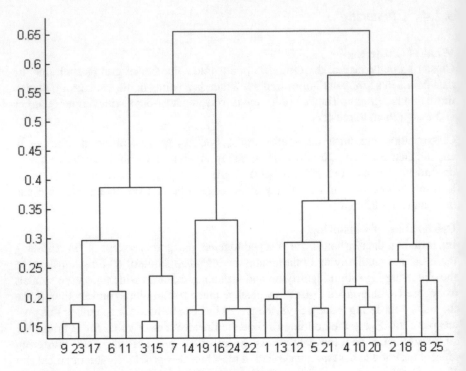

Fig. 3.6 An example of a dendrogram from hierarchical clustering

3.1.5 Optimization

What Is Optimization?

A business-focused definition of optimization can be given as "finding an alternative with the most cost-effective or highest achievable performance under the given constraints, by maximizing desired factors and minimizing undesired ones." Optimizing manufacturing processes, supply chains, or new products is of ultimate interest to businesses. Classical optimization uses numerous linear and nonlinear methods. The linear methods are based on least-squares techniques that guarantee finding analytically the global optimum. The classical nonlinear optimization techniques are based on either direct-search methods (the simplex and Hooke–Jeeves methods) or gradient-based methods (steepest descent, conjugate gradient, sequential quadratic programming, etc.), which can find local optima. An example of a smooth optimization landscape with local and global optima is shown in Fig. 3.8. Usually optimization in business settings is multicriteria and includes many constraints.

Deliverables of Optimization

The deliverable expected from optimization is an optimal or suboptimal set of parameters within the defined constraints. The main sources of profit as a result of successful optimization are as follows: (a) operating plants with the minimum waste

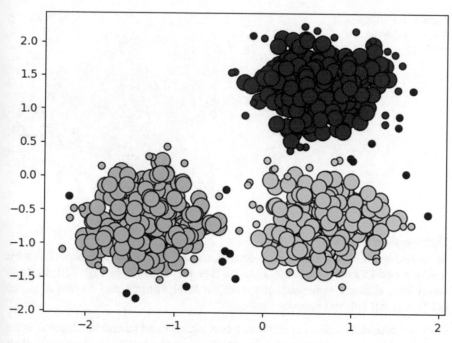

Fig. 3.7   An example of K-means clustering with $K = 3$

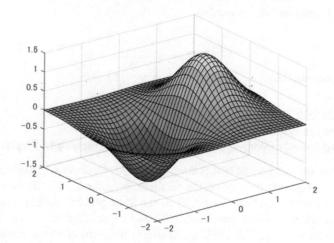

Fig. 3.8   An example of a smooth optimization landscape

of energy and raw materials, (b) distributing products with minimum transport expenses, and (c) designing new products with maximum quality and minimum material losses. An additional factor in favor of optimization is that in most cases the profit gain does not require capital investment.

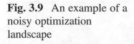

Fig. 3.9 An example of a noisy optimization landscape

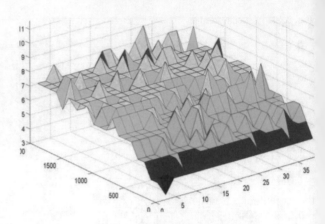

Optimization Methods

Most real-world optimization problems, however, are based on noisy data with multiple optima and complex landscapes (see the example in Fig. 3.9). In these conditions, classical optimizers get stuck in local minima and finding a global solution is difficult and time-consuming.

AI-based methods, such as evolutionary computation and swarm intelligence, offer many opportunities to handle the issues of classical optimizers. Above all, they increase the probability of finding global optima even in very noisy and high-dimensional optimization landscapes.

3.1.6   Association

What Is Association?

Associations are based on the *a priori* principle:

If a set of items is frequent, then all of its subsets must also be frequent.

It is popularly used in market basket analysis, where one checks for combinations of products that frequently co-occur in the database. In general, we write the association rule for "If a person purchases item x, then she or he purchases item y" as $x \to y$.

For example, if a person purchases milk and sugar, then she or he is likely to purchase coffee powder. This could be written in the form of an association rule as: {milk, sugar} $\to$ coffee powder. Association rules are generated after crossing a threshold for support and confidence. The support measure helps prune the number of candidate sets of items to be considered during generation of sets of frequent items.

Deliverables of Association

The expected deliverables are association rules. Association rules find all sets of items (item sets) that have support greater than the minimum support, and then use

Fig. 3.10 Association rule

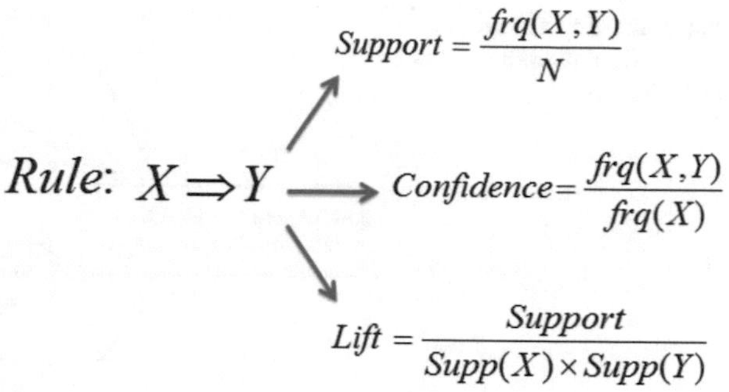

$$Support = \frac{frq(X,Y)}{N}$$

$$Rule: X \Rightarrow Y \longrightarrow Confidence = \frac{frq(X,Y)}{frq(X)}$$

$$Lift = \frac{Support}{Supp(X) \times Supp(Y)}$$

the large item sets to generate the desired rules that have confidence greater than the minimum confidence. The lift of a rule is the ratio of the observed support to that which would be expected if x and y were independent. An example of an association rule is shown in Fig. 3.10.

Association Methods

The most popular association method is called *Apriori*. It proceeds by identifying the frequent individual items in the database and extending them to larger and larger item sets as long as those item sets appear sufficiently often in the database. The frequent item sets determined by *Apriori* can be used to determine association rules which highlight general trends in the database. *Apriori* uses a "bottom-up" approach, where frequent subsets are extended one item at a time (a step known as candidate generation), and groups of candidates are tested against the data. The algorithm terminates when no further successful extensions are found.

3.2 Advanced AI Solutions Related to Data Science

In addition to the typical solutions related to Data Science that can be applied to identified problems in business settings, the AI Olympians are intensively developing several new technologies of a more complex nature. They are using these solutions in their key applications, such as search engines, intelligent assistants, and self-driving vehicles. Most of these technologies require a big investment in infrastructure and training and are works in progress. Probably in the not so distant future they will reach a high maturity level and low cost, but at this development phase they are not ready for mass-scale applications. For that reason, we'll give a very limited overview of these technologies in the book. The interested reader can find many references in the literature and on the Internet. The advanced AI solutions selected for our overview are shown in Fig. 3.11 and discussed below.

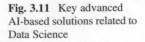

Fig. 3.11 Key advanced AI-based solutions related to Data Science

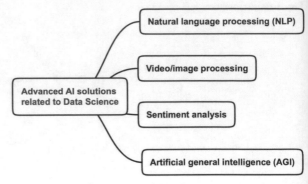

3.2.1 Natural Language Processing

Natural language processing (NLP) is the ability of machines to understand and interpret human language the way it is written or spoken. Natural language processing includes many different techniques for interpreting human language, ranging from statistical and machine learning (recently, deep learning) methods to rules-based and algorithmic approaches. This long list of approaches is needed because text-and voice-based data varies widely, as do the practical applications. In general terms, NLP breaks down language into shorter, elemental pieces, try to understand the relationships between the pieces, and explore how the pieces work together to create meaning.

Natural language generation allows the machine to then communicate results or responses in "plain English" (or any language it is designed to support). Some NLP tools simply perform translation, mapping the words in a command to a dictionary. More sophisticated applications strive for understanding: inferring meaning or intent in order to inform an appropriate action or response. Given the broad variations in dialects, colloquialisms, and individual mannerisms and the rapid evolution of new modes of communication (abbreviations, emoticons), this undertaking is not trivial.

For example, NLP makes it possible for computers to read text, hear speech, interpret it, measure sentiment, and determine which parts are important. Today's machines can analyze more language-based data than humans, without fatigue and in a consistent, unbiased way.

Deep learning, especially recurrent neural network, is having a growing impact on the development of NLP solutions.

3.2.2 Video/Image Processing

In the context of machine vision, image recognition is the capability of software to identify people, places, objects, actions, and writing in images. To achieve image recognition, computers can utilize machine vision technologies in combination with AI-driven methods and a camera.

While it is very easy for human and animal brains to recognize objects, computers have difficulty with the same task. When we look at something like a tree or a car or a cat, we usually don't have to study it consciously before we can tell what it is. However, for a computer, identifying the same objects represents a very difficult problem that is waiting for an effective solution. Recently, deep learning, especially convolution neural networks, is having a growing impact on the development of image recognition solutions.

3.2.3 Sentiment Analysis

According to Wikipedia "A basic task in sentiment analysis is classifying the polarity of a given text at the document, sentence, or feature/aspect level–whether the expressed opinion in a document, a sentence or an entity feature/aspect is positive, negative, or neutral."

Existing approaches to sentiment analysis can be grouped into three main categories: knowledge-based techniques, statistical methods, and hybrid approaches. Knowledge-based techniques classify text by affect categories based on the presence of unambiguous affect words such as "happy"," sad", "afraid", and "bored." Some knowledge bases not only list obvious affect words, but also assign arbitrary words a probable "affinity" to particular emotions. Statistical methods leverage on elements from machine learning. More sophisticated methods try to detect the holder of a sentiment (i.e., the person who maintains that affective state) and the target (i.e., the entity about which the affect is felt). To mine an opinion in context and to find the feature about which the speaker has opined, the grammatical relationships of words are used. Grammatical dependency relations are obtained by deep parsing of the text. Hybrid approaches leverage machine learning and elements of ontologies and semantic networks are used as well.

3.2.4 Artificial General Intelligence

According to Techopedia, "artificial general intelligence (AGI) refers to a type of distinguished artificial intelligence that is broad in the way that human cognitive systems are broad, that can do different kinds of tasks well, and that really simulates the breadth of the human intellect, rather than focusing on more specific or narrower types of tasks." The term is used to distinguish various types of AI from each other— the terms "strong artificial intelligence" and "full artificial intelligence" are also used to discuss broader AI goals. Therefore, "AGI" is closer to the original meaning of "AI", while very different from the current mainstream "AI research", which focuses on domain-specific and problem-specific methods.

3.3   Key AI Methods in a Nutshell

In the same way as human intelligence uses many actions, ideas, and plans to respond reasonably to a changing environment, AI is using many different approaches to imitate humans. These methods are based on diverse scientific bases. The majority of them are founded on the broad field of machine learning, which enables computers to learn by example, analogy, and experience. As a result, the overall performance of a system improves over time and the system adapts to changes. A broad range of computer-driven learning techniques, such as neural networks, deep learning, and random forest have been defined and explored.

A special case of machine learning methods is based on a solid mathematical theory, called statistical learning theory. This gives a theoretical basis for extracting patterns and relationships from a few data points. It is very important in practical applications to have the capability to learn from a small number of data samples. At the core of this learning machine is the balance between performance and complexity of the solutions derived.

Another set of AI methods is driven by bio-inspired computing. In this category are the broad areas of evolutionary computation and swarm intelligence. Evolutionary computation represents in various forms the three major processes in the theory of natural evolution: (1) the existence of a population of individuals that reproduce with inheritance; (2) adaptation to environmental changes by variations in different properties of individuals; and (3) natural selection, where the "best and the brightest" individuals survive and the losers perish. At the center of evolutionary computation is the hypothesis that simulated evolution in a computer will generate individuals which will improve their performance in a similar way to that in which biological species did in millions of years of natural evolution.

Swarm intelligence is based on the social life of different biological species, such as insects, birds, fish, and humans. One way is to reproduce in computer algorithms the collective behavior of some specific forms of social coordination, such as finding food or handling dead bodies in ant colonies. Another way of exploring social interactions is by representing the abilities of humans to process knowledge through sharing information. In both cases, unique algorithms for complex optimization have been developed using even very simple models of collective behavior of both nonhuman species and humans.

From the diverse set of AI-based approaches, we have selected seven: neural networks, deep learning networks, support vector machines, decision trees/random forests, evolutionary computation, swarm intelligence, and intelligent agents (chatbots). The list of technologies with the corresponding icons that will be used in the book is shown in Fig. 3.12.

The selection criteria for AI-based methods are value creation potential, application capacity, implementation availability, algorithmic maturity, and training efforts. For example, neural networks have proven their value creation potential in numerous profitable applications in industry. Their capacity to create predictive, classification, and forecasting models has been demonstrated in various business solutions. Neural

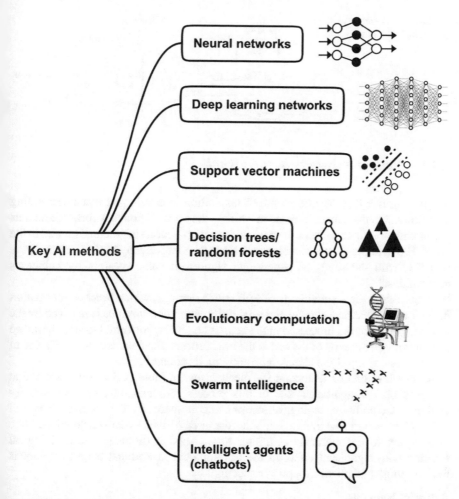

Fig. 3.12 Key AI methods discussed in the book

networks are part of almost every software product related to modeling, analytics, or machine learning. It is a mature technology with many architectures that is user-friendly and has a short learning curve. Similar arguments are valid for the other selected methods.

Each method will be briefly described in this section with an emphasis on its key principles of operation. The advantages and weaknesses, as well as, their applicability, are discussed in Chap. 4.

Fig. 3.13 A schematic view
of a biological neuron

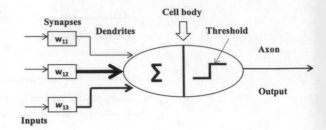

3.3.1   Neural Networks in a Nutshell

It is no surprise that the first source of inspiration in developing machine learning algorithms was the human learning machine—the brain. Neural networks are models that attempt to mimic some of the basic information-processing methods found in the brain. The field has grown tremendously since the first paper by McCulloch and Pitts in 1943, with thousands of papers and significant value creation by numerous applications.[3]

In our brain, there are billions of cells called neurons, which process information in the form of electrical signals. External information or a stimulus is received by the dendrites of a neuron, processed in the neuron cell body, converted to an output, and passed through the axon of the cell to the next neuron. The next neuron can choose to either accept or reject the signal depending on its strength.

Research efforts on analyzing the mechanisms of biological learning show that synapses play a significant role in this process. For example, synaptic activity facilitates the ability of biological neurons to communicate. Thus, a high degree of activity between two neurons at one time, captured by their synapses, could facilitate their ability to communicate at a later time. Another discovery from biological learning with significant effect in the development of artificial neural network is that learning strengthens synaptic connections.

Artificial Neurons
A schematic view of the structure of a biological neuron is shown in Fig. 3.13, where the synaptic strength is represented by the connection weights w_{ij} and visualized by the widths of the arrows. This schematic representation is the basis of the definition of the structure of an artificial neuron, shown in Fig. 3.14.

An artificial neuron, or processing element, emulates the axons and dendrites of its biological counterpart by connections and emulates the synapses by assigning a certain weight or strength to these connections. A processing element has many inputs and one output. Each of the inputs x_j is multiplied by the corresponding weights w_{ij}. The sum of these weighted inputs, u_k, is then used as the input to an activation function $f(u_k)$, which generates the output y_k for that specific processing

---

[3]W. McCulloch and W. Pitts, A logical calculus of ideas immanent in nervous activity, *Bulletin of Mathematical BioPhysics*, **5**, p. 115, 1943.

Fig. 3.14 Structure of an
artificial neuron

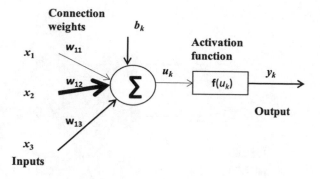

element. The bias term *b* also called a "threshold term," serves as a zero adjustment
for the overall inputs to the artificial neuron. Excitatory and inhibitory connections
are represented by positive and negative connection weights, respectively.

Artificial neurons use an activation function to calculate their activation level as a
function of the total inputs. The feature that is most distinguished between the
existing artificial neurons is based on the activation function selected. The main
popular options are based on a sigmoid function, a radial basis function and recently,
a rectified linear unit function.

Artificial Neural Network

An artificial neural network consists of a collection of processing elements, organized in a network structure. The output of one processing element can be the input
to another processing element. Numerous ways to connect artificial neurons into
different network architectures exist. The most popular and widely applied neural
network structure is the multilayer perceptron. It consists of three types of layer
(input, hidden, and output), shown in Fig. 3.15.

The input layer connects the signals in the incoming pattern and distributes these
signals to the hidden layer. The hidden layer is the key part of the neural network
since its neurons capture the features hidden in the input pattern. The learned patterns
are represented by the weights of the neurons and visualized in Fig. 3.15 by the
widths of the arrows. These weights are used by the output layer to calculate the
prediction. Artificial neural networks with this architecture may have several hidden
layers with different numbers of processing elements, as well as outputs in the output
layer. The majority of applied neural networks use only three layers and are similar
to the network shown in Fig. 3.15.

This type of neural network is a universal approximator (it fits any input to any
output) and can be used to build general-purpose models for both classification and
regression. For example, when a feedforward network is used for classification, the
number of neurons in the output layer is equal to the number of classes. Conceptu-
ally, the output neuron that fires determines the class that the network has predicted.

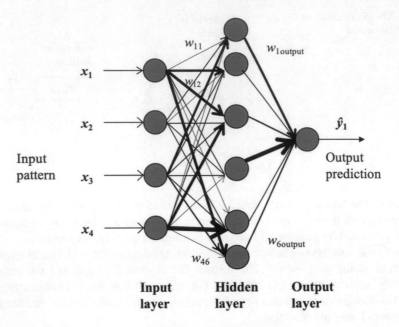

Fig. 3.15 Structure of a multilayer perceptron with one hidden layer

Back Propagation
The key function of neural network learning is based on representing the desired behavior by adjusting the connection weights. Back-propagation, developed by Paul Werbos in the 1970s, is the most popular of the known methods for neural network-based machine learning.[4] The principle behind back-propagation is very simple and is described next.

The performance of a neural network is usually measured by some metric of the error between the desired and the actual output at any given instant. The goal of the back-propagation algorithm is to minimize this error metric in the following way.

First, the weights of the neural network are initialized, usually with random values. The next step is to calculate the outputs y and the associated error metric for that set of weights. Then the derivatives of the error metric are calculated with respect to each of the weights. If increasing a weight results in a larger error then the weight is adjusted by reducing its value. Alternatively, if increasing a weight results in a smaller error then this weight is corrected by increasing its value. Since there are two sets of weights, a set corresponding to the connections between the input and the hidden layer (w_{ij} in Fig. 3.15), and another set between the output and the hidden layer ($w_{ioutput}$ in Fig. 3.15), the error calculated at the output layer is propagated backwards in order to also adjust the first set of weights—hence the name

---

[4]P. Werbos, *Beyond Regression: New Tools for Prediction and Analysis in the Behavioral Sciences*, Ph.D. Thesis, Harvard University, Cambridge, MA, 1974.

Fig. 3.16 Key benefits of
neural networks

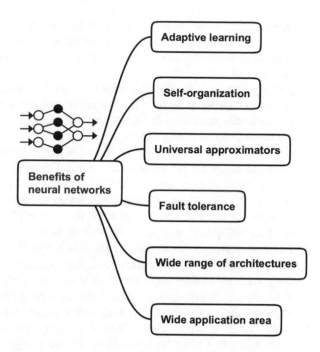

back-propagation. After all the weights have been adjusted in this manner, the whole
process starts over again and continues until the error metric meets some
predetermined threshold.

The basic back-propagation algorithm generally leads to a minimum error.
Problems can occur, however, when local minima are present (which is generally
the case). Another problem with the back-propagation algorithm is its slow speed.
Thousands of iterations might be necessary to reach the final set of weights that
minimize the selected error metric.

Fortunately, these well-known flaws of back-propagation can be overcome by
adding tuning parameters, such as a momentum and a learning rate. The momentum
term causes the weight adjustment to continue in the same general direction. This, in
combination with different learning rates, speeds up the learning process and
minimizes the risk of being entrapped in local minima. As a result, the error back-
propagation process is more efficient and under control. It has to be considered,
however, that biological neurons do not work backwards to adjust the strength of
their synapses, i.e., back-propagation is not a process that emulates the brain activity.

Benefits of Neural Networks

The key benefits of neural networks are captured in the mindmap in Fig. 3.16 and
discussed below:

- *Adaptive learning*. Neural networks have the capability to learn how to perform
  certain tasks and adapt themselves by changing the network parameters in a

surrounding environment. The requirements for successful adaptive learning are as follows: choosing an appropriate architecture, selecting an effective learning algorithm, and supporting model building with representative training, validation, and test data sets.

- *Self-organization.* In the case of unsupervised learning, a neural network is capable of creating its own representations of the available data. The data is automatically structured in clusters by the learning algorithm. As a result, new unknown patterns can be discovered.

- *Universal approximators.* As we have already mentioned, neural networks can represent nonlinear behavior to any desired degree of accuracy. One practical advantage of this property is that they offer a fast test of the hypothesis of a possible functional input–output dependence in a given data set. If it is impossible to build a model with a neural network, there is very little hope that such a model can be developed by any other data-driven method.

- *Fault tolerance.* Since the information in a neural network is distributed over many processing elements (neurons), the overall performance of the network does not degrade drastically when the information in some node is lost or some connections are damaged. In such a case the neural network will repair itself by adjusting the connection weights according to new data.

- *Wide range of architectures.* Neural networks are one of the few modeling approaches that offer high flexibility in the design of different architectures using their basic component—the neuron. In combination with the variety of learning algorithms, such as back-propagation, competitive learning, Hebbian learning, and Hopfield learning, the possibilities for potential solutions are almost infinite. The key architectures from an application point of view are multilayer perceptrons (which deliver nonlinear functional relationships), self-organizing maps (which generate new patterns), and recurrent networks (which capture system dynamics).

- *Wide application areas.* The unique learning capabilities of neural networks combined with the wide range of available architectures allow very broad application opportunities. The main application areas are in building nonlinear empirical models, classification, finding new clusters in data, and forecasting. The application topics vary from process monitoring and control of complex manufacturing plants to online fraud detection in almost any big bank.

Key Messages about Neural Networks

Artificial neural networks are based on the structure and activities of the human brain.

Neural networks learn from examples from historical data or self-organize the data and automatically build empirical models for prediction, classification and forecasting.

Neural networks interpolate well but extrapolate badly outside the known operating conditions.

3.3.2 Deep Learning Networks in a Nutshell

Recently another more complex type of neural network, called deep learning networks, has been gaining momentum in academic research as well as in potential business applications. The fundamental difference between classical machine learning and deep learning is that the latter learns how to learn. The key principles of deep learning, the key structures of deep learning networks, and the benefits of this approach are discussed briefly below.

Principles of Deep Learning

A deep learning neural network can be seen as a set of feature extraction layers with a classification layer on top. The power of deep learning is not in its classification skills, but rather in its feature extraction capabilities. Feature extraction is automatic (without human intervention) and multi layered. Another description of deep learning is as an algorithm that "learns in layers", i.e., it builds a hierarchy of complex concepts out of simpler ones.

We'll illustrate the principles of deep learning with image data. Computers cannot understand the meaning of an image as a collection of pixels. Mappings from a collection of pixels to a complex object are complicated. With deep learning, the problem is broken down into a series of hierarchical mappings, with each mapping described by a specific layer.

The input (representing the variables we actually observe) is presented in pixels at the visible layer. Then a series of hidden layers extracts increasingly abstract features from the input with each layer representing a specific mapping. However, this process is not predefined, i.e., we do not specify what is selected in the layers. For example, from the pixels, the first hidden layer identifies the edges. From the edges, the second hidden layer identifies the corners and contours. From the corners and contours, the third hidden layer identifies parts of objects. Finally, from the parts of objects, the fourth hidden layer identifies whole objects. An example of this principle is illustrated in Fig. 3.17 .[5]

It is possible to implement deep learning systems now because of three reasons: high CPU power (especially with GPU), better algorithms, and the availability of more labeled data.

The two most popular architectures of deep learning neural networks that implement the deep learning algorithms, convolutional neural networks and recurrent neural networks, are described below.

Convolutional Neural Networks

A convolutional neural network (CNN) consists of several specialized layers, the most important of which are the convolution, pooling, and dropout layers. The

---

[5]Based on H. Lee et al., *Convolutional Deep Belief Networks for Scalable Unsupervised Learning of Hierarchical Representations*, ICML, Montreal, 2009.

Fig. 3.17 Hierarchy of features hierarchy based on deep learning

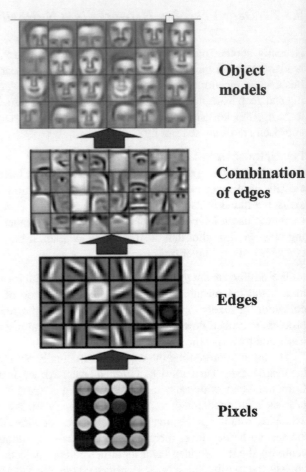

Object models

Combination of edges

Edges

Pixels

convolution layer puts the input images through a set of convolutional filters, each of which activates certain features in the images. The pooling layer simplifies the output by performing nonlinear down sampling, reducing the number of parameters that the network needs to learn about. The drop-out layer randomly drops units along with their connections during training. This helps the network to learn more robust features by reducing complex co-adaptations of units, and alleviates the overfitting issue as well.

CNNs are at the heart of all types of image and video recognition, such as facial recognition, image tagging and recognizing a stop sign from a pedestrian in self-driving cars. They are extremely complex and difficult to train, and while you don't need to specify specific features (a cat has fur, a tail, four legs, etc.), you do need to show a CNN literally millions of examples of cats before it can be successful. The large amount of training data is a huge barrier to mass-scale applications, however.

Fig. 3.18 Structure of an
Elman recurrent neural
network

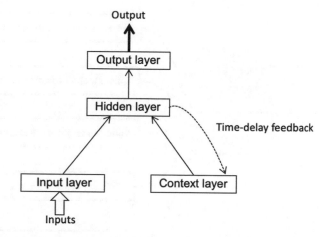

Recurrent Neural Networks

Recurrent neural networks (RNNs) are a type of neural net that maintains an internal memory of the inputs that it has seen before, so it can learn about time-dependent structures in streams of data.

Whereas other types of neural network assume that the inputs are independent of one another, RNNs deal specifically with sequences of data where each data item clearly does have a relationship to the ones before and after. So RNNs learn on any data which can be presented as a sequence.

One key requirement for capturing process dynamics is the availability of short-term and long-term memory. Long-term memory is captured in the neural network by the weights of the static neural network while the short-term memory is represented by using time delays in the input layer of the neural network. In this configuration, the input layer is divided into two parts: one part that receives the regular input signals (which could be delayed in time) and another part made up of context units, which receive as inputs the activation signals of the hidden layer from the previous time step.

A common feature in most recurrent neural networks is the appearance of a so-called context layer, which provides the recurrence. At the current time t, the context units receive some signals resulting from the state of the network at the previous time sample $t-1$. The state of the neural network at a certain moment in time thus depends on previous states due to the ability of the context units to remember certain aspects of the past. As a result, the network can recognize time sequences and capture process dynamics. An example of one of the most popular architectures of a recurrent neural network, the Elman network, is shown in Fig. 3.18.

Recently a new, complex RNN structure, called the long short-term memory (LSTM) network, has been gaining popularity. One common LSTM RNN is composed of a cell, an input gate, an output gate, and a forget gate. The cell is responsible for "remembering" values over arbitrary time intervals, hence the word "memory" in

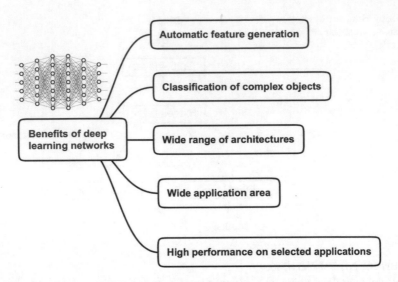

Fig. 3.19 Key benefits of deep learning networks

the name. The expression "long short-term" refers to the fact that LSTM is a model for a short-term memory which can last for a long period of time. An LSTM RNN is well suited to handling time series data, even in the case of time lags of unknown size and unknown duration between important events.

RNNs are at the center of NLP. Unlike CNNs, they process information as a time series in which each piece of data relies in some way on the piece that came before. It may not be obvious, but speech falls into this category since the next character or the next word is logically related to the preceding one. RNNs can work at the character, word, or even long paragraph level, which makes them perfect for providing the expected long-form response to a customer service question. RNNs handle both the understanding of the text of the question and the formulation of complex responses including translation into foreign languages.

Benefits of Deep Learning Networks
The key benefits of deep learning networks are captured in the mindmap in Fig. 3.19 and discussed below:

- *Automatic feature generation.* One of the main benefits of deep learning over various machine learning algorithms is its ability to generate new features from limited series of features located in the training data set. The problem is, however, that most of these automatically created features are very difficult to interpret.
- *Classification of complex objects.* A key benefit of deep learning networks is their capability to correctly classify various complex objects, such as video, audio, images, and text. In fact, this ability opened the door to NLP and image processing in the real world.

- *Wide range of architectures.* Another key benefit of deep learning algorithm is that they allow various structures to be added, some of them with high complexity and tailored to a specific problem. The two key types of structures, CNNs and RNNs, however, are standard for the key application areas of image processing and NLP.
- *Wide application area.* The key reason for the growing interest in and funding for deep learning networks is their impressive application potential. The key target application area is self-driving vehicles, but many other business applications, in face recognition, cybersecurity, healthcare, marketing, automatic translation, etc., will use various architectures of this technology.
- *High performance on selected applications.* With progress in the technology, the performance of specifically designed deep learning networks has increased significantly and is either close to that of or better than humans, especially in the area of image classification. An additional advantage over humans is that this high performance can be kept up continuously.

Key Messages about Deep Learning Networks
Deep learning networks learn how to learn.
Deep learning networks create complex features automatically.
Deep learning networks are the eyes, ears, and voice of AI.

3.3.3 Support Vector Machines in a Nutshell

An implicit assumption in developing and using neural networks is the availability of sufficient data to satisfy the "hunger" of the back-propagation learning algorithm. Unfortunately, in many practical situations we don't have this luxury and here is the place for another machine learning method, called support vector machines (SVMs). It is based on the solid theoretical foundation of statistical learning theory, which can handle effectively statistical estimation with small samples. The initial theoretical work of two Russian scientists, Vapnik and Chervonenkis in the late 1960s, grew into a complete theoretical framework in the 1990s[6] and has triggered numerous real-world applications since the early 2000s.

We need to warn nontechnical readers that understanding SVMs is challenging and requires a mathematical background.

Support Vector Machines for Classification
The key notion of SVMs is the support vector. In classification, the support vectors are the vectors that lie on the margin (i.e., the largest distance between the two closest vectors to either side of a hyperplane), and they represent the input data points that are the most difficult to classify. Above all, the generalization ability of the optimal separating hyperplane is directly related to the number of support

---

[6]V. Vapnik, *Statistical Learning Theory*, Wiley, 1998.

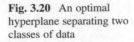
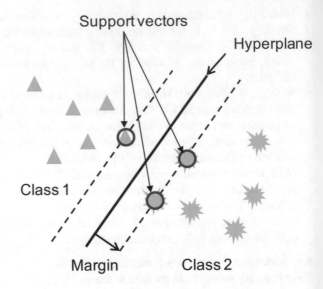

Fig. 3.20 An optimal hyperplane separating two classes of data

vectors, i.e. the separating hyperplane with minimum complexity has the maximum margin. From a mathematical point of view, the support vectors correspond to the positive Lagrangian multipliers of the solution of a quadratic programming optimization problem. Only these vectors and their weights are then used to define the decision rule of the model. Therefore, this learning machine is called a support vector machine. It is important to emphasize that, in contrast to neural networks, the support vectors defined are derived by global optimization and there is no issue of generating inefficient solutions due to local minima.

An example of some support vectors for the classification of two distinct classes of data is shown in Fig. 3.20. The data points from the two classes are represented by stars and triangles. The support vectors that define the hyperplane, or the decision function with the maximum margin of class separation, are encircled. It is clear from Fig. 3.20 that these three support vectors are the critical data points that contain the significant information required to define the hyperplane and separate reliably the two classes. The rest of the data points are irrelevant for the purpose of class separation even if their number is huge.

There is a social analogy that may help in understanding the notion of support vectors.[7] The political realities in the USA are such that there are hard-core supporters of both dominant political parties—the Democratic Party and the Republican Party—who don't change their backing easily. During presidential elections, approximately 40% of the electorate vote for any Democratic candidate and approximately the same percentage vote for any Republican candidate, independently of

---

[7]V. Cherkassky and F. Mulier, *Learning from Data: Concepts, Theory, and Methods*, second edition, Wiley, 2007.

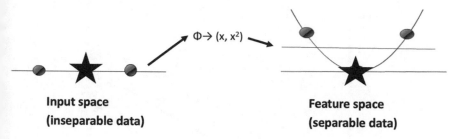

Input space
(inseparable data)

Feature space
(separable data)

Fig. 3.21 Mapping the input space into a high-dimensional feature space

her or his platform. The result from the election depends on the 20% of the undecided voters. These voters are the analogy of social "support vectors."

In general, SVMs use a so-called "kernel trick" to map the input data into a higher-dimensional feature space (see the example in Fig. 3.21). The logic behind this operation is the possibility of a linear solution in the feature space, which can solve the difficult nonlinear classification problem in the input space. For example, the data in the input space in Fig. 3.21 is inseparable. The mapping can be done nonlinearly, and the transformation function is chosen a priori. Usually it is a mathematical function, called the kernel (shown as Φ in Fig. 3.21), that satisfies some mathematical conditions given in the SVM literature, and the selection is problem-specific. By application of the selected kernel Φ, the transformed data in the high-dimensional space can be separated, as is illustrated in Fig. 3.21.

Support Vector Machines for Regression

SVMs were initially developed and successfully applied for classification. The same approach has been used for developing regression models. In contrast to finding the maximum margin of a hyperplane in a classification, a specific loss function with an insensitive zone (called an ε-insensitive zone) is optimized in SVMs for regression. The objective is to fit a tube of radius ε to the data (Fig. 3.22).

It is assumed that any approximation error smaller than ε is due to noise. It is accepted that the method is insensitive to errors inside this zone (visualized by the tube). The data points (vectors) that have approximation errors outside but close to the insensitive zone are the most difficult to predict. Usually the optimization algorithm picks them as support vectors. Therefore, the support vectors in the regression case are those vectors that lie outside the insensitive zone, including the outliers, and as such contain the most important information in the data.

One impressive feature of regression models based on SVMs is their improved extrapolation capabilities relative to those of neural networks. The best effect is achieved by combining the advantages of global and local kernels. It is observed that a global kernel (such as a polynomial kernel) shows better extrapolation ability at lower orders, but requires higher orders for good interpolation. On the other hand, a local kernel (such as a radial basis function kernel) has good interpolation abilities but fails to provide longer-range extrapolation.

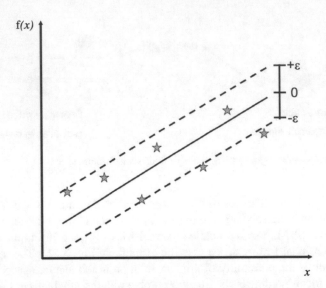

Fig. 3.22 The ε-insensitive zone for support vector regression

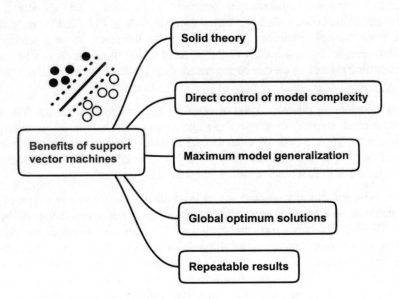

Fig. 3.23 Key benefits of support vector machines

Benefits of Support Vector Machines

The key benefits of SVMs are represented in the mindmap in Fig. 3.23 and discussed below:

- *Solid theory*. One of the very important benefits of SVMs is that the solid theoretical basis of statistical learning theory is closer to the reality of practical applications. Modern problems are high-dimensional, and the linear paradigm needs an exponentially increasing number of terms with the increased dimensionality of the inputs (effect of the so-called "curse of dimensionality"). As a result, the high-dimensional spaces are terrifyingly empty, and only the capability of learning from sparse data offered by statistical learning theory (on which SVMs are based) can derive a viable data-driven solution.

- *Direct control of model complexity.* SVMs provide a method to control complexity independently of the dimensionality of the problem. As was clearly shown in the classification example in Fig. 3.20, only three support vectors (i.e., the right complexity) are sufficient to separate two classes with the maximum margin. In the same way, the modeler has full control over the model complexity of any classification or regression solutions by selecting the relevant number of support vectors. Obviously, the best solution is based on the minimum number of support vectors.

- *Maximization of generalization ability.* In contrast to neural networks, where the weights are adjusted only by using the capability to interpolate training data, SVMs are designed by maximizing the generalization ability. As a result, the performance of the derived models has the best possible generalization ability, i.e., they do not deteriorate dangerously under new operating conditions like neural network models.

- *No local minima.* SVMs are derived based on global optimization methods, such as linear programming or quadratic programming and do not become stuck in local minima, as do the typical back-propagation-based neural networks, and this is a significant advantage.

- *Repeatable results.* The development of SVM models is not based on random initialization, as is the case with most neural network learning algorithms. This guarantees reproducible results.

Key Messages about Support Vector Machines

Support vector machines are based on statistical learning theory, which is a generalization of classical statistics in the case of finite data.

Support vector machine models capture the most informative data in a given data set, called support vectors.

Support vector machine models have optimal complexity and improved generalization ability under new operating conditions, which reduces maintenance cost.

3.3.4 Decision Trees in a Nutshell

The idea behind decision trees is to search for a pair consisting of a variable and a value within the training set and split the set in such a way that this will generate the "best" two child subsets. The goal is to create branches and leaves based on an optimal splitting criterion, a process called tree growing. Specifically, at every branch or node, a conditional statement classifies a data point based on a fixed

Fig. 3.24 An example of a decision tree

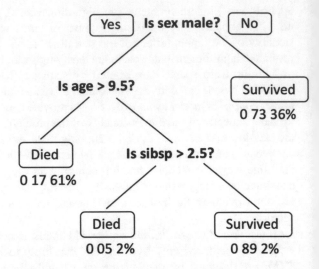

threshold for a specific variable, therefore splitting the data. To make predictions, every new instance starts at the root node (the top of the tree) and moves along the branches until it reaches a leaf node where no further branching is possible.

Decision trees have a high risk of overfitting the training data, as well as of becoming computationally complex if they are not constrained and regularized properly during the growing stage. Therefore, in order to deal with this problem, ensemble learning is used and an ensemble-based approach, called the random forest method has been developed.

Decision Trees

Decision trees are recursive, top-down, divide-and-conquer classifiers. They perform two main tasks: tree induction and tree pruning. Tree induction takes a set of pre classified instances as input, decides which attributes are the best to split on, split the data set, and repeating on the resulting split datasets until all training instances have been categorized. While building the tree, the goal is to split on the attributes which create the purest child nodes possible, which should keep to a minimum the number of splits that need to be made in order to classify all instances in the data set.

A decision tree model generated by induction only can be overly complex, contain unnecessary structure, and be difficult to interpret. Tree pruning is the process of removing the unnecessary structure from a decision tree in order to make it more efficient, more easily readable for humans, and more accurate as well. This increased accuracy is due to the ability of pruning to reduce overfitting.

Models obtained from decision trees are fairly intuitive and easy to explain to businesses. Probability scores are not a direct result but you can use class probabilities assigned to terminal nodes instead. An example of a simple decision tree showing the survival chances of *Titanic* passengers depending on their sex, age, and number of spouses and siblings on board (sibsp) is shown in Fig. 3.24.

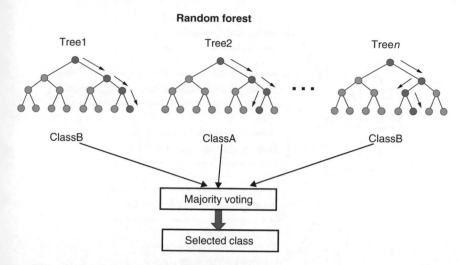

Fig. 3.25 A simplified random forest structure

The numbers under the leaves show the probability of survival and the percentage of observations in the leaf. The decision tree is easily interpreted. Obviously, based on the available data, married females and married males with siblings had a higher probability of survival.

The biggest problem associated with decision trees is that they are a highly biased class of models. You can develop a decision tree model on a training data set that might outperform all other algorithms but it might prove to be a poor predictor on test data. Often this issue cannot be resolved by pruning and cross- validation. A new machine learning algorithm, the random forest algorithm, based on an ensemble of decision trees, delivers the needed solution.

Random Forests

An ensemble method or ensemble learning algorithm, consists of aggregating multiple outputs from a diverse set of predictors to obtain better results. Formally, based on a set of "weak" learners, we are trying to use a "strong" learner for our model. Therefore, the purpose of using ensemble methods is to average out the outcomes of individual predictions by diversifying the set of predictors, thus lowering the variance, to arrive at a powerful prediction model that reduces overfitting of the training set.

In this specific case, the new algorithm, the random forest (a strong learner), is built as an ensemble of decision trees (weak learners) that perform different tasks, such as prediction and classification. The random forest grows many classification trees. Each tree gives a classification, i.e., the tree "votes" for the corresponding class. The forest chooses the classification that has the most votes (over all the trees in the forest). A simplified random forest structure is shown in Fig. 3.25.

Soon after its discovery by Leo Breiman and Adele Cutler in the early 2000s, the random forest method has grown in popularity and has consistently demonstrated

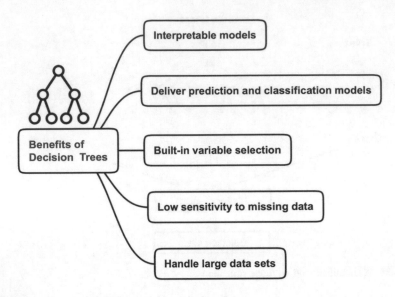

Fig. 3.26 Key benefits of decision trees and random forests

one of the highest classification and prediction accuracies among machine learning algorithms. It runs efficiently on large databases and can handle thousands of input variables without deletion of variables. It gives estimates of what variables are important in the classification and handles variables with missing data.

Unfortunately, random forest models are not interpretable.

Benefits of Decision Trees/Random Forests

The key benefits of decision trees and random forests are represented in the mind map in Fig. 3.26 and discussed below:

- *Interpretable models.* One of the very important benefits of decision trees is their easy interpretability. However, this advantage is limited to relatively simple trees and requires fine tree pruning. As we have discussed, this benefit is lost for random forest models.
- *Delivery of prediction and classification models.* Both decision trees and random forests can be used for developing models that can predict or classify data in a way dependent on an identified business problem. The performance metrics in the two cases are different, however.
- *Built-in variable selection.* The random forest method is one of the few methods that ranks the contribution of the variables to the developed model. This capability is very important in the case of nonlinear variable selection from among thousands of variables.
- *Low sensitivity to missing data.* The random forest method includes an effective method for estimating missing data, and maintains accuracy when a large

proportion of the data is missing. It has methods for balancing errors in class-population-unbalanced data sets.
- *Ability to handle large data sets.* The random forest method can build models from a large number of variables.

Key Messages for Decision Trees/Random Forests
Decision trees generate models as a tree structure of important factors of an identified problem.

A random forest is an ensemble of decision trees with randomly selected structures.

Random forests can generate high-quality predictive or classification models from a large number of variables.

3.3.5   Evolutionary Computation in a Nutshell

Evolutionary computation is a clear result of Darwin's dangerous idea that *evolution* can be understood and represented in an abstract, common terminology as an *algorithmic process*. It uses an analogy with natural evolution to perform a search by *evolving* solutions (equations, electronic circuit schemes, mechanical components, etc.) to problems in the virtual environment of computers. One of the important features of evolutionary computation is that instead of working with one solution at a time in the search space, a large collection, or population, of solutions is considered at once. The better solutions are allowed to "have children" and the worse solutions are quickly eliminated. The "child solutions" inherit their "parents' characteristics" with some random variation, and then the better of these solutions are allowed to "have children" themselves, while the worse ones "die," and so on. This simple procedure causes simulated evolution in the computer. After a number of generations, the computer will contain solutions which are substantially better than their long-dead ancestors that were there at the start. Two evolutionary computation approaches, genetic algorithms (GAs) and genetic programming (GP), have reached a level of maturity and have been applied to many industrial problems.

Genetic Algorithms
A key feature of genetic algorithms is their use of two separate spaces: the search space and the solution space. The search space is a space of coded solutions to the specific problem, and the solution space is the space of the actual solutions. Coded solutions, or genotypes, must be mapped onto actual solutions, or phenotypes, before the quality or fitness of each solution can be evaluated. An example of a mapping in a GA is shown in Fig. 3.27, where the phenotypes are different chairs. The genotypes consist of a sequence of 1 s and 0 s that code the parameters required to generate or describe an individual phenotype.

One of the requirements for a successful GA implementation is the effective definition of the genotype, which depends on the business problem. For example, if

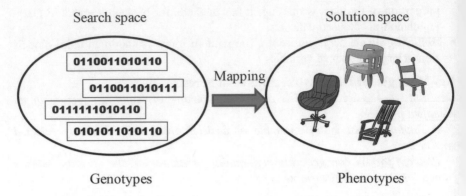

Fig. 3.27 Mapping genotypes in the search space to phenotypes in the solution space

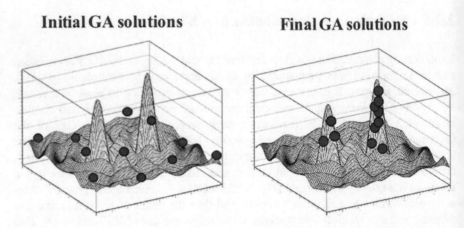

Fig. 3.28 A multimodal fitness landscape, with GA solutions at the start and at the end of simulated evolution

the objective of the GA search is to find a chair with the minimum amount of wood used and the maximum strength, the genotype coding may include the geometrical parameters of the chair, the type of material in each part, and the strength of each material.

In the language of genetic algorithms, a search for a good solution to a problem is equivalent to a search for a particular binary string or chromosome. The universe of all admissible chromosomes can be considered as a fitness landscape. An example of a fitness landscape with multiple optima is shown in Fig. 3.28.

Conventional optimization techniques for exploring such kinds of multimodal landscape will invariably get stuck in a local optimum. Genetic algorithms approach this problem by using a multitude of potential solutions (chromosomes) that sample the search space in many places simultaneously. An example of an initial random distribution of the potential solutions in such a landscape is shown in the left part of Fig. 3.28, where each chromosome is represented by a dark dot. At each step of the

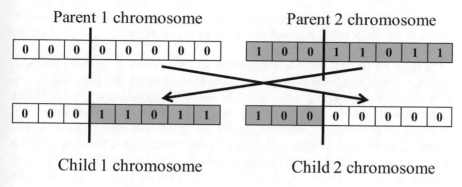

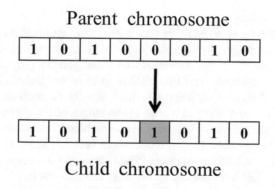

Fig. 3.29 Crossover operator between two GA chromosomes

Fig. 3.30 Mutation operator on a GA chromosome

simulated evolution, the genetic algorithm directs the exploration of the search space into areas with higher elevation than the current position.

This remarkable ability of genetic algorithms to dedicate the most resources to the most promising areas is a direct consequence of the recombination of chromosomes containing partial solutions. First, in successive generations, all strings in the population are evaluated to determine their fitness. Second, the strings with a higher ranking are allowed to reproduce by crossover, i.e., portions of two parent strings are exchanged to produce two offspring. An example of the operation of a crossover operator between two GA chromosomes is shown in Fig. 3.29, where the selected sections of the parent's chromosomes have been swapped in the offspring. These offspring are then used to replace the lower- ranking strings in the population, maintaining a constant size of the overall population.

Third, mutations modify a small fraction of the chromosomes. An example of the operation of a mutation operator on a GA bit string is shown in Fig. 3.30. In a simple GA, mutation is not an important operator, but is mainly used to keep diversity and safeguard the algorithm from running into a uniform population incapable of further evolution.

As a result of the key genetic operations—crossover and mutation—the average fitness of the whole population and above all, the fitness of the best species gradually increase during simulated evolution.

In the final phases, the GA concentrates the population of potential solutions around the most significant optima (see the right plot in Fig. 3.28). However, there is no guarantee that the global optimum will be found. The good news is that according to the *schema theorem*,[8] under general conditions, in the presence of differential selection, crossover and mutation, any bit string solution that provides above-average fitness will grow exponentially. Or, in other words, good bit string solutions (called schemata) will appear more frequently in the population as the genetic search progresses. The "good guys" grab more and more attention and gain influence with the progression of the simulated evolution. As a result, the chance of finding optimal GA solutions increases with time.

Genetic Programming

Genetic programming is an evolutionary computing approach to generating soft structures, such as computer programs, algebraic equations, and electronic circuit schemes. There are three key differences between genetic programming and the classical genetic algorithm. The first difference is in the solution representation. While GAs use strings to represent solutions, the forms evolved by GP are tree structures. The second difference relates to the length of the representation. While standard GAs are based on a fixed-length representation (a chromosome), GP trees can vary in length and size. The third difference between standard GAs and GP is based on the type of alphabets they use. While standard GAs use a binary coding to form a bit string, GP uses coding of varying size and content depending on the solution domain. An example of coding in GP is the representation of mathematical expressions as hierarchical trees, in a manner similar to the taxonomy of the programming language Lisp.

The major distinction between the two evolutionary computation approaches is that the objective of GP is to evolve computer programs (i.e., perform automatic programming) rather than evolve chromosomes of a fixed length. This puts GP at a much higher conceptual level, because suddenly we are no longer working with a fixed structure but with a dynamic structure equivalent to a computer program. GP essentially follows a similar procedure to GAs: there is a population of individual programs that is initialized, evaluated and allowed to reproduce as a function of fitness. Again, crossover and mutation operators are used to generate offspring that replace some or all of the parents.

As with GAs, the average fitness of the population in GP increases from generation to generation. Often, the final high-fitness models are inappropriate for practical applications since they are very complex and difficult to interpret, and crash on even minor changes in operating conditions. Recently, a special version of GP,

---

[8]D. Goldberg, *Genetic Algorithms in Search, Optimization, and Machine Learning*, Addison-Wesley, 1989.

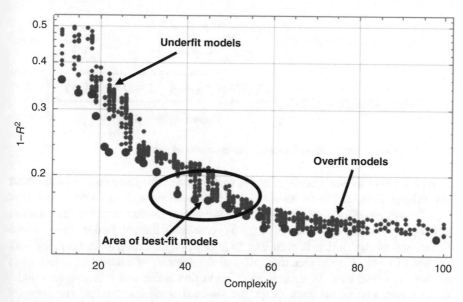

Fig. 3.31 Results from generation of models by Pareto-front GP

called Pareto-front GP, has offered a solution by using multiobjective optimization to direct the simulated evolution toward simple models with sufficient accuracy.[9] This version is especially useful for generating mathematical expressions (symbolic regression derived by GP). An example of some results of Pareto-front GP symbolic regression is given in Fig. 3.31.

In Pareto-front GP, the simulated evolution is based on two criteria—prediction error (for example, based on $1-r^2$) and complexity (for example, based on the number of nodes in the equations). In Fig. 3.31, each point corresponds to a certain model, with the x-coordinate referring to the complexity of the model and the y-coordinate to the error of the model. The large points form the Pareto front of the set of models. Models at the Pareto front are nondominated by any other model with respect to both criteria simultaneously. Compared with other models, elements of the Pareto front are dominant in terms of at least one criterion; they have either lower complexity or lower model error. The Pareto front itself is divided into three areas (see Fig. 3.31). The first area contains the simple, underfit models, which occupy the upper left part of the Pareto front. The second area, of complex overfit models, lies on the bottom right section of the Pareto front. The third and most interesting area, of the best-fit models, is around the tipping point of the Pareto front where the biggest gain in model accuracy for a corresponding model complexity is.

[9]G. Smits and M. Kotanchek, Pareto-front exploitation in symbolic regression, in *Genetic Programming Theory and Practice,* U.-M. O'Reilly, T. Yu, R. Riolo, and B. Worzel (eds.), Springer, pp. 283–300, 2004.

$$GP\_Model1 = A + B\left(\frac{Tray64\_T^4Vapor^3}{Rflx\_flow^2}\right)$$

$$GP\_Model2 = C + D\left(\frac{Feed^3\sqrt{Tray46\_T - Tray56\_T}}{Vapor^2Rflx\_flow^4}\right)$$

Fig. 3.32 Models from a real-world application in inferential sensors

At the end of the simulated evolution, we have an automatic selection of the "best and brightest" models in terms of the trade-off between accuracy and complexity. The "sweet zone" of models with the biggest gain in accuracy for the smallest possible complexity (encircled in Fig. 3.31) contains a limited number of solutions. In the case of Fig. 3.31 it is clear that there is no significant gain in accuracy with models of complexity higher than 50, and the number of interesting solutions is between four and five. As a result, model selection is fast and effective and model development efforts are low. Often the selected nonlinear models are compact analytical functions that are easy for SMEs to interpret. Examples of such models from a real-world application in the area of inferential sensors is given in Fig. 3.32.[10]

where A, B, C, and D are fitting parameters, and all model inputs in the equations are continuous process measurements.

Benefits of Evolutionary Computation

The key benefits of evolutionary computation are represented in the mind map in Fig. 3.33 and discussed below:

- *Reduced assumptions for model development.* The development of models by evolutionary computation can be done with fewer assumptions than with some of the other known methods. For example, there are no assumptions based on the physics of the process as in the case of first-principles modeling. Some statistical models based on the popular least-squares method require assumptions such as independence of variables, multivariate normal distribution, and independent errors with zero mean and constant variance. However, these assumptions are not required for symbolic regression models generated by GP.
- *High-quality empirical models.* The key evolutionary approach for empirical model building is symbolic regression (nonlinear algebraic equations), generated by GP. Pareto-front GP allows the simulated evolution and model selection to be directed toward structures based on an optimal balance between accuracy and expression complexity. A survey of industrial applications in The Dow Chemical

---

[10]E. Jordaan, A. Kordon, G. Smits, and L. Chiang, Robust inferential sensors based on ensemble of predictors generated by genetic programming, in *Proceedings of PPSN 2004*, pp. 522–531, Springer, 2004.

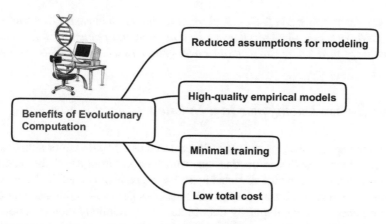

Fig. 3.33 Key benefits of evolutionary computation methods

Company shows that the selected models, generated by Pareto-front GP, are simple[11].

- *Minimal training of the final user.* The explicit mathematical expressions generated by GP are universally acceptable to any user with a mathematical background at high school level. This is not the case either with first-principles models (where specific physical knowledge is required) or with black-box models (where some advanced knowledge of neural networks is a must). In addition, a very important factor in favor of symbolic regression is that process engineers prefer mathematical expressions and very often can find an appropriate physical interpretation.
- *Low total cost of development, deployment, and maintenance.* Evolutionary computation has a clear advantage in terms of marketing the technology to potential users. The scientific principles are easy to explain to almost any audience. We also find that process engineers are much more open to implementing symbolic regression models in manufacturing plants. Most of the alternative approaches are expensive, especially in the area of real-time process monitoring and control systems. In addition, simple symbolic regression models require minimal maintenance. Model redesign is very rare, and most of the models perform with acceptable quality even when they are 20% outside the range of the original development data.

Key Messages about Evolutionary Computation

Evolutionary computation mimics natural evolution in the virtual world of computers.

---

[11]A. Kordon, F. Castillo, G. Smits, and M. Kotanchek, Application issues of genetic programming in industry, in *Genetic Programming Theory and Practice III*, T. Yu, R. Riolo, and B. Worzel (eds.), Springer, pp. 241–258, 2005.

Genetic algorithms can find optimal solutions based on intensive exploration and exploitation of a complex search space by a population of potential solutions.

Genetic programming can generate novel structures which fit a defined objective.

3.3.6   Swarm Intelligence in a Nutshell

Swarm intelligence systems are typically made up of a population of simple agents interacting locally with one another and with their environment. This interaction often leads to the emergence of global behavior which is not coded in the actions of the simple agents. Analyzing the mechanisms of collective intelligence that drives the appearance of new complexity out of interactive simplicity requires knowledge of several research areas, such as biology, physics, computer science, and mathematics.

There are two key directions in the research into swarm intelligence and its application: (1) ant colony optimization (ACO), based on insect swarm intelligence, and (2) particle swarm optimization (PSO), based on social interaction in bird flocks. Both approaches will be discussed in the next two subsections.

Ant Colony Optimization

Individual ants are simple insects that have limited memory and are capable of performing simple actions. However, an ant colony generates a complex collective behavior providing intelligent solutions to problems such as: carrying large items, forming bridges, finding the shortest routes from the nest to a food source, prioritizing food sources based on their distance and ease of access, and sorting corpses. Moreover, in a colony each ant has its prescribed task, but the ants can switch tasks if the collective needs it. For example, if part of the nest is damaged, more ants will do nest maintenance work to repair it.

How can ants manage to find the shortest path? The answer from biology is simple—by applying the stigmergy mechanism of indirect communication, based on pheromone deposition over the path they follow. They follow competing routes, one of which is significantly shorter than the others. This effect is illustrated in Fig. 3.34.

Let's assume that in the initial search phase, equal numbers of ants move on both routes. However, the ants on the short path will complete the journey more times and thereby lay more pheromone over it. The pheromone concentration on the short trail will increase at a higher rate than on the long trail, and in the advanced search phase the ants on the long route will choose to follow the short route (the amount of pheromone is proportional to the thickness of the routes in Fig. 3.34). Since most ants will no longer travel on the long route, and since the pheromone is volatile, the long trail will start evaporating. In the final search phase, only the shortest route remains.

Surprisingly, this simple algorithm is the basis of a method for finding optimal solutions to real problems, called ant colony optimization. Each artificial ant is a probabilistic mechanism that constructs a solution to the problem, using artificial

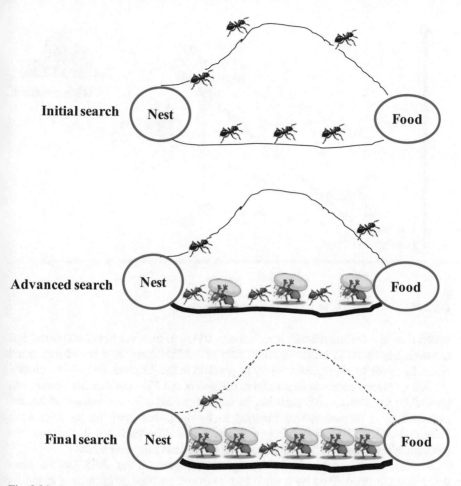

Fig. 3.34 Ant route handling

pheromone deposition, heuristic information about pheromone trails, and a memory of already visited places.

In ACO, a colony of artificial ants gradually constructs solutions to a defined problem, using artificial pheromone trails, which are modified accordingly during the algorithm. In the solution construction phase, each ant builds a problem-specific solution—for example, selection of the next route for the supply chain.

Particle Swarm Optimizers

In contrast to the insect-driven ant colony optimization, the second key direction in swarm intelligence is inspired mostly by the social behavior seen in bird flocking and fish schooling.

In the PSO algorithm, each single solution is like a "bird" in the search space, and is called a "particle." A selected number of solutions (particles) form a flock (swarm)

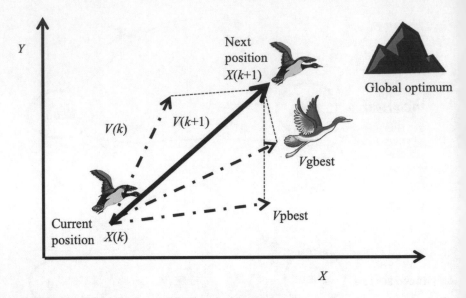

Fig. 3.35 Visualization of particle position update diagram

which flies in a D-dimensional search space, trying to uncover better solutions. For the user, the situation recalls a simulation of bird flocking in a two-dimensional plane. Each particle is represented by its position in the XY plane and by its velocity (V_X is the velocity component parallel to the X-axis and V_Y is the velocity component parallel to the Y-axis). All particles in the swarm have fitness values which are evaluated by a defined fitness function to be optimized, and the particles have velocities which direct the flying of the particles. (The particles fly through the problem space by following the particles with the best solutions so far.)

A visual interpretation of the PSO algorithm is given in Fig. 3.35. Let's assume that a particle (visualized by a bird) has a current position $X(k)$ at time k and its current velocity is represented by the vector $V(k)$. The next position of the bird, at time $k + 1$, is determined by its current position $X(k)$ and the next velocity $V(k + 1)$. The next velocity is a vector blend of the current velocity $V(k)$ with the component of the acceleration toward the swarm-best (represented by the big bird), with a velocity of V_{gbest}, and another component of acceleration toward the particle-best, with a velocity of V_{pbest}. As a result of these velocity adjustments, the next position, X ($k + 1$), is closer to the global optimum.

Benefits of Swarm Intelligence

The value creation capabilities of swarm intelligence based on social interactions may generate tremendous benefits, especially in the area of nontrivial optimization of complex problems. The key benefits of swarm intelligence are represented in the mind map in Fig. 3.36 and discussed below:

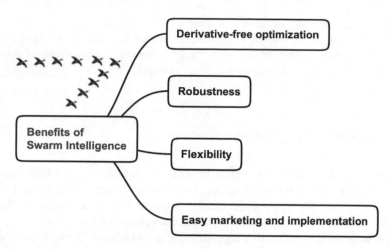

Fig. 3.36 Key benefits of swarm intelligence methods

- *Derivative-free optimization.* The search for an optimal solution in swarm intelligence is based not on functional derivatives but on different mechanisms of social interaction between artificial individuals. In this way the chances of being entrapped in local minima are significantly reduced (but not eliminated!).
- *Robustness.* The population-based ACO and PSO algorithms are more protective toward individual failure. Poor performance of even several members of the swarm is not a danger for the overall performance. The collective behavior compensates for the laggards and the optimum solution is found independently of the variations in individual performance of the members of the swarm.
- *Flexibility.* Probably the biggest benefit from swarm intelligence is its capability to operate in a dynamic environment. The swarm can continuously track even fast-changing optima. In principle, there is no significant difference in functioning of the algorithm between steady-state and dynamic modes. In the case of classical methods, different algorithms and models are required for these two modes.
- *Easy marketing and implementation.* The principles of biology-inspired swarm intelligence are easy to communicate to a broad audience of potential users and there is no need for a heavy math or statistical background. In addition, the implementation of both ACO, and especially, PSO in any software environment is trivial. The tuning parameters are few, and easy to understand and adjust. In some cases, implementing PSO can even be transparent for the final user and be a part of the optimization options of a larger project.

Key Messages for Swarm Intelligence

Swarm intelligence is coherence without choreography and is based on the emergent collective intelligence of simple artificial individuals.

Swarm intelligence is inspired by the social behavior of ants, bees, termites, wasps, birds, fish, sheep, and even humans.

Ant colony optimization uses a colony of artificial ants to construct optimal solutions for a defined problem by digital pheromone deposition and heuristics.

Particle swarm optimization uses a flock of communicating artificial particles searching for optimal solutions of a defined problem.

3.3.7   Intelligent Agents in a Nutshell

Intelligent agents represent complex systems by interactions between different simple entities with a defined structure, called agents. Intelligent agents have the following features: the relationships among the elements are more important than the elements themselves, complex outcomes can emerge from a few simple rules, small changes in the behavior can have big effects on the system, and patterns of order can emerge without control.

The use of agent-based systems is an artificial intelligence technique of bottom-up modeling that represents the behavior of a complex system by the interactions of simple components, called agents. The field is very broad and is related to other areas, such as economics, sociology, and object-oriented programming, to name a few. Recently, a specific type of intelligent agents, called chatbots, have been gaining popularity and has been implemented in several business applications.

The key principles of intelligent agents, agent-based systems, and chatbots are discussed briefly below.

Intelligent Agents
There are probably as many definitions of intelligent agents as there are researchers in this field. There is a consensus, however, that autonomy, i.e. the ability to act without the intervention of humans or other systems, is the most important feature of an agent. Beyond that, different attributes take on different importance based on the domain of the agent. Some of the key generic attributes are summarized below:

- *Autonomy.* An agent is autonomous if its behavior is determined by its own experience (with the ability to learn and adapt). It is assumed that an agent has its own initiative and can act independently. As a result, from the viewpoint of other agents, its behavior is neither fully predictable nor fully controllable.
- *Sociality.* An agent is characterized by its relationships with other agents, and not only by its own properties. Relationships among agents are complex and not reducible. Conflicts are not easily resolvable, and cooperation among agents cannot be taken for granted.
- *Identity.* Agents can be abstract or physical and may be created and terminated. Their boundaries are open and changeable.
- *Rational self-interest.* An agent aims to meet its goals. However, its self-interest exists in a context of social relations. Its rationality is also limited to its perceptions. It is assumed that the right action is the one that will cause the agent to be most successful.

Fig. 3.37 A generic
intelligent agent structure

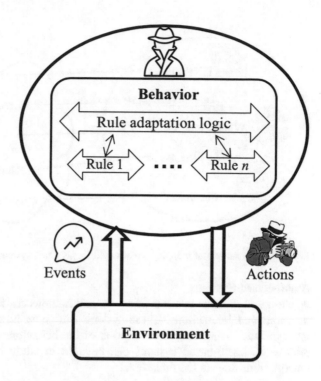

Figure 3.37 illustrates a high-level view of an agent interacting with its environment. An agent receives input as events from its environment and, through a set of actions, reacts to it in order to modify it.

Agents can be defined as the decision-making components in complex adaptive systems. Their decision power is separated into two levels, shown in Fig. 3.37. The first decision level includes the rule base (rule 1, rule 2, and rule n) which specifies the response to a specific event. An example of such a rule for a bidding agent is:

If current profit < last profit then.

Increase our price

Else.

Do nothing

The second-level rules, or the rules to change the rules, provide the adaptive capability of the agent. Different methods and strategies can be used, but we need to be careful not to introduce too much complexity. One of the key issues in the design of intelligent agents is the complexity of the rules. It is not a surprise that defining simple rules is strongly recommended. Above all, simple agent rules allow easy decoupling between the agent's behavior and the interaction mechanisms of the agent-based system. As a result, models are quickly developed, the validation time is shortened, and the chance of use of the model is increased.

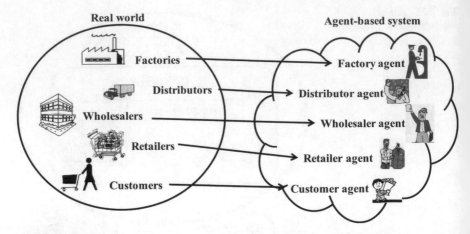

Fig. 3.38 An example of mapping a real-world supply chain system into an agent-based system

Agent-Based Systems

Agent-based systems use sets of agents and frameworks for simulating the agents' decisions and interactions. These systems can show how a system could evolve through time, using only a description of the behaviors of the individual agents. Above all, agent-based systems can be used to study how macrolevel patterns emerge from rules at the microlevel.

Agent-based systems are classic multiagent systems, i.e., a collection of agents and their relationships. Usually the collection of agents reflects the real-world structure of the specific problem under exploration. For example, if the objective of the agent-based system is to develop a supply chain model, the following agent types can be defined: factory, wholesaler, distributor, retailer, and customer. They represent the real five-stage process of the supply chain, which includes a sequence of factories, distributors, wholesalers, and retailers, who respond to customers' demand.[12] The mapping of the real-world supply chain into the agent-based system is illustrated in Fig. 3.38.

Each agent type is defined by its attributes and methods, or rules. For example, the factory agent is represented by the following attributes: the agent's name; the inventory level; the desired inventory level; the amount in the pipeline; the desired amount in the pipeline; the amounts received, shipped, ordered, and demanded; various decision parameters; and the costs incurred of holding inventory or back orders. The values of these variables at any point in time constitute the state of the agent. Examples of agent rules in the case of the factory agent type include a rule for determining how much to order, and from whom at any point in time and a rule for forecasting demand.

---

[12]M. North and C. Macal, *Managing Business Complexity: Discovering Strategic Solutions with Agent-Based Modeling and Simulation*, Oxford University Press, 2007.

Fig. 3.39 Chatbot
architecture

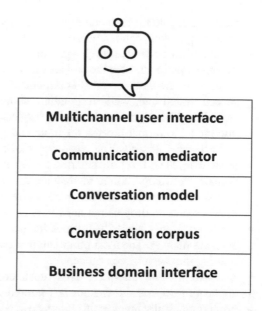

| Multichannel user interface |
| Communication mediator |
| Conversation model |
| Conversation corpus |
| Business domain interface |

In addition to the collection of agents, agent-based systems consist of agent interactions. Each agent relation involves two agents. For example, the factory–distributor relation includes the attributes of the number of items in transit from the factory to the distributor and the order in transit from the distributor to the factory. Agent relations also have methods that operate on them, just as human agents have. For example, "get shipments", "get orders", "get upstream agent", and "get downstream agent" are useful methods for agent interactions.

Chatbots

Recently, a special type of intelligent agent, called chatbots (also known as conversation agents or dialog-based agents) has been an area of intensive research and production in the AI Olympians. A key feature is the use of natural-language conversation models that can understand language, perform some reasoning based on common-sense and business specific knowledge, and communicate results to the user.

The key blocks of a chatbot are shown in Fig. 3.39 and discussed briefly below.[13]

- *Multichannel User Interface*. This chatbot block interacts with the user. The conversation can happen via multichannel interfaces such as phones, laptops, specialized hardware (Amazon Echo), and desktops, and can be text based (messengers) or verbal (Siri, Alexa, etc.). The user interface provides the necessary capabilities through which the user can interact with and use the system. It is

---

[13]https://www.linkedin.com/pulse/anatomy-chatbots-harpreet-sethi/

also responsible for maintaining the user session, keeping track of user activity, managing authentication/authorization, etc.

- *Communication Mediator*. This part of the chatbot is responsible for channelizing the input coming in from the previous module to invoke a service that can execute a query. In the case of verbal communication, it should also be able to convert speech to text and, while responding, should have a text-to-speech generation capability. This layer should also be responsible for selecting the right conversation model to further process the input and generate the appropriate responses.
- *Conversation Models*. The reasoning block of the chatbot is based on several approaches. The level of "smartness" depends upon the availability of data and domain knowledge. Some models are as simple as hard-coded rules, other use predefined templates, and the most sophisticated use machine translation techniques based on deep neural networks.
- *Conversational Corpus*. This block includes the necessary data for the chatbot's decision-making. The most important part consisists of business-specific data for training the conversation models.
- *Business domain interface*. This block connects the chatbot with enterprise applications. Once the chatbot has figured out what the user is looking for, it programmatically invokes this interface to add specifics to the response's product information in a shopping cart, or inventory list or by showing the last orders.

Benefits of Intelligent Agents

The key benefits of intelligent agents are captured in the mind map in Fig. 3.40 and discussed below:

- *Bottom-up modeling*. Agent-based systems begin at the level of the decision-making individual or organization and build upwards from there. Defining an agent's microbehavior for a specific task is relatively easy. Obviously, this

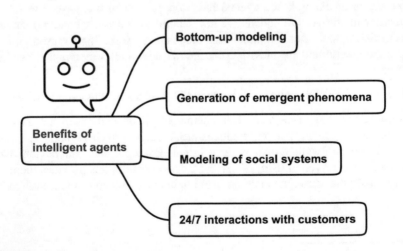

Fig. 3.40 Benefits of intelligent agents

bottom-up modeling is quite different from the existing equation-based methods. Individual behavior is complex. Everything can be done with equations, in principle, but the complexity of differential equations increases exponentially as the complexity of the behavior increases. At some point, describing complex individual behavior with equations becomes intractable and other methods, such as intelligent agents, are needed.

- *Generation of emergent phenomena.* The driving force behind most of the applications of agent-based systems is the assumption that macropatterns will emerge from a given set of agents with specified microbehavior. The important factor here is that the properties of the emergent phenomena are independent of the properties of the agents. As a result, the features of the macropatterns may be difficult to understand and predict. It is also possible that nothing new might emerge from the interactions of agents.

- *Modeling of social systems modeling.* The ability of agent-based systems to mimic human behavior allows the modeling and simulation of social systems in a very natural way. This is probably the approach with the best potential to represent the diverse complexity of the social world. It gives social analysts a proper medium to explore in detail the emerging social phenomena of interest. An agent-based approach may become the economic *E. coli* for analysis of social systems.[14] The idea is taken from biology, where the ability to experiment with animal models such as *E. coli* has led to great advances in understanding the nature of biological phenomena. In the same way, artificial adaptive agents could be the new way to develop and validate new social theories.

- 24/7 *interactions with customers.* A special type of intelligent agent, chatbots, can revolutionize communications with customers. A well-designed chatbot responds instantly to any query without delay, remembers customer's preferences, provides specific information at each question, etc. and is available round the clock. According to Gartner, chatbots will be used in 85% of all customer service interactions by the year 2020 and the average person will have more conversations with them than with their spouse. [15]

Key Messages about Intelligent Agents

An intelligent agent receives inputs from its environment and, through a set of actions, reacts to modify it in its favor.

Agent-based systems generate emerging macropatterns from rules at the microlevel of interacting social agents.

A special type of intelligent agents, chatbots, have been successfully applied in customer machine conversations.

---

[14] J. Miller and S. Page, *Complex Adaptive Systems: An Introduction to Computational Models of Social Life*, Princeton University Press, 2007.

[15] https://chatbotsmagazine.com/chatbot-report-2018-global-trends-and-analysis-4d8bbe4d924b

3.4   Common Mistakes

3.4.1   Obsession with One Method

A well-known secret in the Data Science community is that most data scientists are using their favorite approach without comparing the results with other methods. In most cases the preferred methods are linear and logistic regression. Often the reason for this limited selection is economic—no investment in software that includes a variety of AI-based methods has been made. However, the current trend toward low-cost or even free packages including these algorithms (examples are RapidMiner, and many open source software packages based on Python and R) is removing this obstacle. In order to increase the probability of success in solving business problems, it is recommended to explore solutions derived from different AI approaches and to compare them with the linear benchmark. The selection and integration of methods is discussed in the next chapter.

3.4.2   Focusing on Fashionable Methods

Another observed trend is to demonstrate fast "acceptance" of AI-based methods by blind use of "sexy" approaches, such as deep learning, neural networks, and random forests. Some of these methods are too complex or too costly, and difficult to interpret. Often the desired solution can be accomplished by other simpler and cheaper approaches.

3.4.3   Lack of Knowledge about Broad Options for AI-Based Approaches

The lion's share of references, products, and media attention is focused on a limited set of AI-based approaches, mostly related to machine learning, such as neural networks, deep learning, and random forests. There is some growing interest in chatbots and decision trees. However, the Data Science community at large is not very knowledgeable about the enormous capabilities of support vector machines, evolutionary computation, and swarm intelligence. As a result, the choices are limited for an efficient solution to an identified business problem.

3.4.4   Lack of Knowledge about Cost of Implementation of Methods

A common gap in the knowledge of the Data Science community at large is about the deployment cost of the selected solutions, especially the fashionable machine learning methods. Some methods require specialized run-time licenses and intensive training. Most of the methods that do not deliver robust models require almost continuous maintenance of models to compensate for degrading model performance, which adds asubstantial cost to the project.

3.5   Suggested Reading

- V. Cherkassky and F. Mulier, *Learning from Data: Concepts, Theory, and Methods*, second edition, Wiley, 2007.
- L. de Castro, *Fundamentals of Natural Computing*, Chapman & Hall, 2006.
- A. Eiben and J. Smith, *Introduction to Evolutionary Computing*, second edition, Springer, 2014.
- A. Engelbrecht, *Fundamentals of Computational Swarm Intelligence*, Wiley, 2005.
- I. Goodfellow, Y. Bengio, and A. Courville, *Deep Learning*, MIT Press, 2016.
- T. Hastie, R. Tibshirani, and J. Friedman, *The Elements of Statistical Learning: Data Mining, Inference, and Prediction*, second edition, Springer, 2016.
- S. Heykin, *Neural Networks and Learning Machines*, third edition, Pearson, 2016.
- J. Keller, D. Liu, and D. Fogel, *Fundamentals of Computational Intelligence*, Wiley-IEEE Press, 2016.
- J. Miller and S. Page, *Complex Adaptive Systems: An Introduction to Computational Models of Social Life*, Princeton University Press, 2007.
- M. North and C. Macal, *Managing Business Complexity: Discovering Strategic Solutions with Agent-Based Modeling and Simulation*, Oxford University Press, 2007.

3.6   Questions

Question 1
Give examples of different types of AI-based solutions to business problems.

Question 2
What is the difference between prediction and forecasting?

Question 3

What is the difference between classification and clustering?

Question 4

Select at least two AI methods related to a business problem of your preference.

Question 5

Which AI methods are you familiar with? Which AI methods are unknown to you?

Chapter 4
Integrate and Conquer

Practical sciences proceed by building up; theoretical sciences by resolving into components.

St. Thomas Aquinas

Since value creation capability is the key driving force in business applications, the strategy for applying AI-based Data Science is based on factors such as minimizing modeling cost and maximizing model performance under a broad range of operating conditions. An obvious result of this strategy is the increased effort in robust data-driven model building, which is very often at an economic optimum. Unfortunately, robustness of empirical solutions, i.e., their capability to operate reliably during minor process or business operation changes, is difficult to accomplish with only one modeling method. Very often the nasty reality of real-world problems requires the joint work of several modeling techniques. In order to meet this need, it is necessary to develop a consistent methodology that effectively combines different modeling approaches to deliver high-quality models with minimal effort and maintenance. This statement is based on the almost infinite number of ways to explore the synergies between modeling methods. The result will be an integrated methodology which significantly improves performance and compensates for the disadvantages of the individual methods.

The motto of the chapter is clear: Integrate the modeling methods and conquer the business world. From our experience, this is the winning strategy which opens the door to AI-based Data Science in industry.

The first objective of this chapter is to justify the need for integration of methods to solve business problems, and to clarify the benefits. The second objective of the chapter is to give examples of integration between different AI-based Data Science methods, as well as between these approaches and first-principles and statistical models. The third objective is to help the reader to select the proper combination of AI-based Data Science methods for solving specific business problems based on their capability and applicability.

© Springer Nature Switzerland AG 2020
A. K. Kordon, *Applying Data Science*,
https://doi.org/10.1007/978-3-030-36375-8_4

Fig. 4.1 Main issues in
real-world applications

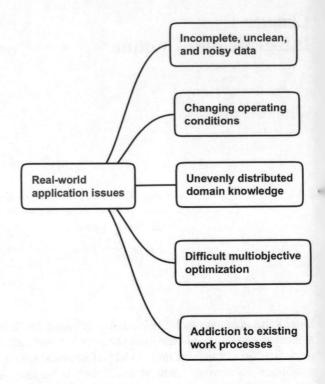

4.1 The Integrate and Conquer Strategy in Applied Data Science

4.1.1 The Nasty Reality of Real-World Applications

In the same way as an abstract spherical cow differs from the real animal, the purely academic versions of a modeling approach are far away from the challenges of business reality. Unfortunately, the reality corrections cannot be done only through intensive computer simulations, as is usually the case. The big issue is that the nasty reality of business applications requires the joint solution of the technical, infrastructural, and people-related components of a given problem. A short list of the main issues in real-world applications, which includes these components, is presented in Fig. 4.1 and discussed below:

- *Unclean, incomplete, and noisy data.* Data quality and availability is at the top of the list of technical real-world application issues. Usually two extreme situations need to be handled. In the first situation, typical of manufacturing, the available data has high dimensionality (up to thousands of variables and millions of records). At the other extreme is the second situation, typical of new product development, where the available data is limited to a few records and the cost of each new record is very high. In both cases, the data may contain gaps, erroneously entered values, units of measurement conversion chaos, and different levels of noise. Handling all of this data mess with effective data preparation is the prerequisite for the success of any data-driven approach, including most AI-based methods.

- *Changing operating conditions.* The real world operates in a dynamic environment. The pace and amplitude of changes, however, vary significantly from microseconds to decades and from small variations to deviations by orders of magnitude. It is extremely difficult to represent this broad dynamic range with one modeling technique only. Even the meaning of a steady-state model has to be interpreted in relation to the current business needs. The use of a model depends strongly on the current demand for the product or service it is linked to. The demand varies with the economic environment and, at some point, the steady-state conditions under which the model was developed become invalid and the model's credibility evaporates. Either parameter readjustments or complete redesign is needed. In both cases, the maintenance cost grows significantly.
- *Unevenly distributed domain knowledge.* Domain knowledge is critical for any real-world application. The presence of subject matter experts is decisive in all phases of model development and deployment—from problem definition to solution selection and use. The popular phrase that a model is as good as the experts involved in its development is not an exaggeration. Unfortunately, domain knowledge has high dynamics (experts change organizations frequently) and is unevenly distributed in an organization. Involving the subject matter experts in driving the real-world application is a challenging but critical task for final success. It has to be considered that some top-level experts have no habits of sharing their knowledge easily.
- *Difficult multiobjective optimization.* The potential solutions to real-world problems depend on a multitude of factors which are related in complex ways and often represented by messy data. The high dimensionality, as well as the complex search space with multiple optima and a noisy surface, is a tough challenge for any individual optimization technique. In addition, the solutions to real-world problems are inherently multiobjective. This implies active participation of the final user in the trade-off process of solution selection.
- *Addiction to existing work processes.* The working environment in a business is based on the established infrastructure, work processes, and behavior patterns (or culture). Any successful real-world application must adapt to this environment and not try to push the opposite, i.e. to adapt the existing work processes to the technical needs of the application. Most people are addicted to their working environment and hate significant changes, especially in well-established work processes. Understanding how the solution to real-world problem will fit into the existing working infrastructure is one of the most critical (and most ignored) factors for success.

In summary, when we solve real-world problems, we realize that such systems are typically ill-defined and difficult to model, and possess large solution spaces with multiple optima. The data we may use is messy and noisy, the operating conditions change, and the domain knowledge is difficult to extract. In addition, we need to understand and integrate the solution into the existing work infrastructure and to find support for, not opposition to, the application.

4.1.2   Why Integration of Methods Is Critical for Real-World Applications

The challenges of industrial reality and the tough criteria for the success of real-world applications set a very high standard for any modeling method. It is much more difficult for a new emerging technology, such as AI, to satisfy this standard. Each AI-based method has its strengths and weaknesses, which will be analyzed in this chapter. The problem is that their individual application potential is very limited and the probability of a credibility fiasco is high.

At the same time, the synergetic potential between AI-based methods is enormous. Almost any weakness of a given approach can be compensated for by another method. A clear example is when a neural network model optimizes its structure using a genetic algorithm. Exploring the integration potential of AI-based methods gives tremendous application opportunities at a relatively low price and only a slightly more complicated model development process.

Benefits of Integration
First, we'll focus on the key benefits from the synergetic effects between the various AI-based methods, shown in Fig. 4.2:

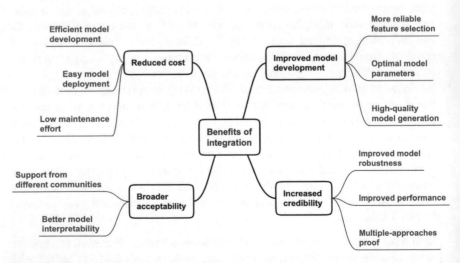

Fig. 4.2 Key benefits of integrating different modeling approaches for successful real-world applications

- *Improved model development.* The combination of different methods leads to the following gains during model development:

  - *More reliable feature selection.* Feature or variable selection is critical for effective model development. However, different AI-based methods estimate the contributions of variables using different metrics and algorithms. Selecting the most important variables based not on one but on several methods increases the probability of a credible choice.
  - *Optimal parameters.* A typical case is the use of GAs or PSO for parametric or structural optimization of neural networks or SVM-based models.
  - *High-quality model generation.* Model quality depends on structure and parameters. The critical factor for configuration of the structure is the selection of variables. The critical factor for selection of parameters is their optimization. Both of these criteria benefit from integration of methods and the final result is generation of high-quality models with most of the available AI-based methods.

- *Increased credibility.* An obvious benefit from the synergy between AI-based methods is an increase in the critical factor for success of an application—its credibility. This is based on the following specific factors:

  - *Improved model robustness.* The final result of integrated model development is generation of models with an optimal trade-off between accuracy and complexity. Such relatively simple solutions are more likely to operate reliably during minor process or business condition changes.
  - *Improved performance.* The final result of the integrated methodology is an application with the best trade-off with respect to accurate prediction, robustness, and extrapolation crisis handling.
  - *Multiple approaches proof.* The fact that solutions with similar performance are generated by methods with an entirely different scientific basis increases the level of trust in the proposed solution. The users of the technology are really impressed by its capability to deliver multiple proofs of the expected results in different algorithmic forms without a substantial increase in the development cost.

- *Broader acceptability.* Using several methods in model development increases the support for the application, especially if the opportunities for synergies with first-principles methods and statistics are explored. The sources of this broader acceptability are as follows:

  - *Support from different communities.* Integration not only optimizes the technical capabilities of methods but also builds bridges between different AI and Data Science communities. Of critical importance for introducing AI into large organizations with advanced R&D activities is to have on board the first-principles modelers and the statisticians. The best way to accomplish this is by integrating AI with their approaches.

- *Better interpretability*. As a result of integration, the methods which are more appropriate for interpretation, such as genetic programming or decision trees, can be used for final delivery.

- *Reduced cost*. All components of the total cost of ownership are positively affected by integration, as discussed briefly below:

  - *Efficient model development*. The optimal development process significantly reduces the development time and increases the quality of the developed models.
  - *Easy model deployment*. The final models are from technologies such as symbolic regression or statistics that have minimal run-time software requirements and can easily be integrated into the existing infrastructure.
  - *Low maintenance efforts*. The increased robustness and potential for handling extrapolation crises reduce significantly the maintenance cost.

The Price of Integration

Integration adds another level of complexity over the existing methods. Having in mind that the theoretical foundations of most AI-based approaches are still under development, the analytical basis of an integrated system is nonexistent. However, the practical benefits of integration have already been demonstrated and explored before developing a solid theory.

Integration also requires more complex software and broader knowledge about the methods discussed in this book. In comparison with the benefits, however, the price for integrating AI-based methods is relatively low.

4.2   Integration Opportunities

The good news about building an integrated system is that the ways to explore the synergistic capabilities of its components are practically unlimited. The bad news is that it is practically unrealistic to devote theoretical efforts to analyzing all of them. As a result, most of the integrated systems than have been designed and used lack a solid theoretical basis. However, the experiences of several big multinational companies, such as GE,[1] Ford,[2] and Dow Chemical,[3] demonstrate the broad application of integrated systems in different areas of manufacturing, new product design, and financial operations.

---

[1]P. Bonissone, Y. Chen, K. Goebel, and P. Khedkar, Hybrid soft computing systems: Industrial and commercial applications, *Proceedings of the IEEE*, **87**, no. 9, pp. 1641–1667, 1999.

[2]O. Gusikhin, N. Rychtyckyj, and D. Filev, Intelligent systems in the automotive industry: Applications and trends, *Knowledge and Information Systems*, **12**, 2, pp. 147–168, 2007.

[3]A. Kordon, Hybrid intelligent systems for industrial data analysis, *International Journal of Intelligent Systems*, **19**, pp. 367–383, 2004.

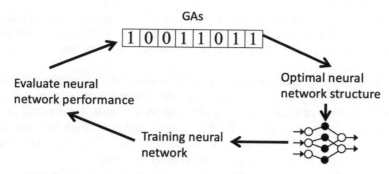

Fig. 4.3   An example of an evolutionary neural network

We'll focus on four opportunities for integration. The first opportunity includes the synergy between the AI methods themselves, the second opportunity integrates AI methods with first-principles models, and the third opportunity explores the bridges between AI and statistics. Finally, we'll discuss the fourth opportunity, of building an ensemble of models generated by different methods.

4.2.1   Integration Between AI-Based Methods

Usually an integrated system between AI-based methods is called a hybrid intelligent system and it combines at least two AI-based technologies, for example neural networks and particle swarm optimizers. The key topic in the field of hybrid intelligent systems is the methodology of integration. The best way to illustrate the direction of potential integration is by borrowing a social analogy from a famous quote from the inventor of fuzzy logic, Lotfi Zadeh, about a good and a bad hybrid system: "A good hybrid system is when we have British Police, German Mechanics, French Cuisine, Swiss Banking and Italian Love. But a hybrid of British Cuisine, German Police, French Mechanics, Italian Banking, and Swiss Love is a disaster." In the same way, a hybrid intelligent system could be a tremendous success or a complete fiasco depending on the effectiveness of a component's integration. As is well-known, each AI-based technology has its strengths and weaknesses. Obviously, the synergistic potential is based on the compensation of the weaknesses of one approach with the strengths of another.

A clear example is the combining of the machine learning capabilities of neural networks with the optimization power of GAs to develop evolutionary neural networks. As a result, the marriage between evolutionary computation and neural networks can lead to structures with optimal topology and parameters that can improve significantly the machine learning process. An example of an evolutionary neural network using genetic algorithm is shown in Fig. 4.3.

The principle of integration between genetic algorithms and neural networks is simple. The problem domain of the neural network that needs to be optimized—

represented by its structure or weights—is encoded as a genotype chromosome. Then a population of genotypes evolves using genetic operators until the performance of the corresponding phenotype neural network reaches some specified optimal fitness.

Recently, the idea of integrating neural networks and evolutionary algorithms has had a revival for deep neural networks. It is expected that neuroevolution might be able to evolve the best structure for a very complex convolutional neural network intended for training with stochastic gradient descent. Another area where neuroevolution is making a comeback is in reinforcement learning, which focuses on control and decision-making problems in deep learning.

The same idea of optimizing the structure of different AI-methods with powerful AI-based optimizers such as GAs and PSO has been implemented in integrated systems such as Neuro-PSO, evolutionary SVMs, and SVM-PSO. With their diverse features and capabilities, AI-based methods create many opportunities for integration. This broadens the range of options for using high-quality solutions based on these technologies. It is strongly recommended to explore them when solving business problems.

4.2.2   Integration with First-Principles Models

One of the key issues in industrial model development, especially in the chemical industry and biotechnology, is reducing the total cost of ownership of first-principles modeling. This is a niche that AI-based Data Science can fill very effectively with various ways of integration of its methods into the first-principles model building process. We'll focus on four potential ways of integrating AI-based methods and mechanistic models: (1) joint integration, (2) parallel operation, (3) accelerating first-principles model development by use of empirical proto-models, and (4) representing fundamental models by empirical substitutes for online optimization.

The first integration scheme for fundamental model and empirical models generated by AI-based methods, is shown in Fig. 4.4.

Fig. 4.4 An example of joint integration of AI-based methods and first-principles models

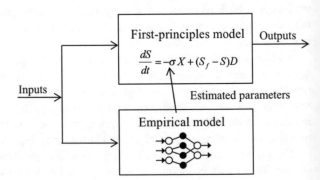

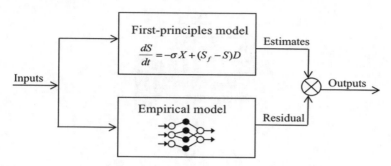

Fig. 4.5 An example of parallel operation of AI-based methods and first-principles models

The idea is to combine a simplified parametric first-principles model with some empirical estimator of the unknown parameters. A typical case is a mass balance hybrid model of a fed-batch bioreactor, in which an artificial neural network estimates specific kinetic rates, such as biomass growth and substrate consumption.

In the second integration scheme, shown in Fig. 4.5, the simple fundamental model and the empirical model (a neural network in this example) operate in parallel to determine the model output. The first-principles model serves as a physically sound but not detailed representation of the process. The deficiency in the model, however, is compensated for by the parallel empirical counterpart. The neural network is trained on the residual between the real response and the fundamental model-based estimates. As a result, the disagreement between the models is minimized and the combined prediction is more accurate.

The third scheme for integrating fundamental and AI-based models, shown in Fig. 4.6, is not based on direct or indirect model interaction but on using empirical methods during first-principles modeling. The idea is to reduce the cost of one of the most expensive and unpredictable phases in the development of fundamental models—hypothesis search. Usually this phase is based on a limited amount of data and a large number of potential factors and physical mechanisms. The current process of finding the winning hypothesis and mechanisms based only on the available data can be time-consuming and costly. However, if the data is used first to generate symbolic regression models, generated by genetic programming, the process could be significantly shortened. In this case, the starting position for hypothesis search is based on a set of empirical proto-models, which may help the first-principles modeler with ideas for appropriate mechanisms.

The fourth scheme for integrating fundamental and AI-based models, shown in Fig. 4.7, links the two paradigms with data generated by design of experiments (DOE). The first-principles model (the face in Fig. 4.7) generates high-quality data offline, which is the basis for deriving an empirical model that emulates the high-fidelity model's performance online (the mask in Fig. 4.7). This form of integration opens the door to using complex first-principles models for online optimization and control. Very often, the execution speed of fundamental models is too slow for

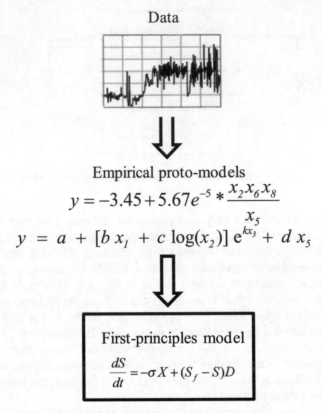

Data

Empirical proto-models

$$y = -3.45 + 5.67e^{-5} * \frac{x_2 x_6 x_8}{x_5}$$

$$y = a + [b\,x_1 + c\,\log(x_2)]\,e^{kx_3} + d\,x_5$$

First-principles model

$$\frac{dS}{dt} = -\sigma X + (S_f - S)D$$

Fig. 4.6 A generic scheme for using empirical proto-models in development of first-principles models

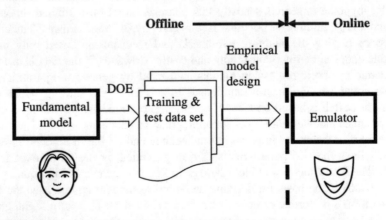

Fig. 4.7 Representing first-principles models as empirical models or emulators for online optimization

real-time operations and this limitation reduces the frequency and efficiency of process optimization.

Empirical emulators are one of the rare cases where neural networks can be reliably used, since we have full control over the ranges of data generation and the potential for extrapolation is minimal.

4.2.3   Integration with Statistical Models

Surprisingly, the least explored area of integration is between AI-based methods and statistics, even though the benefits could be used by a large army of statisticians. Of special interest is the synergy between evolutionary computation, which is population-based, and statistical modeling. We'll focus on the integration opportunities between genetic programming, as one of the most applicable evolutionary computation methods, and statistics.[4]

There are several economic benefits from the synergy between GP and statistical model building. The most obvious is the elimination of additional experimental runs to address model lack of fit (LOF).

The capability of a model to represent the data can often be assessed through a formal LOF test when experimental replicates are available. Significant LOF in the model indicates a regression function that is not linear in the inputs, i.e., the polynomial initially considered is not adequate. However, the fitting of a higher-order polynomial may be impractical because experiments are very expensive or technically infeasible due to extreme experimental conditions. This problem can be handled if appropriate input transformations are used, provided that the basic statistical assumption that errors are uncorrelated and normally distributed with zero mean and constant variance are satisfied.

Fortunately, GP-generated symbolic regression provides a unique opportunity to rapidly develop and test these transformations. This multiplicity of models, derived by GP, with different analytical expressions, provides a rich set of possible transformations of the inputs, which has the potential to solve the LOF issue. The original variables are then transformed according to the functional form of these equations. Then a linearized regression model, called a transformed linear model, is fitted to the data using the transformed variables. The adequacy of the transformed model is initially analyzed considering LOF and r^2. Then the error structure of the models not showing significant LOF is considered and the correlation among model parameters is evaluated.

[4]F. Castillo, A. Kordon, J. Sweeney, and W. Zirk, Using genetic programming in industrial statistical model building, *Genetic Programming Theory and Practice III.*, U.-M. O'Reilly, T. Yu, R. Riolo, and B. Worzel (eds.), Springer, pp. 31–48, 2004.

This integration methodology has been successfully applied in industrial settings, with significant cost reduction due to savings from reducing the number of designed experiments.[5]

Another benefit of integrating statistics with GP is that it allows to tailor a statistical dress to the selected symbolic regression model. The idea is very simple, and, as such, very effective: decompose the nonlinear symbolic regression model into linearized transforms and compose a statistical model which is linear in the transformed variables. In this case we have the best of both worlds. On the one hand, the nonlinear behavior is captured by the transformed variables, derived automatically from the selected GP equation. On the other hand, the final linearized solution is linear in the parameters and statistically correct, and has all related measures, such as confidence limits and variance inflation factors (VIFs).[6] Recent analysis demonstrates that these types of linearized models, with transforms derived by GP, have reduced multicollinearity and do not introduce bias into the parameter estimates.[7]

However, selecting linearized transforms is not automatic and trivial. It requires experience in statistical interpretation of the models derived. For example, looking at the error structure of the linearized models is a must. If the error structure does not have constant variance and shows patterns, models with other transforms need to be selected. Another test for acceptance of the derived linearized models is to calculate the VIFs and validate the lack of multicollinearity between the transformed variables. As the statisticians like to say, the linearized model must make "statistical sense."

4.2.4 Integration by Ensembles of Models

Probably the most popular and obvious way to achieve integration is to develope different types of ensembles of appropriate methods. An example of such an ensemble that integrates several prediction models derived by various AI-based approaches, such as genetic programming, neural networks, support vector machines for regression, and random forests for regression is shown in Fig. 4.8.

The ensemble is based on the individual predictions of the corresponding methods. All the methods use the same data set and their tuning parameters (such as the population size for GP, the number of neurons in the hidden layer for neural networks, the kernel for SVMs, and the number of trees for random forests) are appropriately selected. An important recommendation for developing these types of

---

[5]F. Castillo, K. Marshall, J. Green, and A. Kordon, Symbolic regression in design of experiments: A case study with linearizing transformations, *Proceedings of GECCO 2002*, New York, pp. 1043–1048, 2002.

[6](VIF) is a statistical measure of collinearity between input variables.

[7]F. Castillo and C. Villa, Symbolic regression in multicollinearity problems, *Proceedings of GECCO 2005*, Washington, DC, pp. 2207–2208, 2005.

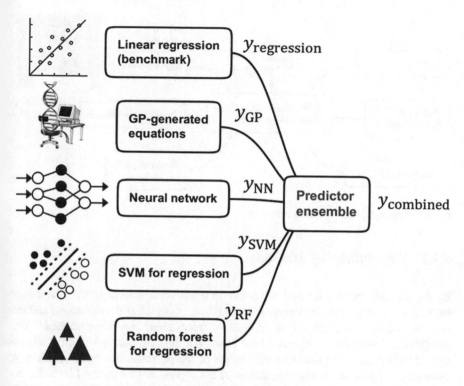

Fig. 4.8 An example of an ensemble of predictors based on different AI-driven Data Science methods and a linear regression model as benchmark

ensembles is to include a statistical model as a benchmark for comparison to find out if the other, nonlinear options can deliver a better solution. There are many options for combining the predictions of the individual models in the ensemble, and the most popular are discussed in Chap. 10. In most cases, the ensemble solution is better than that obtained from any of the individual methods.

Classification ensembles can be developed in a similar way, using the corresponding AI-based classification methods and logistic regression as a statistical benchmark.

4.3   How to Select the Best Solutions for the Business Problem

The selection is based on two criteria: (1) The capabilities of AI-based methods and (2) its potential for different business applications.

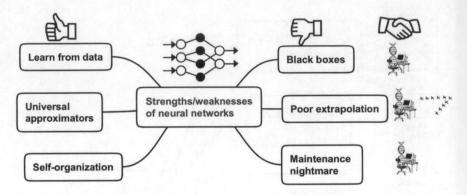

Fig. 4.9 Key capabilities of neural networks

4.3.1   Capabilities of Methods

By the capabilities of a method we mean an overall assessment of the features of each AI-based approach, focusing on the top three strengths and weaknesses and the options to compensate for the drawbacks by integration with other methods. The strengths and weaknesses discussed are based not only on purely technical advantages/disadvantages but also on assessments of the potential for value creation. A mind map template for the representation of the capability of a method includes the blocks with the strengths and weaknesses and icons for corresponding approaches that can counteract the flaws.

Capabilities of Neural Networks

The key feature of neural networks is their ability to learn from examples through repeated adjustments of their parameters (weights). The capability mind map of neural networks is shown in Fig. 4.9 and discussed briefly below.

Key advantages of neural networks are:

- *Learning from data*. Neural networks are the most well-known and broadly applied machine learning approach. They are capable of defining unknown patterns and dependencies from available data. The process of learning is relatively fast and does not require deep fundamental knowledge of the subject. The learning algorithms are universal and applicable to a broad range of practical problems. The required tuning parameters are few and relatively easy to understand and select. In cases where patterns are to be learned from many variables, neural networks outperform humans.
- *Universal approximators*. It has been theoretically proven that neural networks can approximate any continuous nonlinear mathematical function, provided that there is sufficient data and a sufficient number of neurons.
- *Self-organization*. A big advantage of neural networks is their capability to handle unlabeled data by means of self-organized structures, such as Kohonen

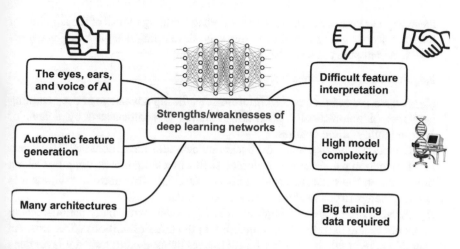

Fig. 4.10 Key capabilities of deep learning networks

self-organized maps. This allows clustering of the data into a two-dimensional feature map that is a basis for further data analysis.[8]

The key disadvantages of neural networks, namely black-box models, poor extrapolation, and the maintenance nightmare, have already been discussed in Sect. 1.4.

Capabilities of Deep Learning Networks

The key characteristic of deep learning networks is their ability to learn how to learn features. The capability mind map of deep learning networks is shown in Fig. 4.10 and discussed briefly below.

Key advantages of deep learning networks are:

- *They are the eyes, ears, and voice of AI.* Deep learning networks are the core technology that allows AI to communicate with humans. Image processing and object identification are the algorithmic eyes of AI. Natural language processing algorithms are the ears of AI, and natural language generation methods allow the machine to communicate in plain English.
- *Automatic feature generation.* The wide range of application opportunities of deep learning networks is based on their unique capability to automatically generate a hierarchy of features with increasing complexity. In this way, image recognition is made more efficient.
- *Many network architectures.* Another factor that contributes to the success of deep learning networks is the broad range for options of different neural net architectures—convolutional neural networks, recurrent neural networks, and long short-term memory neural networks to name the most popular. In fact,

---

[8]S. Heykin, *Neural Networks and Learning Machines*, third edition, Pearson, 2016.

these popular network architectures are used as building blocks of the structure of the final deep learning network. In this way, developing deep learning networks with different structures can be done very flexibly.

Key disadvantages of deep learning networks are:

- *Difficult feature interpretation.* Unfortunately, the big advantage of deep learning networks of automatically generating features is compensated by a lack of interpretability of the features.
- *High model complexity.* Some deep learning networks have more than 100 layers and 100 million parameters. Developing such a modeling monster requires a long time even on the most powerful hardware. One area for future development is deriving better structures based on neuroevolution.
- *Big training data required.* High-complexity models with many parameters are data-hungry and need big data, sometimes in the order of millions of records. An example of ten million records with cat images being used to train a 9-layer deep neural network was given by Le et al.[9] The training was performed on a cluster of 1000 machines (16,000 cores) for 3 days! Labeling these records can increase significantly the cost.

Capabilities of Support Vector Machines
In the real world, learning is the process of estimating an unknown relationship or structure using a limited number of observations. SVMs derive solutions with an optimal balance between accurately representing the existing data and dealing with unknown data. In contrast to neural networks, the results from statistical machine learning have optimal complexity for the given learning data set and may have some generalization capability if proper parameters are selected. SVMs represent the knowledge learned by means of the most informative data points, called support vectors. The capability mind map of SVM is shown in Fig. 4.11 and discussed briefly below.

Key advantages of support vector machines are:

- *Learning from small data records.* In contrast to neural networks, which require significant amounts of data for model development, SVMs can derive solutions from a small number of records. This capability is critical in new product development, where each data point could be very costly.
- *Model complexity control.* SVMs allow explicit control over the complexity of the derived models by tuning some parameters. In addition, statistical learning theory, on which SVMs are based, defines a direct measure of model complexity. According to this theory, an optimal complexity exists. This is based on the best balance between the interpolation capability of the model on known data and its generalization abilities on unknown data. As a result, derived SVM models with

---

[9]Q. Le et al., Building high-level features using large scale unsupervised learning, *Proceedings of the 29th International Conference on Machine Learning*, Edinburgh, UK, 2012.

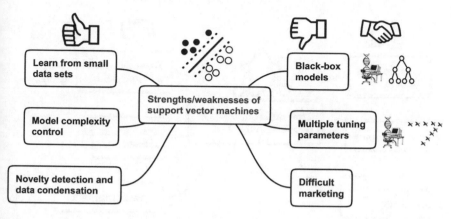

Fig. 4.11 Key capabilities of support vector machines

optimal complexity have the potential for reliable operation during process changes.

- *Novelty detection and data condensation.* At the foundation of SVM models are information-rich data points, called support vectors. This feature gives two unique opportunities. The first opportunity is to compare the informational content of each new data point with the existing support vectors and to define novelty if some criteria are met. The second opportunity is to condense a large number of data records into a small number of support vectors with high information content. Very often, this data condensation could be significant and high-quality models could be built with only 10–15% of the original data records.

Key disadvantages of support vector machines are:

- *Black-box models.* For a nonexpert, the interpretability of SVM models is an even bigger challenge than understanding neural networks. The main flaw, however, is the black-box nature of these models, which provokes the same negative response from users, as that already discussed for neural networks. Even the better generalization capability of SVMs relative to neural networks cannot change this attitude. One option for improvement is to avoid direct implementation of SVMs as final models but to use some of their advantages, such as data condensation, in the development of models, based on methods with higher interpretability, such as genetic programming or decision trees.
- *Multiple tuning parameters.* Developing SVM models requires more specialized parameter selection, including complex mathematical transformations, called kernels. The prerequisite for understanding the latter is some specialized knowledge of math. Using a GA or PSO for optimal parameter selection is strongly recommended
- *Difficult marketing.* Explaining SVMs and their basis, statistical learning theory, includes very heavy math and is a challenge even to an audience of experienced data scientists. In addition, the black-box nature of the models creates significant

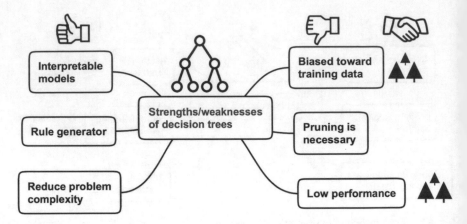

Fig. 4.12 Key capabilities of decision trees

obstacles in marketing the approach to a broad class of users. SVMs are the most difficult AI-based approach to sell.

Capabilities of Decision Trees

The key feature of decision trees is their ability to learn tree-based structures from data. The capability mind-map of decision trees is shown in Fig. 4.12 and discussed briefly below.

Key advantages of decision trees are:

- *Interpretable models.* Decision tree models are very easy to understand even for people from nonanalytical background. They do not require any statistical knowledge to read and interpret them. An additional advantage is that they identify the most significant variables and relations between multiple variables. Decision trees are excellent tools for helping the user to choose between several courses of action.
- *Rules generator.* Some decision trees can be represented directly as generated rules. Many users prefer this form of representation.
- *Reducing problem complexity.* Derived decision tree models reduce the complexity of the problem by the automatic selection of variable for the most relevant inputs/factors. An additional, unique option for reducing complexity provided by this approach is the capability of pruning the trees.

Key disadvantages of decision trees are:

- *They are biased toward training data.* The original decision tree models fit very well the training data but are unreliable predictors or classifiers on different data sets. The problem can be resolved by using a more powerful version of this technology, random forests, which can deliver high-quality predictive and classification models.

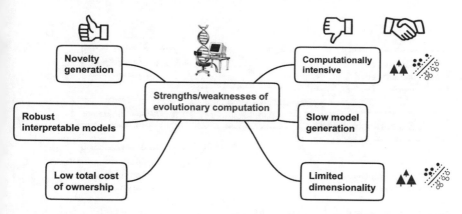

Fig. 4.13 Key capabilities of evolutionary computation

- *Pruning is necessary.* Often, interpretable decision tree models require additional pruning to be done in an interactive way by the user.
- *Low performance.* On average, interpretable decision trees have lower performance as predictors and classifiers relative to other methods. The best way to address this issue is to use random forests as the final modeling technique for obtaining solutions to business problems.

Capabilities of Evolutionary Computation

Evolutionary computation creates novelty by means of automatically generated solutions of a given problem with a defined fitness. The capability mind map of evolutionary computation is shown in Fig. 4.13 and discussed briefly below.

Key advantages of evolutionary computation are:

- *Novelty generation.* The most admirable feature of evolutionary computation is its unlimited potential to generate novel solutions for almost any practical problem. When the fitness function is known in advance and can be quantified, the novel solutions are selected after a prescribed number of evolutionary phases.
- *Robust models.* If complexity has to be taken into account, evolutionary computation can generate gray-box models with an optimal balance between accuracy and complexity. This is a best-case scenario for the final user, since she/he can implement simple empirical models which can be interpreted and are robust toward minor process changes. Several examples of robust models generated by evolutionary computation are given in different chapters in the book.
- *Low total cost of ownership.* A very important feature of evolutionary computation is that the total development, implementation, and maintenance cost is often at an economic optimum. The development cost is minimal, since robust models are generated automatically with minimal human intervention. In the case of a multiobjective fitness function, the model selection is fast and evaluating the best trade-off solutions between the different criteria takes a very short time. Since the resulting empirical solutions are explicit mathematical functions, they can be

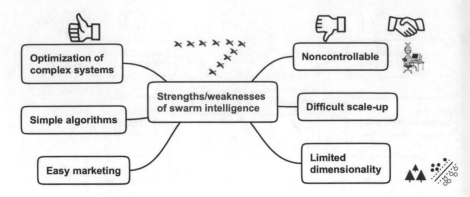

Fig. 4.14 Key capabilities of swarm intelligence

implemented in any software environment, including Excel. As a result, the implementation cost is minimal. The robustness of the selected models also reduces the need for model readjustment or redesign during minor process changes. This minimizes the third component of the total cost of ownership, maintenance cost.

Key disadvantages of evolutionary computation are:

- *They are computationally intensive.* Simulated evolution requires substantial number-crunching power. Fortunately, the continuous growth of computational power, according to Moore's law, is gradually resolving this issue. Additional gains in productivity are being made by use of improved algorithms. The third way to reduce computational time is to shrink down the dimensionality of the data. For example, the most informative data can be selected for simulated evolution by other methods such as random forests (by variable selection) and support vector machines (by record selection).
- *Time-consuming solution generation.* An inevitable effect of the computationally intensive simulated evolution is slow model generation. However, the slow speed of model generation does not significantly raise the development cost. The lion's share of this cost is taken by the model developer's time, which is relatively small and limited to that required for a small number of potential models on the Pareto front for selection and validation.
- *Limited dimensionality.* Evolutionary computation is not very efficient when the search space is extremely large. It is preferable to limit the number of variables to up to one thousand and the number of records to up to one million in order to get results in an acceptable time. It is strongly recommended to reduce the dimensionality before using evolutionary computation.

Capabilities of Swarm Intelligence

Swarm intelligence offers new ways of optimizing and classifying complex systems in real time by mimicking social interactions in animal and human societies. The capability mind map of swarm intelligence is shown in Fig. 4.14 and discussed briefly below.

Key advantages of swarm intelligence are:

- *Optimization of complex systems.* Swarm intelligence algorithms are complementary to classical optimizers and are very useful in cases with static and dynamic combinatorial optimization problems, distributed systems, and very noisy and complex search spaces. In all of these cases, classical optimizers cannot be efficiently applied.
- *Simple algorithms.* In contrast to classical optimizers, most swarm intelligence algorithms are simple and easy to understand and implement. However, they still need the setting of several tuning parameters, most of them problem-specific.
- *Easy marketing.* The principles of swarm intelligence systems are well understood by potential users. Most of them are willing to take the risk of applying the approach. The methods do not have special software or hardware requirements.

Key disadvantages of swarm intelligence are:

- *They are noncontrollable.* There is no authority in charge. Guiding a swarm system can only be done as a shepherd would drive a flock by applying force at crucial leverage points, and by subverting the natural tendencies of the system to new ends. One way to resolve this issue is either to optimize the tuning parameters by evolutionary computation (in particular by genetic algorithms) or to use genetic algorithms as an alternative optimization method for solving the business problem.
- *Difficult scale-up.* Applying swarm intelligence solutions to similar problems on a larger scale is not reproducible. This increases the development cost and reduces productivity.
- *Limited dimensionality.* Swarm intelligence methods are not very effective with big data. It is recommended, before applying them to reduce the dimensionality of the data by variable selection using random forests and records selection using SVMs.

Capabilities of Intelligent Agents
Intelligent agents can generate emergent behavior by mimicking human interactions, such as negotiation, coordination, cooperation, and teamwork. Chatbots are a popular agent-based application for effective human–machine communication. The capability mind map of intelligent agents and chatbots is shown in Fig. 4.15 and discussed briefly below.

Key advantages of intelligent agents are:

- *Bottom-up modeling.* In contrast to first-principles modeling, based on the laws of nature, or empirical modeling, based on different techniques of data analysis, intelligent agents are based on the emergent behavior of their constituents—the different classes of agent. At the foundation of agent-based modeling is the hypothesis that even a simple characterization of the key agents can generate

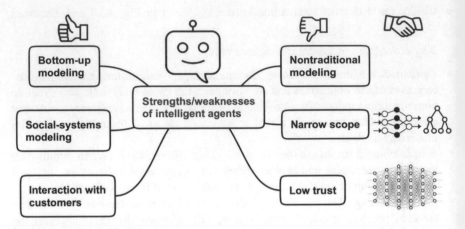

Fig. 4.15 Key capabilities of intelligent agents

complex behavior of the system. Such types of system cannot be described by mathematical methods. A classic example is analyzing the complex behavior of supply chain systems based on a bottom-up description of the key types of agent—customer, retailer, wholesaler, distributor, and factory.

- *Social-systems modeling.* Intelligent agents are one of the few methods available to simulate social systems. It is our strong belief that simulations of social response will be the most significant addition to the current modeling techniques in Data Science. The fast market penetration in a globalized world with diverse cultures of potential customers requires modeling efforts for accurate assessment of customers' acceptance of future products. Intelligent agents can help to fill this gap.
- *Interaction with customers.* Chatbots and digital assistants can communicate with humans 24/7. They combine communication capabilities with learning customer behavior from available data and reasoning algorithms.

Key disadvantages of intelligent agents are:

- *Nontraditional modeling.* Modeling with intelligent agents is quite unusual in comparison with other Data Science methods. One cannot expect results in the form of tangible models such as equations or defined patterns and relationships. Usually the results from the modeling are visual, and the new emergent phenomena are represented by several graphs. In some cases, it is possible that the simulation outcome can be verbally condensed into a set of rules or a document. However, the qualitative results may confuse some of the users with a classical modeling mindset and the credibility of the recommendations could be questioned. Validating agent-based models is also difficult, since the knowledge about the emerging phenomena is nonexistent.
- *Narrow scope.* Chatbots are limited to performing an exact task, or maybe a few similar tasks, but no more. In the case of customer requests beyond the range of

the chatbot's knowledge, there could be a misunderstanding if there is no human consultant to intervene. A possible solution to this issue is a better reasoning algorithm, with more choices based on decision trees and improved learning based on neural networks.

- Low trust. Chatbots are not mature enough to build solid trust in customers. Mistakes in speech recognition and NLP still happen rather frequently—and, as a result, customer instructions are not carried out properly. There are chatbots which are used for sending out spammy promotional content. Most of the chatbots ask customers to sign up for automatic updates to keep them posted with relevant information, but there are chatbots which consider an initial conversation as permission to send regular updates, which are unwanted information. In a nutshell, the quality of service based on chatbots still requires considerable improvement, probably with better NLP capabilities.

4.3.2 Applicability of Methods

The other important factor in making an appropriate selection of an AI-based method is its record of application to relevant business solutions. Here, we give the reader a condensed view of the six key application area of each AI method selected for discussion in the book. The visual part includes a mind map with the names and icons or clipart images[10] of the application areas. It is combined with a short description of the applications, and in many cases specific company names are given.

Applicability of Neural Networks
Neural networks have been applied in thousands of industrial applications over the last 20 years. Among the interesting cases are: using an electronic nose for the classification of honey, applying neural networks to predict beer flavors from chemical analysis, and using an on-boat navigator to make better decisions in real time during competitions. The selected application areas of neural networks are shown in Fig. 4.16 and discussed briefly next:

- *Marketing.* Neural networks are used in a wide range of marketing applications by companies such as Procter & Gamble, Sears, and Experian. Typical applications include: targeting customers by learning critical attributes of customers in large databases, developing models of aggregate behavior that can suggest future consumption trends, and making real-time price adjustments as incentives to customers with discovered patterns in purchase behavior.
- *Nonlinear process control.* With their unique capability to capture nonlinearity from available process data, neural networks are at the core of nonlinear process control systems. The two key vendors of these systems—Aspen Technology and

---

[10]The icons in the book are based on those available at www.thenounproject.com. The clipart images are based on those available at www.clipart.com

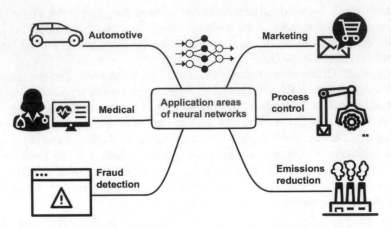

Fig. 4.16 Key application areas of neural networks

Rockwell Automation—have implemented hundreds of nonlinear control systems in companies such as Fonterra Cooperative, Eastman Chemical, and Sterling Chemicals, with total savings of hundreds of millions of dollars. The sources of savings are consistently operating units in their optimal range through cost-effective debottlenecking and identification of optimal operating conditions, transition strategies and new grades, minimizing inventory through accurate demand forecasting, optimum planning and scheduling of grades, and agile manufacturing.

- *Emissions reduction.* Environmental limits are one of the significant constraints in optimal control. Unfortunately, most hardware emissions sensors are very expensive (in the range of several hundred thousand dollars). An alternative software sensor, based on a neural network, is a much cheaper solution and has been applied in companies such as The Dow Chemical Company, Eastman Chemical, and Michigan Ethanol Company. The economic impact of continuous NOx, CO, or CO_2 emissions estimation is a significant reduction in compliance cost, increasing production by 2–8%, reduced energy cost per ton by up to 10%, and reduced waste and raw materials by reducing process variability by 50–90%.
- *Fraud detection.* One of the most popular applications of neural networks is in fraud detection. For example, the London-based HSBC, one of the world's largest financial organizations, uses neural network solutions to protect more than 450 million active payment card accounts worldwide. An empirical model calculates a fraud score based on transaction data at the cardholder level as well as the transaction and fraud history at the merchant level. The combined model significantly improves the percentage of fraud detection.
- *Medical.* Hospitals and health management organizations use neural networks to monitor the effectiveness of treatment regimens, including physicians' performance and drug efficacy. Another application area is in identifying high-risk members of health plans and providing them with appropriate interventions at the right time to facilitate behavior change. For example, Nashville-based

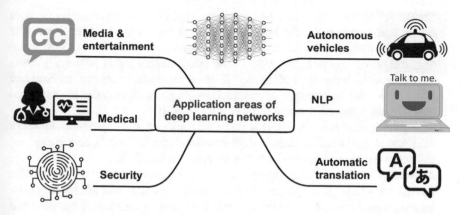

Fig. 4.17 Key application areas of deep learning networks

Healthways, a health provider to more than two million customers in all 50 states of the USA, is building predictive models that assess patient risk for certain outcomes and establish starting points for providing services.

- *Automotive*. Neural networks have a presence in our cars. In the Aston Martin DB9, for example, Ford has used a neural network to detect misfires in the car's V12 engine. To discover the misfire patterns, the development team drove a vehicle under every imaginable condition. They then forced the vehicle into every conceivable type of misfire. The data collected during the test drives was returned to the lab and fed into a training program. By examining the patterns, the neural network was trained in a simulation to distinguish the normal patterns from the misfires. The final result determines a critical threshold number of misfires in the system. Once the neural network detects the critical number, the misfiring cylinder is shut down to avoid damage such as a melted catalytic converter. Neural networks were the only cost-effective way Ford could make the DB9 meet the demanding Californian emissions requirements.

Applicability of Deep Learning Networks

Deep learning is at the core of image recognition and automatic image labeling, text to speech, speech to text, automatic translation, and sentiment analysis. It also is beginning to be used in entertainment recommenders and for detection of anomalies such as transaction fraud. Even though the technology is still in the R&D domain, many companies regard deep learning as a critical application area. The selected application areas of deep neural networks are shown in Fig. 4.17 and discussed briefly next:

- *Autonomous vehicles*. Autonomous vehicle (AV) technology offers the possibility of fundamentally changing transportation. Equipping cars and light vehicles with this technology will likely reduce crashes, energy consumption, and pollution–and also reduce the costs of congestion. AV technology will also increase mobility for those who are currently unable or unwilling to drive.

Level 4 AV technology, when the vehicle does not require a human driver, would enable transportation for the blind, disabled, or those too young to drive. The benefits for these groups would include independence, reduction in social isolation, and access to essential services. The competition for this potentially enormous market includes most of the established car manufacturers such as Ford and GM, Tesla, and technology giants such as Apple and Google. Deep neural networks are a key component in most AV systems.

- *Natural language processing*. This powerful capability of deep neural networks has many potential applications in digital assistants (online retail), search engines, and many solutions that require a human–machine interface.
- *Automatic translation*. One application area with growing potential is machine translation. Although big players such as Google Translate and Microsoft Translator offer near-accurate, real-time translations, some professional domains and industries call for highly specific training data related to a particular domain in order to improve accuracy and relevancy. Here, generic translators would not be of much help, as their machine learning models are trained on generic data. This gap gives opportunities to specialized machine translation businesses that use domain-specific corpuses to train their deep learning networks and deliver high-quality translations.
- *Security*. Face recognition is another popular application of deep learning networks and is a critical component of many systems for security.
- *Medical*. The enormous image-processing capabilities of deep learning networks have begun to penetrate the medical area, following the foot steps of neural networks. One specific type of application is medical diagnostics.
- *Media and entertainment*. The unique combination of video, image, and natural language processing creates many opportunities in the media and entertainment industry. Some examples are video captioning, video search, and computer games.

Applicability of Support Vector Machines

In general, SVMs deliver high-quality models from a small data set. Unfortunately, the method is not very popular in industry due to the difficulty of explaining its principles and the black-box nature of the models generated.

The selected application areas of support vector machines are shown in Fig. 4.18 and discussed briefly next:

- *Medical*. SVMs have been used in several medical diagnostics applications, such as detection of leukemia and prostate and colon cancer, as well as in image analysis of mammograms and tumor cell imaging.
- *Bioinformatics*. Recently bioinformatics has appeared as one of the most promising application areas of SVMs. Examples are gene-expression-based tissue classification, protein function and structure classification, and protein subcellular localization.

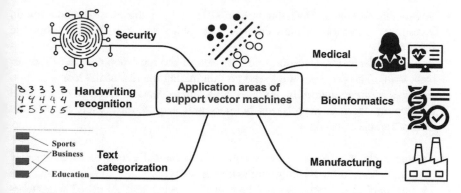

Fig. 4.18 Key application areas of support vector machines

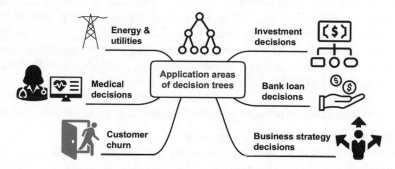

Fig. 4.19 Key application areas of decision trees

- *Manufacturing.* SVMs have been applied in two key areas of manufacturing—inferential sensors and advanced process control.
- *Text categorization.* A typical application of SVMs is text and hypertext categorization. This uses training data to classify documents into different categories such as news articles, e-mails, and web pages.
- *Handwriting recognition.* The most popular application of SVMs is in recognizing hand-written characters used in data entry and validating signatures on documents.
- *Security.* SVMs can be used as a low-cost solution for face recognition.

Applicability of Decision Trees

With their high interpretability, decision trees are one of the most popular AI-based approaches. Most of the applications are in the financial industry, where interpretability of models is expected.

The selected application areas of decision trees are shown in Fig. 4.19 and discussed briefly next:

- *Investment decisions.* Decision tree models suggest the best decision for an investment based on defined criteria, historical data, and selected economic factors.
- *Bank loan decisions.* Decision trees are one of the most popular classifiers for bank loans. The decisions are based on minimizing the risk of the loan.
- *Business strategy decisions.* Another area where decision trees are used is in developing decision models for mergers and acquisitions based on relevant factors and historical data.

- *Customer churn.* Decision trees have been successfully applied to analysis of customer churn in the telecommunication industry.
- *Medical decisions.* Decision trees are one of the acceptable AI-based approaches in healthcare. A well-established application area is in analyzing the impact of multiple risk factors on a disease. A typical case is an analysis of risk factors related to major depressive disorder. The decision tree model developed identified the most important risk factors from a pool of 17 potential risk factors, including gender, age, smoking, hypertension, education, employment, and life events.
- *Energy and utilities.* A major power distribution organization with over one million residential and business customers is using decision trees to configure work orders for asset remediation, including the labor, materials, services, and plant/tool resources required to perform the construction and maintenance work.

Applicability of Evolutionary Computation

The business potential of evolutionary computing in the area of engineering design was rapidly recognized in the early 1990s by companies such as General Electric, Rolls Royce, and BAE Systems. In a few years, many industries, such as the aerospace, power, and chemical industries, transformed their interest in evolutionary computing into practical solutions. Evolutionary computing entered industry in a revolutionary rather than evolutionary way!

The selected application areas of evolutionary computation are shown in Fig. 4.20 and discussed briefly next.

- *Optimal design.* This was one of the first and is one of the most popular implementation areas of evolutionary computation. A typical example is the application of genetic algorithms to the preliminary design of gas turbine blade-cooling hole geometries at Rolls Royce.[11] The objective was to minimize the fuel consumption of the gas turbine. Another example of a real-world application of a genetic algorithm is a built-in optimizer in a program called ColorPro at Dow Chemical. This program is used to optimize mixtures of polymer and colorants to match a specific color. The optimizer allows simultaneous optimization of

---

[11]I. Parmee, *Evolutionary and Adaptive Computing in Engineering Design*, Springer, 2001.

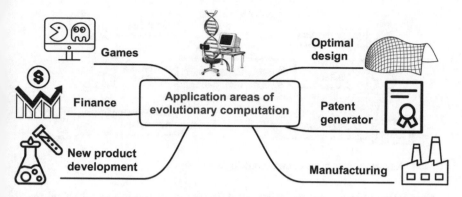

Fig. 4.20 Key application areas of evolutionary computation

multiple objectives (such as color match for a specific light source, transmittance, mixing, and cost).

- *Patent generator*. Genetic programming, with its unique capability to generate novel structures, can be used as a method for automatically synthesizing patentable structures of physical devices. This is an application area that was explored very intensively by the founding father of this approach John Koza.[12] The generation of novel structure was demonstrated for two specific physical systems—analog circuits and optical systems. The analog circuits that were successfully synthesized by GP included both passive components (capacitors, wires, resistors, etc.) and active components (transistors). The target system had to satisfy a specific frequency response curve. First, the credibility of GP was validated by duplicating the functionality of six current patents in the area of analog circuit design. In the next step, the technology was used to create new patentable structures.

- *Manufacturing*. One area of manufacturing with tremendous potential for evolutionary computation and especially for symbolic regression, generated by GP, is inferential or soft sensors. The current solutions on the market, based on neural networks, require frequent retraining and specialized run-time software. The alternative solution of simple algebraic equations is better accepted by process engineers and applied in many industrial processes. More details are given in Chap. 14.

- *New product development*. GP-generated models can reduce the cost of new product development by shortening the development time of fundamental models. For example, symbolic regression proto-models can significantly reduce the hypothesis search space for potential physical/chemical mechanisms. As a result, effort in new product development could be considerably reduced by

---

[12]J. Koza, et al., *Genetic Programming IV: Routine Human-Competitive Machine Intelligence,* Kluwer, 2003.

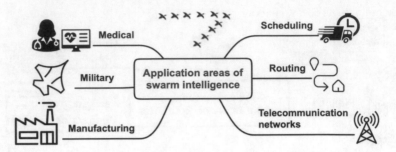

Fig. 4.21   Key application areas of swarm intelligence

eliminating unimportant variables, enabling rapid testing of new physical mechanisms, and reducing the number of experiments for model validation.

- *Finance*. Evolutionary computation, and especially GAs and symbolic regression
  are widely used in various financial activities. For example, State Street Global
  Advisors is using GP for derivation of stock selection models for a low-active-
  risk investment style. Historical simulation results indicate that portfolios based
  on GP models outperform the benchmark and portfolios based on traditional
  models. In addition, GP models are more robust in accommodating various
  market regimes and have more consistent performance than the traditional
  models.[13]

- *Games*. One of the best examples of the use of evolutionary computation in
  games is the famous game[14] Blondie24. It is based on evolving neural
  networks. The game begins with 15 parents, with neural network weights randomly initialized. During simulated evolution, each parent generates one offspring and all 30 players compete with five players randomly selected from the
  population. Fifteen players with the greatest total of points are selected as parents
  for the next generation, and so on for 100 generations.

Applicability of Swarm Intelligence

The application record of swarm intelligence is not as impressive as that of the more
established AI-based methods. However, the speed of adoption in real-world applications, especially of particle swarm optimization, is growing. The selected application areas of swarm intelligence are shown in Fig. 4.21 and discussed briefly next:

- *Scheduling*. The unique features of ant colony optimization for task allocation
  and job scheduling have been successfully used in several industrial applications.
  One of the most impressive applications is a joint ACO and genetic algorithm for
  optimal scheduled deliveries of liquid gas to 8000 customers at American Air

---

[13]Y. Becker, P. Fei, and A. Lester, Stock selection: An innovative application of genetic programming methodology, In R. Riolo, T. Soule, and B. Worzel (eds.), *Genetic Programming Theory and
Practice IV*, pp. 315–335, Springer, 2007.

[14]www.digenetics.com

Liquide.[15] The cost savings and operational efficiencies of this application for one of the company's 40 plants are more than $six million per year. Another application of swarm intelligence-based methods, to scheduling of truck painting at General Motors, claims at least $three million per annum. An interesting industrial scheduling problem in an Alcan aluminum casting center was successfully resolved by using an ACO method which can generate 60 optimal schedules in less than 40 seconds.[16]

- *Routing.* One of the first successful ACO applications was in finding optimal routs for Southwest Airlines' cargo operations. The derived solutions looked strange, since it was suggested that cargo should be left on a plane headed initially in the wrong direction. However, implementing the algorithm resulted in significant cutback on cargo storage facilities and reduced wage costs. The estimated annual gain is more than $ten million.[17]

  Another fruitful application area is optimal vehicle routing. Several impressive applications have been implemented by the Swiss company AntOptima.[18] Examples are DyvOil, for the management and optimization of heating oil distribution and OptiMilk, for improving the milk supply process.

- *Telecommunication networks.* Optimizing telecommunication network traffic is a special case of routing with tremendous value creation potential due to the large volume. This is a difficult optimization problem because the traffic load and network topology vary with time in unpredictable ways and there is a lack of central coordination. All of these features suggest that ACO could be an appropriate solution for this type of problem. A special routing algorithm, called AntNet, has been developed and tested on different networks under different traffic patterns. It has proved to be very robust and, in most cases, better than the competitive solutions[19]. ACO has been used by leading telecommunication companies such as France Telecom, British Telecom, and the former MCI WorldCom.

- *Manufacturing.* Recently PSO has been applied in several process optimization problems in manufacturing. Some applications in The Dow Chemical Company include using PSO for optimal color matching, estimation of the optimal parameters of acoustic foams, estimation of the optimal parameters of crystallization kinetics, and selection of optimal neural network structures for day-ahead

---

[15]C. Harper and L. Davis, Evolutionary computation at American Air Liquide, *SIGEVO newsletter*, **1**, 1, 2006.

[16]M. Gravel, W. Price, and C. Cagne, Scheduling continuous casting of aluminum using a multiple objective ant colony optimization metaheuristic, *European Journal of Operational Research*, **143**, pp. 218–229, 2002.

[17]E. Bonabeau and C. Meyer, Swarm intelligence: A whole new way to think about business, *Harvard Business Review*, May 2001.

[18]www.antoptima.com

[19]M. Dorigo, M. Birattari, and T. Stützle, Ant colony optimization: Artificial ants as a computational intelligence technique, *IEEE Computational Intelligence Magazine*, **1**, pp. 28–39, 2006.

forecasting of electricity prices.[20] Examples of other interesting applications in this area include optimization of numerically controlled milling reactive power and voltage control, estimation of the state of charge of battery packs, and optimization of cracking furnaces.

- *Military.* The most well-known military application of swarm intelligence is in developing a "swarm" of small unmanned aerial vehicles (UAVs) with capabilities to carry out key reconnaissance and other missions at low cost. For example, a swarm of surveillance UAVs could keep watch over a convoy, taking turns to land on one of the trucks for refueling.

  A different approach is the "cooperative hunters" concept, where a swarm of UAVs search for one or more "smart targets," moving in a predefined area while trying to avoid detection. By arranging themselves into an efficient flight configuration, the UAVs optimize their combined sensing and are thus capable of searching larger territories than a group of uncooperative UAVs. Swarm control algorithms can optimize flying patterns over familiar terrain and introduce fault tolerance to improve coverage of unfamiliar and difficult terrain.[21]

- *Medical.* One of the first medical applications of PSO was for successful classification of human tremors, related to Parkinson's disease. A hybrid clustering approach based on self-organizing maps and PSO has been applied in several different cases for gene clustering of microarrays. Another idea, of using a swarm of nanoparticles to fight cancer cells, has come close to reality. A research team led by scientists at the University of Texas M.D. Anderson Cancer Center and Rice University has shown in preclinical experiments that cancer cells treated with carbon nanotubes can be destroyed by noninvasive radio waves that heat up the nanotubes while sparing untreated tissue. The technique completely destroyed liver tumors in rabbits without side effects.[22]

Applicability of Intelligent Agents

Intelligent agents are still knocking at the door of industry. The application record is short and the value created is low relative to the other AI-based methods. There is a growing tendency, though, and we believe in the big potential for value creation of this technology in both of its directions, agent-based systems and chatbots.

The selected application areas of intelligent agents are shown in Fig. 4.22 and discussed briefly next:

- *Sales.* Gradually, chatbots have become an alternative sales channel, attracting more customers with their improved communication capabilities and 24/7

---

[20]A. Kalos, Automated neural network structure determination via discrete particle swarm optimization (for nonlinear time series models), *Proceedings of fifth WSEAS International Conference on Simulation, Modeling and Optimization*, Corfu, Greece, 2005.

[21]Smart Weapons for UAVs, *Defense Update*, January 2007.

[22]C. Gannon et al. Carbon nanotube-enhanced thermal destruction of cancer cells in a noninvasive radiofrequency field, *Cancer*, **110**, pp. 2654–2665, 2007.

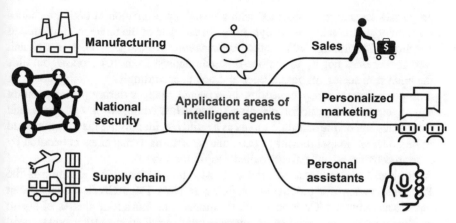

Fig. 4.22 Key application areas of intelligent agents

presence. Ecommerce brands such as H&M, eBay shopBot, and Tommy Hilfiger are selling many products using chatbots.

- *Personalized marketing.* Many companies, including Kik, Uber, Sephora, and more, are now using chatbots to interact with their customers and drive engagement. The trend is to build customized models based on individual customers' preferences.
- *Personal assistants.* A growing number of personal assistants for narrowly focused activities, such as finance handling, control of health activities, and daily scheduling, are gaining in popularity. Their integration and personalization are expected.
- *Supply chain.* An important niche for the application of intelligent agents is dynamic supply chain scheduling. An example is the system applied at Tankers International, which operates one of the largest oil tanker pools in the world. [23] The objective is to dynamically schedule the most profitable deployment of ships to cargo for its very large crude carrier fleet. An agent-based optimizer, Ocean i-Scheduler, was developed by Magenta Technology for use in real-time planning of the assignment of cargo to vessels in the fleet. The system can dynamically adapt plans in response to unexpected changes, such as transportation cost fluctuations or changes to vessels, ports or cargo.
- *National security.* An unexpected application area of agent-based simulations is in analyzing the formation and dynamic evolution of terrorist networks. An example is the agent-based system NetBreaker,[24] developed at Argonne National Laboratory. This considers both the social and the resource aspects of terrorist networks, providing a view of possible network dynamics. As the simulation

---

[23]M. Luck, P. McBurney, O. Shehory, and S. Willmott, *Agent Technology Roadmap*, AgentLink III, 2005.

[24]http://www.anl.gov/National_Security/docs/factsheet_Netbreaker.pdf

progresses an analyst is provided with a visual representation of both the shapes
the network could take and its dynamics, an estimate of the threat, and quantified
questions illustrating what new information would be most beneficial.
NetBreaker does not try to remove a human analyst from this process, but aids
the analyst in seeing all possibilities and acting accordingly.

- *Manufacturing.* In many industries, unexpected changes during plant operation
  can invalidate the production planning and schedule within one or two hours, and
  from that point on the schedule serves as a build list, identifying the jobs that need
  to be done, while assignment of parts and operations to machines is handled by
  other mechanisms, collectively called shop-floor control.
  Agent-based systems can be used to handle this problem. An example is the
  ProPlantT system for production planning at Tesla-TV, a Czech company that
  constructs radio and TV broadcasting stations. It includes four classes of agent:
  production planning, production management, production, and a meta-agent,
  which monitors the performance of the agent community. Chrysler has also
  explored the capabilities of agent-based shop-floor control and scheduling.[25]

4.3.3 One Method Is Not Enough

An important conclusion that can be drawn after discussing the capabilities and
applicability of AI-based methods is that solving a specific business problem is not
limited to one method. Several alternatives are possible, and it is strongly
recommended that at least some of them are explored. If this is done, there is a
high probability that the final solution will be based on the best selection or
combination of the most appropriate algorithms. The steps for the selection of
methods is shown in Fig. 4.23 and discussed below.

The first step in this process is dictated by the definition of the business problem,
where the type of problem and specific requirements for the solutions are defined.
For example, if the business problem is defined as developing an inferential sensor
for emissions, the key type of application is manufacturing, and a more specific
category is enhanced process monitoring. An important piece of information for the
selection of methods is the specific requirement that the generated models are not
black boxes.

The second step in the process of selection of AI-methods is looking at the
applicability of the available methods for the defined problem. In this specific
example, three AI-based algorithms—neural networks, support vector machines,
and evolutionary computation—have been applied to similar problems. Of special
importance are applications to obtain reduced emissions by the use of neural

---

[25]V. Parunak, A practitioner's review of industrial agent applications, *Autonomous Agents and Multi-Agent Systems*, **3**, pp. 389–407, 2000.

Fig. 4.23 Key steps for selection of AI-based methods to solve a defined business problem

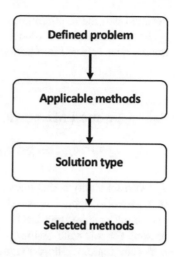

Table 4.1 The most appropriate AI-based methods for key types of solution

Solution type	NN	DLN	SVM	DT	EC	SI	IA
Prediction	x		x	x	x		
Forecasting	x		x		x		
Classification	x	x	x	x			x
Clustering	x	x	x				x
Optimization					x	x	
Association				x			x

networks and inferential sensors based on symbolic regression generated by genetic programming.

The third step in the process of selection of AI-methods is selecting the type of solution for the defined problem. In our example, the type of solution is prediction, because we want to predict the emissions based on other chemical-process measurements.

The final and most important step is actual selection of the AI-methods. Usually the first broad selection is based on the availability of methods for the selected solution types. A very generic mapping between the six key solution types that, we have discussed in the book and the seven selected AI-based methods is given in Table 4.1. The abbreviations in the table have the following meanings: NN, neural networks; DLN, deep learning networks; SVM, support vector machines; DT, decision trees; EC, evolutionary computation; SI, swarm intelligence, and IA, intelligent agents.

For our example, the broad selection for predictive emissions models includes four AI-based methods—neural networks, support vector machines, decision trees, and evolutionary computation. In order to make the final selection, however, a more detailed analysis of the capabilities of each method is needed. As a result of this analysis, two of the methods from the broad selection—neural networks and support vector machines—were removed from consideration. The key reason for this decision was the requirement that the delivered models were not black boxes. Two AI-based methods for model development—genetic programming and decision trees—were selected in addition to the benchmark linear regression model.

4.4   Common Mistakes

4.4.1   Ignoring Integration of Methods

The current reality is that data scientists are not well-prepared for the big challenges of real-world applications. They don't understand the power of integrating the appropriate methods to deliver the best solution to the defined problem. One of the reasons for this state of things is the lack of training and good references on that important topic. We hope that this book will give the reader an initial start in gaining experience in integrating various methods into their business solutions.

4.4.2   Lack of Knowledge of Strengths and Weaknesses of Methods

Another observed issue in data scientists' decision-making processes is their biased attitude towards selected modeling methods. Usually the advantages of their favorite methods are exaggerated at the expense of their weaknesses. As a result, no attempts to compensate some of these disadvantages with integration of other methods are made. A more balanced evaluation of the most important strengths and weaknesses (considering all related factors, not focusing only on technical performance) is needed. The discussed advantages and disadvantages in this book are a good starting step in this direction. It is recommended, however, to look for broad feedback about applying these methods from other references as well.

4.4.3   Lack of Knowledge about Selecting the Most Appropriate Methods for the Business Problem

In most cases the selection of methods is not systematic but based on subjective factors, such as previous experience and knowledge. Usually this selection is limited to one or two methods, which reduces significantly the efficiency of the proposed solution. The systematic approach to selection of methods based on their applicability and capabilities proposed in the book increases the chance of finding the most appropriate combination of methods for the defined business problem.

4.5   Suggested Reading

- H. Martinez and R. Urraca, *Hybrid Artificial Intelligent Systems*, Springer, 2017.
- L. Medsker, *Hybrid Intelligent Systems*, Kluwer, 1995.
- Z. Michalewicz, M. Schmidt, M. Michalewicz, and C. Chiriac, *Adaptive Business Intelligence*, Springer, 2007.
- D. Ruan (editor), *Intelligent Hybrid Systems: Fuzzy Logic, Neural Networks, and Genetic Algorithms*, Kluwer, 1997.

4.6   Questions

Question 1
Give examples of issues that may arise in applying Data Science in business settings.

Question 2
What are the advantages of integrating different modeling methods?

Question 3
Give examples of real-world applications of your favorite AI-based method.

Question 4
What are the key disadvantages of deep learning networks?

Question 5
Give examples of AI-based methods suited for classification.

Chapter 5
The Lost-in-Translation Trap

Good explanations are like bathing suits; they are meant to reveal everything by covering only what is necessary.

E. L. Konigsburg

The AI-based Data Science methods discussed above and the numerous techniques of their integration create enormous opportunities for diverse business applications. However, a lot of business opportunities are lost due to the slow speed of implementing these technologies in industry. One of the key factors in the slow introduction of AI-driven approaches in business is a lack of knowledge about their capabilities. An obvious solution to this problem is training potential users in the basics of the technologies. Another option is to gradually increase knowledge about the methods discussed by applying them to solving business problems. The key for success in this case is good communication between domain experts and data scientists. They must speak the same language and avoid the lost-in-translation trap (not understanding each other due to technical jargon). One way to accomplish this goal is to define a special role—a business translator who will own and communicate to data scientists the domain knowledge in an organization.[1]

A more realistic solution is suggested in this book. It is assumed that data scientists are the key drivers of the introduction of AI-driven technologies into a business, and, as such, take the initiative and leadership in this process. Part of their responsibility is to establish an effective dialog with business users with maximum clarity. Data scientists can improve the communication on both sides. On the business side, they can help domain experts to communicate business problems in a systematic and understandable way. On the Data Science side, they can educate domain experts with basic knowledge about applied AI-driven methods.

The key approaches to avoiding the lost-in-translation trap between business-savvy domain experts and analytics-savvy data scientists are described in this chapter and illustrated with useful examples. The first objective of the chapter is to give guidance on how to translate a defined business problem to data scientists. The

---

[1]McKinsey Global Institute, *The Age of Analytics: Competing in a Data-Driven World,* 2016.

© Springer Nature Switzerland AG 2020
A. K. Kordon, *Applying Data Science,*
https://doi.org/10.1007/978-3-030-36375-8_5

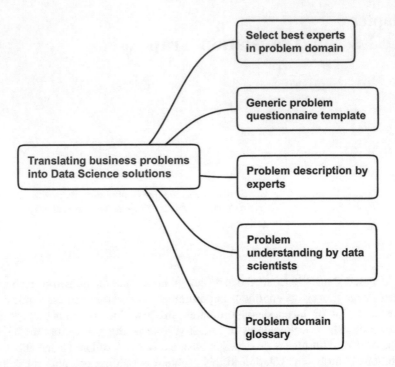

Fig. 5.1 Key steps in translating business problems into Data Science solutions

second is to prepare data scientists for an effective dialog with domain experts by translating related AI-based methods in plain English. The third is to suggest examples of questionnaires, glossaries, and explanation slides to be used to avoid the lost-in-translation trap.

5.1 Translation from Business Problems to Data Science Solutions

First, we'll focus on clarifying the problem domain language and make it understandable to data scientists. The key steps in this process that data scientists should follow are shown in Fig. 5.1 and discussed below.

5.1.1 Select Best Experts in Problem Domain

The first step is the most critical one—finding and engaging the best SMEs with problem domain knowledge. Often the list may include not only employees in the organization which requested the project but SMEs with unique historical experience of the problem who have moved to other departments or have retired. In this case management support is needed to guarantee the SMEs engagement.

It is also recommended to include SMEs with broader areas of expertise than the narrow technical domain according to the problem definition. Typical examples are financial and legal experts. The former are needed to estimate and explain the financial impact of solutions that might be applied and the latter are necessary to estimate and explain potential legal issues and intellectual property protection.

We'll illustrate the process of avoiding the lost-in-translation trap with the example of the emissions estimation project defined in Chap. 2. The team of experts included two technical SMEs working in the unit and responsible for emissions control (who would use the inferential sensor for emissions), one technical expert with 25 years' experience in unit operations (currently working in another department), and a legal SME.

5.1.2   Generic Problem Questionnaire Template

In order to increase the information content and communication efficiency, it is recommended that data scientists take the lead and navigate the dialog. A document that can contribute to a systematic start of the discussion contains generic questions about the nature of the problem. The objective of the questionnaire is to give an overall view of the already defined business problem. The list of questions to SMEs should include business objectives, a technical description, key hypothesis, data sources, knowledge sources, computer infrastructure, and software infrastructure. An example of a questionnaire is given in Sect. 5.4.1.

5.1.3   Problem Description by Domain Experts

The key document for starting the dialog that needs translation is the problem description. There is a high chance that the original explanation of the problem will be full of technical jargon. An example of a description of the emissions estimation problem by process SMEs is given below:

> Current emissions from the unit are measured by lab samples taken every 8 h and entered in a PV tag in the IP21 system. This value is used as a constraint to production rate control. This constraint is linked to the SP tag of the cascade PID controller of the rate. However, the control loop operates every minute and an estimate of the emissions at that frequency is needed for better control. An important condition for using the emissions estimator is that its performance must satisfy the quarterly RATA tests.

5.1.4   Problem Understanding by Data Scientists

An understanding of the business problem by the data scientists by means of a clear explanation from the domain SMEs is the key objective of this translation. Usually this includes a more detailed description of the manufacturing or business process and takes several meetings of the team. It is part of the problem knowledge acquisition step of the Data Science work process, described in detail in Chap. 7.

In our example of emissions estimation project data scientists in the team must understand the principles of the chemical process that is the source of the emissions, the related control system, the existing emissions measurement process, and the environmental regulations in the specific state of the country concerened.

5.1.5   Create a Problem-Related Glossary

An inevitable part of the understanding of business problems by data scientists is translating the business jargon, abbreviations, key concepts, etc. A recommended solution is to create a problem-specific glossary. An example of some terms in such a glossary based on the description of the emissions estimation problem in Sect. 5.1.3 is given below:

- *Cascade process control.* A cascade control loop consists of a primary loop and a secondary loop. These loops might also be referred to as the outer loop and inner loop, respectively. The primary loop provides the secondary loop with a setpoint, or target, for a process related to the primary control objective. The primary loop is sometimes known as the master loop because it provides a setpoint that the secondary loop, or slave loop, must follow.

    The primary purpose of using cascade control as a process control strategy is to allow the secondary loop to control disturbances before they can affect the primary control objective. For this control system to work effectively, the process dynamics of the secondary loop must be much faster than the dynamics of the primary loop. As a general rule, the process dynamics of the secondary loop must be at least four times faster than the process dynamics of the primary loop.
- *IP21 system.* This is the process historian of the plant with full name Aspen Infoplus21 by AspenTech.
- *PID controller.* A proportional–integral–derivative (PID) controller or three-term controller) is a control loop feedback mechanism widely used in industrial control systems. A PID controller continuously calculates an error value $e(t)$ as the difference between a desired setpoint (SP) and a measured process variable (PV) and applies a correction based on proportional, integral, and derivative terms (denoted P, I, and D, respectively) which give the controller its name.
- *Process variable (PV) tag.* The measured value of a process variable.
- *RATA test.* Relative Accuracy Test Audit. The relative accuracy is calculated as the absolute mean difference between the estimated emissions and a value

determined by the measured emissions from a reference measurement of a series of tests divided by the mean of the reference measurement tests. The RATA must be conducted at least once every four calendar quarters.

- *Setpoint (SP) tag.* The desired value of a process variable.

5.2 Translation from Data Science Solutions to Business Problems

Translating the technical language related to selected AI-based methods into plain English, understandable to problem domain experts is the topic of this section. The key steps in this process that data scientists should follow are shown in Fig. 5.2 and discussed below.

5.2.1 Explain Data Science Work Process

The first step in establishing the dialog with the business domain SMEs is to explain the Data Science work process, described in detail in Chap. 6. It is very important that the business members of the team understand that Data Science is not an obscure magic box but a logical sequence of actions from problem knowledge acquisition and data collection and preparation to data analysis and model development and

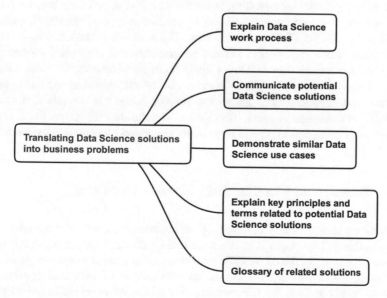

Fig. 5.2 Key steps in translating Data Science solutions into defined business problems

deployment. Of special significance is the need to emphasize the role of data quality in the expected performance of the potential solutions. Another objective of explaining the Data Science work process is to show the steps, such as problem knowledge acquisition, variable selection, and model exploration, where domain expertise is needed for decision-making. It is also recommended that the very important, and often forgotten, step of model deployment and maintenance should be discussed at the beginning. The business users of the Data Science solutions must understand that after model deployment the ownership of the solutions is transferred from the data scientists to them. They have to be prepared for the necessary action items, such as user training, organizing long-term support for the solutions and the corresponding budget.

5.2.2   Communicate Potential Data Science Solutions

Data scientists make decisions about potential solutions based on problem domain knowledge. In this book, the solutions are selected from the list of AI-based technologies, described in Chap. 3. The first step in selection of solutions is choosing the type of solution, described in Sect. 3.1. In the example of the emissions estimation problem, the appropriate solution type is prediction. The defined objective of the problem is to estimate the emissions from available process measurements. This is a typical prediction solution, with the target variable being the emissions and the explanatory variables being the process variables recommended by SMEs.

The next step in decision-making about the selection of solutions is to look at the broad options for technologies related to prediction-type applications, given in Table 4.1. The list includes neural networks, SVMs, decision trees, and evolutionary computation (more specifically, genetic programming). For prediction-type solutions, it is recommended to develop a linear regression benchmark model, i.e., this method is included by default. However, two of the potential methods, neural networks and support vector machines, deliver black-box models that are not acceptable to the business users. This limits the solution methods to linear regression, decision trees, and symbolic regression, generated by genetic programming.

5.2.3   Demonstrate Similar Data Science Use Cases

Of significant importance in convincing the business members of the project team about the selected solutions is to illustrate them with use cases of resolved similar business problems. In the case of our emissions estimation problem, the problem is in the category of inferential sensors, with thousands of industrial applications, discussed briefly in Sect. 2.2.1. However, most of the use cases are based on neural networks, which are not a desirable solution for the users. One potential solution that can satisfy most of the project objectives is the growing application of robust inferential sensors, based on symbolic regression, generated by genetic programming.[2]

5.2.4   *Explain Key Principles Related to Potential Data Science Solutions*

It is recommended that data scientists describe the key principles and benefits of the selected AI-based approaches as well as communicate their strengths and weaknesses. In the case of the emissions estimation problem, Sects. 3.3.4 and 3.3.5 can be used to explain the nature of the two selected approaches—decision trees and evolutionary computation, with an emphasis on genetic programming. An assessment of the capabilities of the methods can be found in Sect. 4.3.1.

5.2.5   *Create a Solution-Related Glossary*

In addition to explanation of the methods, it is recommended that data scientists create a glossary of the key terms used in the terminology of the selected methods. Examples of such glossaries for each of the AI-based methods discussed in the book are given in Sect. 5.4.2. In our example of the emissions estimation project, the solution-related glossary may include the terms for decision trees and evolutionary computation.

5.3   Typical Lost-in-Translation Cases

To some extent, the lost-in-translation trap happens at the beginning of every project, especially in organizations with little experience in data analysis, analytics, or modeling. The most typical cases are described and analyzed below.

5.3.1   *Inexperienced Data Scientists*

A common example of the lost-in-translation trap is when the data scientist responsible for the project is a freshly graduated student with a lack of communication and

---

[2]For this technology, see the recent survey paper A. Kordon, L. Chiang, Z. Stefanov, and I. Castillo, Consider robust inferential sensors, *Chemical Processing*, September 2014.

social skills and only basic business knowledge. The typical behavior in this case is a long process of aligning the business jargon with the Python-R vocabulary of the data scientist. In order to avoid this waste of energy, it is recommended that only socially experienced data scientists lead projects. Another recommendation is that new data scientists take training in communication and social skills and understand the value of avoiding the lost-in-translation trap. Following the action items in this chapter is a good starting point for resolving this problem.

5.3.2   Resistance from Experts

Another typical lost-in-translation trap scenario is driven by domain SMEs. Some of them are not interested in sharing their expertise, due to fear of losing their job if there is a successful AI-driven solution. One strategy they use is to give a very complicated problem description in a jargon-rich language. This behavior continues during all steps of the project, with no attempts to improve the explanations. The other strategy they use is to slow down the process of understanding the selected AI-based technologies. Usually several iterations of technology explanation are requested.

The best way to correct this behavior is to address explicitly the experts' concerns. The management sponsoring the project should give guarantees of job security. It is also recommended to link the success of the project with career performance and bonuses.

5.3.3   Improper Problem Definition

Often the lost-in-translation trap begins with a problem definition full of business and technical jargon. It is preferable to focus on clarifying the message and polishing the style of this important document until it is well understood by all members of the team. It is recommended to follow the suggestions in Sect. 2.4.

5.3.4   Management Intervention

The worst-case scenario of the lost-in-translation trap is when top management intervenes explicitly in project development. This adds an additional communication line with slightly different perceptions and language. Usually management-type language is a rich bullshit generator. There is a high chance that even very clear communication of the business domain problem and the selected AI-driven solutions will be distorted by this intervention. Be prepared for surprises, such as replacing the correct technical terms with strange non-technical phrases or inflated expectations of expected results.

The only solution for avoiding this worst-case scenario is to patiently explain to the management the problem and the selected AI-methods in plain English and to prepare a short summary that can be used in communication with top-management.

5.4 How to Avoid the-Lost-in-Translation Trap

The key objective of avoiding the lost-in-translation trap is to clarify communication among all members of the project team by reducing language ambiguity. The purpose is not to make data scientists problem domain experts and vice versa. The goal is to share the bare minimum of knowledge required for understanding the problem and the selected AI-based methods. In this way, the communication language is jargon-free and a lot of wasted energy due to misunderstanding is saved.

5.4.1 Translators

Several documents, such as questionnaires, glossaries, and visually rich presentations, can be used in the translation. Some examples are given below.

Questionnaires

Most potential business users of AI-based Data Science do not know what kind of information about the business problem they need to communicate to data scientists. It is suggested that the latter should help the users by means of an appropriate questionnaire that requests the most important information to start the dialog. A generic example of such a questionnaire, organized in sections, is given below.

Questionnaire for Starting a New Data Science Project
Business Objectives

- What is the business impact (e.g., increased sales, reduced energy cost, increased profit)?
- How can the business impact be measured and documented (e.g., percentage of sales relative to previous period, documented for the last 3 years)?

Technical Description

- What is the technical problem (e.g., increased steam consumption in one of the units)?
- What is the technical objective (e.g., reduce steam consumption by x%)?

Key Hypothesis

- How is the key hypothesis defined (e.g., steam consumption depends in a nonlinear fashion on measurable process variables)?
- How is the key hypothesis supported (e.g., some patterns between steam consumption and selected process variables have been observed)?

Knowledge Sources

- Who are the SMEs (e.g., inside and outside the organization)?
- What documents are needed to understand the defined problem (e.g., process diagrams and description)?

Data Sources

- What are the data requirements for solving the problem (e.g., data frequency-every minute, hourly, etc.), handling missing data, handling outliers?
- Do we have sufficient data (e.g., length of historical data, changes in operation conditions)?
- Do we need external data (e.g., macroeconomic and microeconomic data from outside the organization)?

Hardware Infrastructure

- On what hardware will the expected solution be deployed (e.g., laptop, cloud, process monitoring system)?

Software Infrastructure

- On what operating system will the expected solution be deployed?
- What is the current database system?
- Is there any statistical, analytical, or math modeling tool used in the business?

Special Requirements

- Are there any regulations to be considered (e.g., emissions model test requirements)?
- Are there any specific business requirements (e.g., no black-box models)?

Glossaries

In the same way as dictionaries are absolutely necessary in translation between different languages, glossaries are recommended tools for improving significantly the understanding between business experts and data scientists. Usually problem-related glossaries are specific to the targeted project and should be requested from the team SMEs. There are many glossaries about AI-based methods, however, and a selected set is given in this book. This set can be used in preparing a selected glossary with the definitions needed for the defined project.

Explanations of Business Problems and AI Methods

The key part of the translation process is an explanation of the business problem from the experts and the AI-based methods from the data scientists both of them as

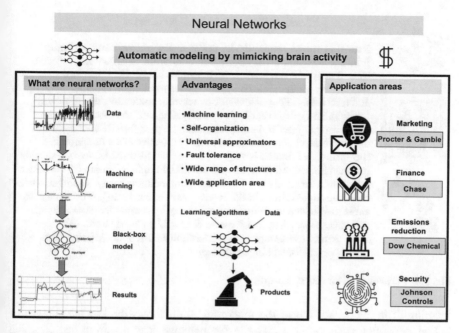

Fig. 5.3 Marketing slide for neural networks

clear as possible. The business explanation usually includes description of the current work process and an explanation of the key buzzwords.

We'll focus in the book on recommendations for delivering a good explanation of the selected AI-based methods by data scientists. This may include the descriptions of the methods with their benefits (given in Sect. 3.3), an assessment of their strengths and weaknesses, and the key application areas (both given in Sect. 4.3). To introduce each method, we recommend that data scientists should include three documents: (1) a marketing slide, (2) an elevator pitch, and (3) a glossary. Examples of these "translators" for all AI-based technologies discussed are given in the next section.

5.4.2   *Examples of Translators for AI-Methods*

Neural Networks Translators

The purpose of the marketing slide is to deliver a very condensed message in a one-page PowerPoint slide about the technology. At the top of the slide is the key slogan that captures the essence of the value creation capability of the technology. In the case of neural networks the slogan is "Automatic modeling by mimicking brain activity" (Fig. 5.3).

The marketing slide is divided into three sections. The left section demonstrates in a very simplified and graphical way how the technology works. In the case of neural networks, this is transforming data into black-box models by machine

Elevator Pitch for Neural Networks

Neural networks automatically discover new patterns and dependencies from available data. They imitate human brain structure and activity. Neural networks can learn from examples in historical data and successfully build data-driven models for prediction, classification, and forecasting. They can even extract information by self-organizing the data. The value creation is by analyzing data and discovering the unknown patterns and models that can be used for effective decision-making. Neural networks are especially effective when developing first-principles models is very expensive and historical data on a specific problem is available. The development and implementation cost of neural networks is relatively low but their maintenance is costly. There have been numerous applications in the last 20 years in marketing, process control, emissions reduction, finance, and security by companies like Eastman Chemical, Dow Chemical, Procter & Gamble, Chase, Ford, etc. It's a mature technology with well-established vendors offering high-quality products, training, and support.

Fig. 5.4 Elevator pitch for neural networks

learning. The middle section of the marketing slide shows the key advantages of neural networks, such as the capability for machine learning from samples, the potential for self-organizing the data when no knowledge is available, and fitting the data to any nonlinear dependence to any desired degree of accuracy. The key application areas of neural networks are shown in the right section. Only the most popular application areas, such as marketing, finance, emissions reduction, and security with the corresponding industrial leaders, such as Procter & Gamble, Chase, Dow Chemical, and Johnson Controls, are selected.

This organization of the marketing slide allows a nontechnical person to get a rough idea about the approach, its unique capabilities, and its application record in around 30 s.

Another popular way to present a condensed view of a technology in plain English is the elevator pitch. An example of an elevator pitch for neural networks is given in Fig. 5.4.

The third translator for AI-based methods is the glossary, with specific terms related to the particular technology. A full glossary is available at the end of this book, and the reader can use it in explaining selected terms. The list of recommended terms for neural networks is given below:

Recommended terms for neural networks
activation function
backpropagation
black-box model
epoch
global/local minimum
hidden layer
machine learning

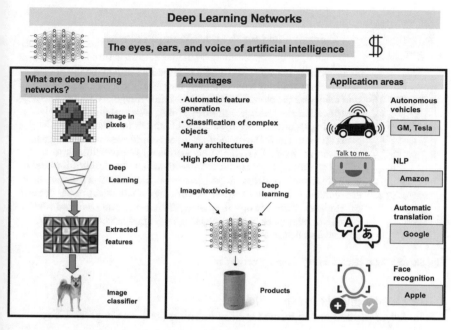

Fig. 5.5 Marketing slide for deep learning networks

multilayer perceptron
recurrent neural network (RNN)
self-organization
supervised learning
universal approximation
unsupervised learning

Translators for Deep Learning Networks
The marketing slide for deep learning networks is shown in Fig. 5.5.

The slogan of this fashionable approach, "The eyes, ears, and voice of AI," captures the leading role of deep learning in the current state of the art in AI. The left section of the marketing slide demonstrates the principle of deep learning networks in the case of image processing. It includes the key step of automatic feature extraction from the original image in pixels, performed by the deep learning algorithm. These features are presented the basis of successful image classification.

The middle section of the marketing slide emphasizes the key advantages of deep learning networks, such as automatic feature generation, classification of complex objects, a rich set of network architectures, and high performance comparable to that of humans.

The key application areas of deep neural networks are shown in the right section of the marketing slide in Fig. 5.5. This shows the most popular applications in autonomous vehicles, natural language processing, automatic translation, and face recognition, in companies such as GM, Tesla, Amazon, Google, and Apple.

Elevator Pitch for Deep Learning Networks

 Deep learning networks learn how to learn and automatically discover new complex features from available data. They are the core technology that allows AI to communicate with humans. Image processing and object identification are the algorithmic eyes of AI. Natural language processing algorithms are the ears of AI, and natural language generation methods allow machines to communicate in plain English. However, the models developed are very complex, are difficult to Interpret, and require a lot of training data. The key application areas are autonomous vehicles, digital assistants, automatic translation, face recognition, etc. Even though the technology is still in the research and development phase, many companies, like Google, Amazon, and NVIDIA, regard deep learning networks as a critical application area of their current and future products.

Fig. 5.6 Elevator pitch for deep learning networks

The elevator pitch is shown in Fig. 5.6.

The list of recommended terms for deep learning networks is given below:

Recommended terms for deep neural networks
autonomous vehicle
convolutional neural network
deep learning
feature
graphics processing unit (GPU)
hidden layer
long short-term memory (LSTM)
natural language processing (NLP)
recurrent neural network (RNN)

Translators for Support Vector Machines
Marketing SVMs is a real challenge, because explaining the approach in plain English without the mathematical vocabulary of statistical learning theory is very difficult. The proposed marketing slide for SVMs is shown on Fig. 5.7.

The purpose of the suggested slogan for SVMs, "Automatic modeling by extracting the best from the data" communicates the unique feature of the approach that it can build models based on data with the highest information content. The left section of the marketing slide represents a simplified version of the nature of SVMs as a method that develops models with the right complexity from the data by optimal learning from the most informative data points, the support vectors. The middle section of the marketing slide focuses on the obvious advantages of SVMs, such as the solid theoretical basis of statistical learning theory, and explicit control of the complexity of the model which leads to the most important advantage of SVMs, the best generalization capability of derived models under unknown process conditions. Unfortunately, the right section of the marketing slide for SVMs, presenting

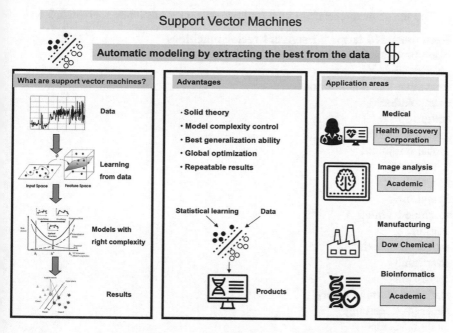

Fig. 5.7 Marketing slide for support vector machines

Elevator Pitch for Support Vector Machines

Support vector machines automatically learn new patterns and dependencies from finite data. The name is too academic and there is a lot of math behind the scenes. The derived models have optimal complexity and improved generalization ability in unknown operating conditions, which reduces maintenance cost. The key idea is simple: we don't need all the data to build a model but only the most informative data points, called support vectors. The value creation is by extracting the "golden" data into unknown patterns and models that can be used for effective decision-making. Support vector machines are especially effective when developing mathematical models is very expensive and only finite historical data on a specific problem is available.
There are applications in the areas of bioinformatics, medical analysis, process monitoring, text categorization, and security.

Fig. 5.8 Elevator pitch for support vector machines

applications, is not as impressive as in the case of neural networks. Only two application areas (medical and manufacturing) are supported by industrial applications at Health Discovery Corporation and Dow Chemical, respectively.

The elevator pitch for SVMs is shown in Fig. 5.8 and requires a lot of effort to grab the attention of any manager.

The list of recommended terms for support vector machines is given below:

Recommended terms for support vector machines
black-box model
classification
derivative-free optimization
extrapolation
feature space
generalization ability
global minimum
hyperplane
kernel
kernel trick
machine learning
margin
model complexity
overfitting
regression
statistical learning theory
supervised learning
support vectors
Vapnik–Chervonenkis (VC) dimension

Translators for Decision Trees
The marketing slide for decision trees is shown in Fig. 5.9.

The motto of the approach, "Translate data into tree-like structures," gives a condensed description of its essence. The marketing slide illustrates the two technologies, classic decision trees and random forests that are part of this family. The left part of the marketing slide emphasizes that the translation of data into tree-like structures can be done in both prediction and classification types of solutions. The key benefits of decision trees, such as interpretability of the delivered models and reduced problem complexity (in the case of classic decision trees), and built-in variable selection (in the case of random forests), are listed in the middle section. The most popular application areas of decision trees, such as in making investment decisions and developing business strategy, are shown in the right section.

The elevator pitch for the decision trees is shown in Fig. 5.10.

The list of recommended terms for decision trees is given below:

Recommended terms for decision trees
branches
ensemble learning
leaf nodes
tree induction
tree pruning

Translators for Evolutionary Computation
The generic concept of evolutionary computation is easy to explain. We suggest defining different marketing strategies for each of the specific evolutionary

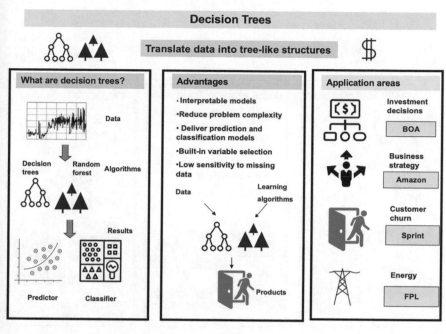

Fig. 5.9 Marketing slide for decision trees

Elevator Pitch for Decision Trees

Decision trees automatically translate data into a tree structure of important factors of the identified problem. Decision tree models are fairly intuitive and easy to explain to business. However, they are highly biased toward training data and a poor predictor on test data. This issue is overcome by a new machine learning algorithm, the random forest algorithm, based on an ensemble of decision trees with randomly selected structures. Each tree "votes" for a specific class but the forest chooses the classification having the most votes out of all the trees in the forest. The random forest algorithm has consistently demonstrated one of the highest classification and prediction accuracies among machine learning algorithms. It runs efficiently on large data sets with thousands of variables and has a built-in variable selection. The key application areas are in the financial sector when investment and bank loan decisions are made, in strategic business decisions, such as mergers and acquisitions, and the area of customer churn.

Fig. 5.10 Elevator pitch for decision trees

computation techniques. An example of a marketing slide for the use of symbolic regression generated by genetic programming is given in Fig. 5.11.

The key slogan of symbolic regression: "Translate data into profitable equations," captures the essence of the value creation basis of the approach. The left section of

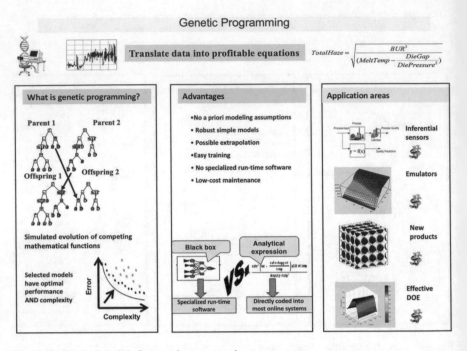

Fig. 5.11 Marketing slide for genetic programming

the marketing slide represents a very simplified view of the nature of genetic programming, the method that generates symbolic regression, which is a simulated evolution of competing mathematical functions. The presentation also emphasizes that the selected models have optimal accuracy and complexity, which makes them robust under changing operating conditions.

The middle section of the marketing slide represents the key advantages of GP-generated symbolic regression, such as automatically derived robust, simple models with the potential for extrapolation with no a priori assumptions, easy training, no need for specialized run-time software, and low maintenance cost. A direct comparison with neural-network-based models is shown, with emphasis on the model representation (black-box vs. mathematical expression), the need for run-time software, and the level of extrapolation. These clear advantages of symbolic regression are visually demonstrated. The key application areas of GP are shown in the right section of the marketing slide in Fig. 5.11. The slide includes the most valuable application areas of symbolic regression of Dow Chemical, such as inferential sensors, emulators of first-principles models, accelerated new product development, and effective design of experiments.

The proposed elevator pitch for communicating evolutionary computation to managers is shown in Fig. 5.12.

The list of recommended terms for evolutionary computation is given below:

Elevator Pitch for Evolutionary Computation

Evolutionary computation automatically generates novelty from available data and knowledge. Solutions to many practical problems can be found by simulating natural evolution in a computer. However, instead of biological species, artificial entities like mathematical expressions, electronic circuits, or antenna schemes are created. The novelty is generated by the competition for high fitness of the artificial entities during the simulated evolution. The value creation is by translating data into newly discovered empirical models that are interpretable analytical functions. Evolutionary computation is especially effective when developing mathematical models is very expensive and there is some historical data and knowledge on a specific problem. The development and implementation cost of evolutionary computation is relatively low. There are numerous applications in optimal design, inferential sensors, new product invention, finance, scheduling, games, etc. by companies like Dow Chemical, GE, Boeing, Air Liquide, etc.

Fig. 5.12 Evolutionary computation elevator pitch

Recommended terms for evolutionary computation

chromosome
convergence
crossover
expressional complexity
extrapolation
fitness landscape
generation
genotype
global/local minimum
model interpretability
multiobjective optimization
mutation
Pareto-front genetic programming
phenotype
population
symbolic regression

Translators for Swarm Intelligence

As in the case of evolutionary computation, the generic concept of swarm intelligence is easy to explain. However, one potential source of confusion could be the existence of two different approaches—ant colony optimization and particle swarm optimization. It is recommended to emphasize and demonstrate the common basis of both methods—social interaction. An example of a marketing slide for swarm intelligence, organized according to this philosophy, is given in Fig. 5.13.

The motto of swarm intelligence, "Translate social interactions into value" captures the essence of the value creation basis of the approach. The left section of the marketing slide represents the generic view of the approach and presents the two

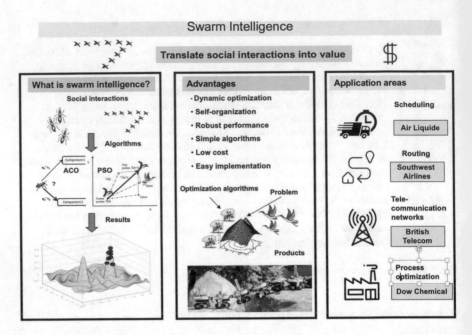

Fig. 5.13 Marketing slide for swarm intelligence

key methods (ACO and PSO), inspired by ants and bird flocks. It is focused on the three key phases of swarm intelligence: (1) analyses of social interactions in biology (represented by ants and birds); (2) derived algorithms for optimization; and (3) the results of finding the optimal solution.

The middle section of the marketing slide represents the key advantages of swarm intelligence, such as derivative-free optimization, self-organization, robust performance, simple algorithms, low cost, and easy implementation. The visualization section represents the problem as a fitness landscape with optima, the two optimization algorithms as ants and birds targeting the optimum, and a line of swarming mini-robots, as products.

The key application areas of swarm intelligence are shown in the right section of the marketing slide in Fig. 5.13. The slide includes the most valuable application areas of swarm intelligence in scheduling, routing, telecommunication networks, and process optimization. Examples of leading companies in the specific application areas are given.

The proposed elevator pitch for inspiring managers about the great capabilities of swarm intelligence is shown in Fig. 5.14.

The list of recommended terms for swarm intelligence is given below:

Recommended terms for swarm intelligence
ant colony optimization (ACO)
agent-based system
bottom-up modeling
classical optimizers

Elevator Pitch for Swarm Intelligence

Surprisingly, we can improve the way we schedule, optimize, or classify by learning from social behavior of ants, termites, birds, and fish. Swarm intelligence captures the wisdom emerging from the social interactions of these biological species and translates it into powerful algorithms. For example, ant colony optimization uses a colony of virtual ants to construct optimal solutions by digital pheromone deposition. Another powerful algorithm, particle swarm optimization, mimics a flock of interacting particles to find optimal solutions in difficult conditions of a rugged fitness landscape with multiple optima. The swarm intelligence is especially effective when operating in a dynamic environment. The total Cost of ownership of swarm intelligence algorithms is relatively low due to their simplicity. Swarm intelligence is a new technology but with growing interest in industry. It has been successfully applied in scheduling, telecommunication and data network routing, and process optimization by companies like Air Liquide, Dow Chemical, Southwest Airlines, France Telecom, and British Telecom.

Fig. 5.14 Elevator pitch for swarm intelligence

derivative-free optimization
emergent phenomena
ontology
particle swarm optimization (PSO)
pheromone
premature convergence
search space
stigmergy

Translators for Intelligent Agents

The concept of intelligent agents is easy to explain and communicate. The deliverables and the value creation mechanisms, however, are confusing to the user and more marketing efforts than for the other AI-based methods are needed. It is very important to explain to the user upfront that there are no guarantees of the appearance of emergent patterns, which could lead to more insightful understanding and eventual improvement of the system. Marketing the value of one specific type of agent, chatbots, is more obvious. An example of a marketing slide for intelligent agents is given in Fig. 5.15.

The main slogan for intelligent agents, "Translate emergent patterns from microbehavior into value," communicates the source of the value creation basis of the approach. However, the message may be confusing to a nontechnical user and needs to be clarified with the other sections of the slide. The left section of the marketing slide tries to play this role and represents a generic view of intelligent agents. The key phases of agent-based modeling systems are illustrated: (1) defining the agents' microbehavior by rules and actions, (2) defining the multiagent system by the agents' interactions, (3) identifying emergent macropatterns from simulations, and (4) using the discovered macrobehavior for system improvement.

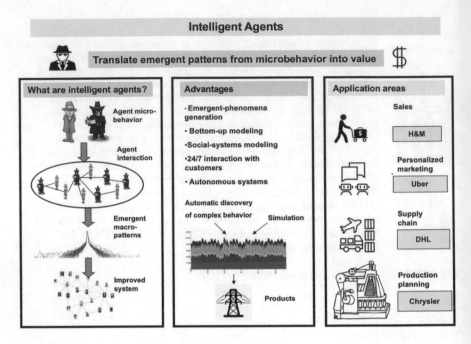

Fig. 5.15 Marketing slide for intelligent agents

The middle section of the marketing slide represents the key advantages of intelligent systems and chatbots, such as generation of emergent phenomena, bottom-up modeling, system integration, social-systems modeling, 24/7 interaction with customers, and autonomous system design. The visualization section represents a simulation as an illustration of the key value source—automatic discovery of complex behavior (in this case, bidding patterns for electricity prices), and a power grid, as an example of a product where an agent-based system can be applied.

The key application areas of intelligent agents are shown in the right section of the marketing slide in Fig. 5.15. The slide includes the most valuable application areas of agent-based systems and chatbots in sales, personalized marketing, supply chains, and production planning. Examples of leading companies in the specific application areas are given.

The proposed elevator pitch representing intelligent agents is shown in Fig. 5.16.

The list of recommended terms for support vector machines is given below:

Recommended terms for intelligent agents
agent-based modeling
bot
bottom-up modeling
chatbot
complex system
conversational corpus

Elevator Pitch for Intelligent Agents

Recently, there has been a growing need for estimating social responses to key business decisions. One of the few available technologies that is capable of modeling social behavior is intelligent agents. Agent-based systems are a method for bottom-up modeling where the key social components are defined as artificial agents. Their microbehavior is captured by actions and rules they follow in order to achieve their individual goals. Intelligent agents interact in different ways and can negotiate. Previously unknown social responses emerge in a computer simulation from the microbehavior of the interacting agents. For example, different price-bidding strategies emerge in simulations of generation, transmission, distribution, and demand in companies in the electricity market. Another big application areas of intelligent agents are chatbots and digital assistant. Agent-based systems have been successfully applied in sales, personalized marketing, supply chains, planning, etc., by companies like H&M, DHL, Uber, and Chrysler.

Fig. 5.16 Elevator pitch for intelligent agents

emergent phenomena
intelligent agent
machine learning
natural language processing (NLP)
ontology
rule base
self-organization

5.5   Common Mistakes

5.5.1   *Ignoring the Dialog Between Domain Experts and Data Scientists*

Many inexperienced data scientists are focused on coding and learning the details of the technologies and do not pay attention and devote time to improve the dialog with their partners on the business side. Some domain experts also ignore the importance of explaining the business problem with clarity to data scientists and do not want to spend time and efforts on understanding the proposed AI-based solutions. They think that this is "rocket science" that is impossible to understand without a big effort on their side. As a result of this behavior, the lost-in-translation effect is in full power and a lot of energy and time are wasted at the critical problem definition phase.

The techniques discussed in this chapter can help data scientists and domain SMEs to overcome the initial communication issues and to gradually improve the dialog to an acceptable level of mutual understanding.

5.5.2   Ignoring the People Factor

A potential root cause of not paying sufficient attention to clear communication with business partners is ignoring the people factor in the Data Science community at large. Most of the attention is focused on the methods and infrastructure, especially the software. Usually training classes in this discipline offer machine learning, deep learning, Python, and R but not courses in project management, business communication, or team building. As a result, newly graduated data scientists enter industry with a lack of the background needed to develop good people and communication skills.

5.5.3   Ignoring Team Building

Another potential root cause of dropping into the lost-in-translation trap is the lack of a team-building skillset. Taking appropriate training can compensate for this gap to some extent.

5.6   Suggested Reading

- D. Hardoon and G. Shmueli, *Getting Started with Business Analytics*, CRC Press, 2013.
- D. Pyle, *Business Modeling and Data Mining*, Elsevier, 2003.

5.7   Questions

Question 1
Give examples of the lost-in-translation trap.

Question 2
Suggest action items for how to improve understanding of business problems.

Question 3
Suggest a questionnaire for your business SMEs.

Question 4

Which AI-based method is the most difficult to explain, and how can the description be improved?

Question 5

Give examples of explanations of AI-based methods to domain SMEs.

Part II
The AI-Based Data Science Toolbox

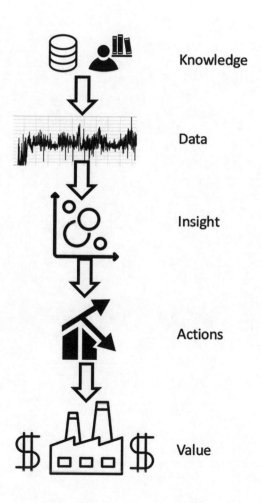

Knowledge

Data

Insight

Actions

Value

Chapter 6
The AI-Based Data Science Workflow

If you fail to plan you plan to fail.
Anonymous

While the focus of the first part of the book was on introducing AI-based Data Science for solving business problems, the objective of this second part is to discuss how this powerful technology can be applied in industrial settings. It includes six chapters. Chapter 6 gives an overview of the work process and the project organization step. The remaining five chapters are focused on the details of the key steps of the process. Chapter 7 discusses the important and neglected step of problem knowledge acquisition. Chapter 8 gives details of the most time-consuming step, data preparation, while Chap. 9 clarifies the different methods for analyzing the data. The topic of model development and validation is discussed in Chap. 10 and model deployment and maintenance issues are covered in Chap. 11.

Often when we think of a Data Science project, the main thing that comes to mind is the algorithmic technique that needs to be applied. While that is crucially important, there are many other steps in a typical Data Science solution that require equal attention. The first objective of this chapter is to explain the need for a well-defined workflow to accomplish efficiently a Data Science project. The second is to give an overview of all necessary steps and to clarify their importance. The third is to discuss the project organization step.

6.1 Overview of Workflow

There is no doubt that the key factor that differentiates a good restaurant from a terrible one is the food preparation in the kitchen. Above all, the food quality depends on the appropriate recipe with clear steps for how to transform the raw food into tasty dishes. A great chef designs a very detailed recipe and pays attention to the proportions of the ingredients used. Often, she/he applies amazing imagination to deliver not a simple dish but an attraction on the plate. At the other extreme is the

© Springer Nature Switzerland AG 2020
A. K. Kordon, *Applying Data Science*,
https://doi.org/10.1007/978-3-030-36375-8_6

automated process of junk-food preparation in the kitchens of the big fast-food chains.

Similar effects are observed when the efficiency of the Data Science "kitchen" is analyzed. The best data scientists use well-structured work processes and pay attention to the details of the steps of these processes. Often the critical feature and model selection steps are driven by their imagination. The final result is a high-quality solution, accepted with enthusiasm and immediately used by the clients. At the other extreme is the attempts by inexperienced citizen data scientists to use a shortened version of the workflow. They are the target audience of the current "One-Click Data-In-Model-Out" movement toward fully automating the process. The result is similar to the junk-food equivalent—creating "burger models" with low robustness towards process changes that become a potential maintenance nightmare if applied.

6.1.1   Why we Need an Effective AI-Based Data Science Workflow

Trying to deliver a useful solution for a defined business problem without a sequence of logical steps is the same as hoping to prepare a tasty dish without a recipe. Defining the optimal "recipe" for the complex AI-based Data Science "dish," however, is not a trivial task. It requires a good balance of phases that integrate business problem and solution space issues in a reasonable number of steps. Here is the difference between a good and a standard Data Science workflow. The standard one usually starts with the data and is focused on the steps of modeling based on machine learning. The good one starts with a detailed definition of the business problem, uses experts' knowledge as much as possible in all project phases, and ends with a well-defined maintenance plan for long-term support of the derived models. Such a workflow is proposed in the book.

The progression of the proposed workflow from the classical scientific method and the dominant CRISP-DM method is described briefly below.

6.1.2   Why Is the Classical Scientific Process Not Enough?

The scientific method encompasses an ongoing process with the following steps: formulate a hypothesis—test it with an experiment—analyze the results—prove it or reformulate the hypothesis. Such a way of proceeding has been in use for centuries and is basically accepted as the most reliable means to produce robust knowledge.

The AI-based Data Science work process is centered on the scientific method. There are some nuances, though. First, the formulation and testing of the hypothesis are done on two levels. In addition to the classical level of hypothesis definition by experts, some AI-based approaches, such as machine/deep learning, evolutionary computation, and intelligent agents generate new patterns and relationships, which inspire the experts to define a new set of hypotheses.

Second, the experimental testing of the hypothesis cannot be organized in a clean way, as in an R&D lab. The business conditions are much more complicated with many physical and financial constraints. Here is one of the key limitations of the scientific method when applied to solving real-world problems. It doesn't consider the full complexity of business problems, especially the different objective of value creation. Additional steps are needed to address these issues.

6.1.3 Comparison with CRISP-DM

Importance of CRISP-DM
Since its creation in 1997, CRISP-DM has been the leading approach for managing data mining, predictive analytic, and Data Science projects, accepted by the business world at large. CRISP-DM is made up of the following six steps:

- business understanding;
- data understanding;
- data preparation;
- modeling;
- evaluation;
- deployment.

The steps reflect the standard data mining and modeling process: we start by asking a question or looking for insight into some particular phenomenon, we need some data to examine, the data must be inspected or prepared in some manner, and the data is used to create and validate an appropriate model, which can be deployed or communicated.

Suggested Improvement
The growing experience in implementing data-driven modeling in business settings has shown some gaps in CRISP-DM, though. For example, the role of effective project management in solving complex industrial problems is underestimated. As a result, many projects are not handled in a very professional way.

A very important area for improvement of the classic CRISP-DM process is integrating existing knowledge into the whole process. Knowledge acquisition includes not only better understanding of the data but also using all available expertise related to the defined problem.

Probably the most significant gap in the classic CRISP-DM process is the neglected issue of maintenance and support for the deployed solutions. The long-term survival of the derived models is the real criteron for success in any modeling project, and this issue must be a part of the whole model development process.

6.1.4   AI-Based Data Science Workflow Sequence

The suggested steps of an AI-based Data Science work process that includes the improvement outlined above are shown in Fig. 6.1 and discussed below.

The key expected deliverable from each phase is shown below each step in Fig. 6.1. For example, preparing a project charter is the expected result of the first step in the AI-based Data Science work process—problem definition. It triggers project organization, which is expected to deliver the infrastructure of people, software, and teams needed for accomplishing the objectives of the project. The deliverable of the critical problem knowledge acquisition step is the knowledge captured from the domain experts.

Translating business problems into Data Science solutions is an iterative process. The proposed workflow includes two loops—a model development cycle and a model deployment life cycle. The inner, and shorter in time, model development loop includes an iterative sequence of data preparation with corresponding analysis that can lead to model generation and validation.

The outer, and much longer in time, model deployment life cycle includes the necessary actions for the maintenance and support of the deployed models. In the

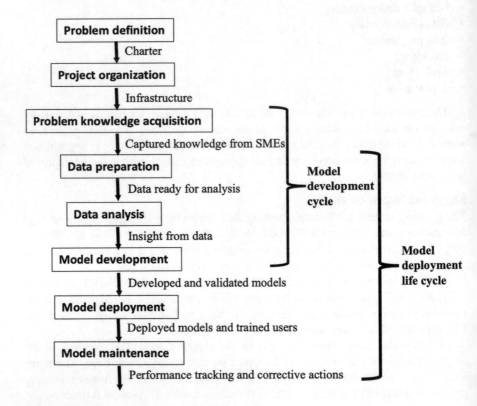

Fig. 6.1 AI-based Data Science work process

case of consistent poor performance of the deployed model, a new model development cycle is triggered. Often it requires a problem knowledge update to reflect changing operating conditions after model deployment.

6.2 Key Steps of AI-Based Data Science Workflow

A short description of the key steps of the proposed methodology, with an emphasis on their importance, is given below. Detailed descriptions are provided in the corresponding chapters of the book.

6.2.1 Problem Definition

This first step in the workflow builds the foundation for the future business-problem-solving project. It includes the problem description, the targeted financial and technical metrics used to measure success over the baseline (doing nothing), the expected deliverables, commitment to funding, and a well-defined scope and boundaries.

The expected deliverable from this step is the initial project charter. This step is described in Chap. 2.

Problem definition is critical for the success of the project. The more specific the information it includes, the better the chances of success of the other phases. The alternative is a possible collapse of the "modeling building" created on the shaky foundation of ambiguous project objectives and deliverables.

6.2.2 Project Organization

The second step covers the key organizational issues that need to be dealt with before starting the technical work, such as building the team, allocating the necessary hardware and software resources, a preliminary project schedule proposal, and an initial cost estimate.

The expected deliverable from this step is the organization of the infrastructure in terms of hardware, software, and people. This step is described in this chapter.

Solving business problems requires a project structure that needs to be organized to accomplish the defined objectives, i.e., the importance of this step of the workflow is obvious.

6.2.3   Problem Knowledge Acquisition

The purpose of the third step is to capture and document the available problem domain knowledge. It includes identifying the SMEs, collecting process diagrams and descriptions, and searching for publicly available references and patents. Knowledge acquisition usually takes several brainstorming sessions facilitated by a moderator and attended by selected SMEs.

The expected deliverable from this step is the documented knowledge, captured from internal and external sources. This step is described in Chap. 7.

This step is missing in the classic CRISP-DM method where the model development process begins from data understanding and preparation. In the real world, however, starting any solution of a problem begins not from the data but from the existing knowledge about the defined problem. Using this experience increases significantly the efficiency of the whole modeling process.

6.2.4   Data Preparation

The fourth step of the proposed AI-based Data Science workflow includes all necessary procedures to collect, integrate, preprocess, and balance the available data. In reality, data preparation is time-consuming, nontrivial, and difficult to fully automate.

The expected deliverable from this step is a data set ready for analysis (i.e., with missing data and outliers handled). This step is described in Chap. 8.

This step is an absolute must in any data analysis or model development effort.

6.2.5   Data Analysis

The fifth step describes the key transformation of the data into insight. It includes variable selection, features definition, multivariate view, and potential pattern discovery by data visualization.

The expected deliverable from this step is the insight from the data in terms of identified key factors, correlations, and discovered patterns. This step is described in Chap. 9.

This is an important step, where the first results from a standard AI-based Data Science project are summarized.

6.2.6   Model Development

The sixth step is focused on the key transformation of insight into action. In most cases, the key actionable item that solves the defined business problem is a mathematical model. This step includes the important issues of fighting overfitting, generating models by multiple approaches, design of ensembles of models, and model validation. Model development is a cycle that usually takes three to five iterations until the target model quality is achieved.

The expected deliverable from this step is a set of high-quality developed and validated models that satisfy the defined performance criteria. This step is described in Chap. 10.

This is the most important step, in that it delivers the results for solving the defined problem. Unfortunately, it is the final step of most of the Data Science projects that are started, i.e., the models are communicated to the final users but are not deployed.

6.2.7   Model Deployment

The seventh step covers the final transformation of actions into value. It includes the necessary sequence of transferring the derived models from the development to the production environment. It is assumed that beyond this phase, the final users will take ownership of the models.

The expected deliverable from this step is deployed models, and trained final users, ready to use them. The step is described in Chap. 11.

This is the key step to beginning to deliver value, by using the derived models to solve the defined problem.

6.2.8   Model Maintenance

The last step of the proposed AI-based Data Science workflow is ignored in almost all known modeling processes. It includes the tracking of financial and technical performance, model supervision, and corrective actions that guarantee a long-term maintenance and support for the deployed solutions.

The expected deliverable from this step is a defined and implemented method for tracking the performance of models with corresponding corrective actions to improve the functioning of the deployed models. This step is also described in Chap. 11.

This is the key step to guaranteeing a long life cycle for the derived solution that will deliver the expected return of investment (ROI) for a long period of time.

6.2.9   Automation of AI-Based Data Science Workflow

There is a recent trend, driven by some vendors, toward automating a significant part of the work process. Some steps of the process, such as problem definition, project organization, knowledge acquisition, and model maintenance, however, are very difficult to automate. Even some operations in the data preparation, such as missing-data imputation and outlier handling, are too risky to be blindly executed by built-in algorithms. This could lead to distorted data and lost important information that could reduce the quality of the data analysis and model development.

However, some operations in the data preparation, data analysis, and model development steps could benefit from automation. A typical example is normalizing the data for specific methods (e.g., for neural networks) in the data preparation steps. Another example is automated feature creation in the data analysis (e.g., generation of transforms from genetic programming). The most used case of automation in model development is hyperparameter optimization, which has become part of many software tools. It is usually combined with automatic model selection by running several approaches in parallel or developing an appropriate ensemble.

The key principle of successful automation of the AI-based Data Science work process is to increase data scientists' productivity, not to replace them. The automation can free the data scientists from nasty repetitive tasks and allow them to focus on extracting insight, making decisions about selection of the final model, and communicating the results to the business users. And, above all, for looking for new business problems!

6.3   Project Organization

The key difference between talking about solving a business problem and making it happen occurs when the project charter triggers a real project. Businesses use diverse project management systems, but the following important steps are needed to start a well-run modeling project: building the project team, allocating the needed resources, initial scheduling, and estimating the funding. These steps are discussed briefly below.

6.3.1   Organizing Project Teams

Identifying appropriate stakeholders is one of the most important substeps to take to guarantee the success of Data Science projects. In the case of a typical middle-to-large-scale business project for applying AI-based Data Science, the following stakeholders are recommended as members of the project team:

- the management sponsor, who provides project funding;

- the project owner, who has the authority to allow changes in the existing business process;
- the project leader, who coordinates all project activities;
- the model developers, who develop, deploy, and maintain the models;
- the subject matter experts, who know the business process and the data;
- the final users, who use the deployed models on a regular basis and deliver the value.

Usually the team is organized in two layers—a decision layer and an execution layer. The decision layer includes the key project leadership that makes the key decisions, provides the funding, allocates needed resources, and performs project management. The role of this layer is critical in two cases—starting the project and organizing model deployment. Often, without intervention from the decision layer it may be challenging to navigate the organizational change necessary to deploy the solution.

The execution layer includes the data scientists, the domain SMEs, the IT specialists, and the final users. This part of the team runs the technical steps of the project according to the proposed AI-based Data Science workflow. It is recommended that all team members actively participate in all phases of the project. Of special importance is the involvement of the final users from the beginning. In this case the difficult process of transferring ownership to them after model deployment will be smooth and require minimal training.

It is important that every decision-maker and stakeholder understands the value creation objectives of the project and how the team will achieve them. Having technical members in the team such as IT or data management staff, can help mitigate concerns about data access or security, software deployment, and infrastructure requirements.

6.3.2   Resources Allocation

One of the first task that the project team needs to fulfill is to define, agree, and deliver the needed resources. They can be outlined in the following categories: specialized hardware needed, development and deployment software, required data sources, and additional expertise.

Specialized Hardware
Some AI-based methods, such as deep learning and evolutionary computation, require more number-crunching power than a standard laptop or desktop. These specialized hardware requirements should be analyzed and properly allocated to the corresponding members of the team. Fortunately, with the growing capabilities and decreased cost of cloud services, this issue can easily be resolved.

Development and Deployment Software

In most cases the business has already made its investment in a software platform for model development and the project team will use its capabilities. If not, many commercial enterprise-wide (SAS, IBM SPSS, etc.) and open source (R, Python, etc.) options are available. The choice has to be made very carefully, however, because each option has strengths and weaknesses. Open source software is free and has all the latest features, but may not be user-friendly, may not scaleup well, or may not be as maintainable as the commercial tools. Another factor in favor of open source options is that the new generation of data scientists prefer to develop models in this way.

Commercial enterprise-wide software is very expensive, but includes tested algorithms with known quality, is user-friendly, and is more readily accepted by IT managers than open source tools are. Another advantage of commercial platforms is that they include model deployment environment, including tools for tracking model performance, which is critical for the maintenance and support of solutions.

If an open source option is selected, the issue of implementing the models on the deployment platform has to be solved by the project team. Often, this platform is limited by the existing software used by the client. This might be Microsoft Office, Web-based software, Tableau, etc. In many manufacturing applications, the deployment environment is the existing process monitoring system. In both cases, additional resources for software transfer are needed. They are usually outside the model development group and their commitment should be guaranteed at the beginning of the project.

Data Sources

Allocating resources to supply the team with the necessary data is another important step in project organization. A key consideration is data availability and access. What data is available (storage format, quantity, number of tables, etc.) to support the project and how will the project team have access to it? Who owns the data, and what is the security around it? At this early phase of the project we don't need to expect perfect data, but it is important to understand whether or not sufficient data exists and how much effort will be required to collect the data so that it is usable for accomplishing the project objectives.

Some projects require external data that is purchased from another organization. The project management should provide funding and organize consistent data delivery in this early phase of project development.

Additional Expertise

Often the business doesn't have the internal resources to solve efficiently the defined problem. This is a realistic scenario when a new AI-based technology is introduced. The recommended solution is to transfer knowledge from external sources and to use proper training. The knowledge gap, however, should be identified and funded at this step of the workflow.

6.3.3   Project Scheduling

Project organization involves creating and maintaining a schedule for the project. A project schedule is a list of milestones, activities, and deliverables, usually with intended start and finish dates. Developing a project schedule requires a combination of activities, resources, and activity performance sequences that gives the project leader the greatest chance of meeting the stakeholders' expectations with the least amount of risk. An advantage of a well-defined and detailed workflow is that it is easier to do the scheduling of individual steps.

Project scheduling is an iterative process, starting with an initial estimate of the length of time required based on the problem definition and previous experience. After data collection, when a reality correction happens, a major update with more realistic time estimates is needed.

6.3.4   Project Funding

The amount of funding for a project depends on the estimated cost of the project. It is strongly recommended that this activity should be done professionally by a financial expert. Some guidelines for estimating the key components of the cost are discussed briefly below.

Labor Cost
The lion's share of the cost is based on compensating the time of the project team members. The estimate of this cost is proportional to the assessment of this time in man-hours. It is easier and more accurate to estimate the time for each individual step than to estimate the total time for the project. An example of such an assessment for the project for emissions estimation discussed earlier is shown in Table 6.1.

This table includes three columns related to the time duration of the steps discussed above in the AI-based Data Science workflow. The first column gives the initial estimate for each step in man-hours. This estimate is made by the project leader based on her/his previous experience. The second column is filled in after project completion and includes the actual time spent on each step. The third column calculates the correction factor, i.e., the ratio between spent and estimated time. This is an important piece of information for analyzing the accuracy of the initial assessment. As is shown, in the example in Table 6.1, some steps, such as knowledge acquisition, model deployment, and maintenance, were underestimated. Their assessed time duration was 2–2.8 times shorter than the actual time spent. Another steps, model development, was overestimated. In this case, the actual time spent was 0.8 times the initial assessment. The total time spent on the project was 1.7 times longer than estimated. It is assumed that this warning message will help the project leader to analyze her/his assessments and continually improve them, and reach correction factors close to one in future projects.

Table 6.1 Example of time estimate for project steps

Project Step	Time estimate in hours	Time spent in hours	Correction factor (spent/ estimated)
1. Problem definition	20	25	1.25
2. Project organization	40	75	1.88
3. Knowledge acquisition	10	28	2.80
4. Data preparation	40	56	1.40
5. Data analysis	60	86	1.43
6. Model development	60	48	0.80
7. Model deployment	20	42	2.10
8. Model maintenance	60	122	2.03
Total estimate in hours	310	482	1.71

Infrastructure Cost

Usually this includes investment in hardware and software if the business has not already spent on data analytics infrastructure. In our example of the emissions estimation project, investing in two powerful desktops for the two data scientists and specialized software for generating symbolic regression models by genetic programming was needed and was added to the project cost estimation.

Data Collection Cost

For internal data collection, this cost is included in the time spent by the team members responsible for this task. Collecting external data, however, requires subscription to the corresponding data delivery services. The estimated project cost for this subscription should cover the estimated solution maintenance life cycle, i.e., at least 3–5 years.

Maintenance Cost

This component of the cost is difficult to estimate because it depends on the performance of the deployed solutions. In the worst-case scenario of fully degrading models, development of new models will need to be done, which can require a significant amount of time and money.

As a recommended rule, it is suggested to plan for 12–24 man-hours of annual cost for at least 3 years for tracking the performance of each deployed model. This rule was used in the example of the emissions estimation project.

Training Cost

Usually this includes the cost of attending training courses for new software products and the time required to train the final users when the ownership of the models is transferred to them after deployment.

6.4   Common Mistakes

6.4.1   *Ignoring a Detailed Workflow*

An important litmus test for a mature Data Science group is whether or not project development is based on a comprehensive work process such as the one proposed in the book. Inexperienced data scientists avoid following a strict sequence of steps and solve the business problem in an improvised way, driven by their personal preferences. Often their starting step is jumping to data analysis without doing the necessary data preparation. In many cases the initial phase of the project is even jumping directly to model development! As a result of this "project management by gut" style, the delivered solutions have low quality and robustness. The argument that the objective of this "fast-track" model development mode is to please the client is not very solid. The business clients are more interested in solutions with high performance than on fast delivery of crappy models.

6.4.2   *Ignoring some Steps in the Workflow*

A more common mistake is skipping selected steps in the proposed AI-based Data Science work process. At the top of the list is the knowledge acquisition step. The fashionable "data first" mindset contributes to the attitude that there is no need to spend time on collection of knowledge about the problem. The result is inefficient, context-free data analysis and model building.

Second on the list of ignored steps is model maintenance. Many data scientists think that the project is completed with model deployment and ownership transfer to the final user. This is a deeply wrong impression. In fact, the most important part of the project—delivering value, created by the applied solution of the problem, − has not been accomplished. Usually achieving the financial objectives of the project requires long-term use of the deployed models. In order to make this possible, a well-developed maintenance and support program has to be implemented.

Two other steps, problem definition and project organization, might not be totally disregarded but might be partly fulfilled without the needed details. Of special concern is neglecting the preparation of a problem definition without clear objectives and deliverables, as was discussed in Chap. 2.

6.4.3   *Insufficient Efforts on Cost Estimates*

Often project cost estimation is done very unprofessionally when financial experts are not part of the project team. Some generic guidelines on how to make an assessment of the key components of the cost—labor, infrastructure, data collection, maintenance, and training—have been given in this chapter.

6.4.4   *Not Documenting the Deliverables*

Another observed mistake made by data scientists is saving time by skipping the stage of documenting the deliverables from the different project steps. This is often the case after finishing the important data preparation step, when critical decisions about data imputation, transformation, and outlier handling are not documented. As a result, the quality of data preprocessing remains unknown and the credibility of the corresponding data analysis is questionable. Another negative consequence of undocumented steps is that it becomes very difficult to transfer the project to another data scientist if necessary. The same issue exists in the case of leveraging the project to a similar business problem.

6.5   Suggested Reading

- D. Pyle, *Business Modeling and Data Mining*, Elsevier Science, 2003.
- W. Verbeke, B. Baesens, and C. Bravo, *Profit Driven Business Analytics*, Wiley, 2017.
- S. Williams and N. Williams, *The Profit Impact of Business Intelligence*, Elsevier, 2007.

6.6   Questions

Question 1
Give arguments why the scientific method is not sufficient for solving business problems.

Question 2
What workflow is used in your organization to develop Data Science solutions?

Question 3
What steps are not used in your Data Science workflow?

Question 4
Why do we need a model maintenance step?

Question 5
How can labor cost be estimated?

Chapter 7
Problem Knowledge Acquisition

Beware of false knowledge; it is more dangerous than ignorance.

George Bernard Shaw

The logical step after defining the business problem and organizing the project is to collect the available internal (inside the business) and external (in publicly available references) information about the topic. Problem knowledge acquisition and organization is the activity of gleaning and arranging existing knowledge from domain experts, documents, papers, patents, etc. in a form suitable for effective use by the project team. Examples of deliverables from knowledge acquisition are a problem description, an explaination of known features of the problem, defined hypotheses, the recommended data, to name a few. They are used in different steps of the Data Science workflow and contribute to more efficient data analysis and model development.

Knowledge acquisition is different from another similar activity—knowledge management. Knowledge management is the process of building expert systems that require a significant amount of human expertise to solve a class of domain-specific problems. However, it is too time-consuming to scarch for and organize stored knowledge, and the approach has been of little use recently.

The first objective of this chapter is to clarify the importance of problem knowledge across the Data Science work process. The second is to discuss the key sources of problem knowledge and the methods for dealing with them while the third is to review the different options for integrating this knowledge. The fourth is to explain the key items in defining a solution strategy for the problem.

7.1 Importance of Problem Knowledge

Knowledge of the domain in which a problem lies is immensely valuable and irreplaceable. It provides an in-depth understanding the context of the available data and the factors influencing the project objectives. Often domain knowledge is

© Springer Nature Switzerland AG 2020
A. K. Kordon, *Applying Data Science*,
https://doi.org/10.1007/978-3-030-36375-8_7

one of the key contributors to a Data Science team's success. It influences the whole Data Science work process: how we engineer and select features, impute data, choose an algorithm, determine success, etc. The specific importance of problem knowledge in the individual steps of the Data Science workflow is discussed below.

7.1.1 Problem Knowledge in Problem Definition

Problem knowledge is the key factor in describing and defining the business problem. An idea about the problem is spawned by some issues based on existing knowledge. Often, data is used in support of the problem description but the detection and formulation of the problem, triggered by domain experts, is based on available knowledge.

Another issue that is critical in problem definition is outlining the scope of the proposed project. In a more detailed version, this includes defining the assumption space of targeted solutions. A clear example is a quantitative measure of a normal operating regime in manufacturing, needed in many use cases. These definitions are formulated by domain experts based on their business knowledge.

Problem knowledge is needed also in another task at the beginning of the project—to prevent the "reinventing the wheel" effect when looking for solutions to the problem in publicly available references.

7.1.2 Problem Knowledge in Data Preparation

Adding context to the data is one of the key contributions of the domain experts. The description of metadata depends strongly on process knowledge as well. The initial selection of data is one of the key steps in the Data Science workflow and is based on problem knowledge. It is also not recommended to blindly remove or impute missing data but to discuss these cases with domain experts before making decisions. A similar strategy is suggested for handling outliers.

7.1.3 Problem Knowledge in Data Analysis

The key task of data analysis is to deliver insight about the business problem based on the data used and methods applied. This requires identification and interpretation of patterns discovered in the data and should be done in close collaboration with problem domain SMEs. Their experience is very important for identifying these patterns. General, the more the SMEs understand the nuances of the business, the better the insights they will find.

7.1.4   Problem Knowledge in Model Development

Problem domain knowledge is needed at the beginning and at the end of model development. The SMEs' expectations about the behavior of the model are a critical piece of information for defining the initial structure of the model. For example, nonlinear relationships may be assumed when models for some chemical processes are developed, based on the nonlinear Arrhenius law. In this case, planning model development based on linear methods is not recommended.

Even more important is the SMEs' knowledge in the last phases of model development. The selection of the final models is based on their behavior according to the existing domain knowledge. Statistical performance metrics are a good starting step for exploring the potential modeling solutions, but the domain experts make the final choice by carefully exploring the behavior of the models. The SMEs' interpretation of the models is also critical for their acceptance. In addition to the for the models expressions (their structure), this interpretation is based on a set of "what-if" scenarios of responses of the models. There is no way that a model will be accepted by the domain SMEs if its behavior deviates significantly from existing knowledge.

7.1.5   Problem Knowledge in Model Deployment

In order to accomplish model deployment, domain knowledge is needed in two key steps—(1) designing the user interface and (2) defining intangible performance metrics.

One of the critical factors for the success of applied Data Science solutions is their acceptance and long-term use by the final users. An intuitive user-friendly design contributes a lot toward accomplishing this objective. The best-case scenario is to actively include the final users in shaping this design and validating the final version of the interface.

Of special importance is the case where model deployment will require significant changes in the current business process. The proposed new process should be carefully designed in coordination with the domain SMEs. It is also recommended that they actively participate in communicating the changes and training the final users.

The other step that is needed in model deployment is defining intangible performance metrics for solutions applied. Performance criteria are problem- and business-specific. A typical example is the requirement for ease of use, which has different qualitative components for each specific application.

7.2   Sources of Problem Knowledge

Domain knowledge sources can be classified into two categories: internal and external. The first category includes the experts and the problem-related documents, such as process diagrams, manuals, and business structure documents inside the organization where the targeted solution will be applied. The second category involves all problem knowledge resources external to the organization, such as outside SMEs and publicly available references such as papers, reports, patents, and presentations. Some details of these knowledge sources are given below.

7.2.1   Subject Matter Experts

Domain experts are one of the most critical factors for the success of AI-based Data Science projects. An SME has usually become an expert by both education and experience in a domain. This is a person who can define a framework for a Data Science project, as she/he will know what the current challenges are and how they must be answered to be practically useful given the state of the domain in the organization. The SME can judge what data is available and how good it is. These experts can use and apply the deliverables of a Data Science project in the real world. Most importantly, this person can communicate with the intended users of the project's outcome. This is crucial, as many projects end up being shelved because the conclusions are either not actionable or not acted upon.

In addition to the SMEs directly involved in the project it is recommended to have a good network of contacts in the business who can answer questions related the defined problem. The more expertise is used, the better. It is preferable that AI-based methods complement natural intelligence than to try to find a solution based only on available data.

How to Attract SMEs

Good experts are too busy and sometimes difficult to deal with. Often they don't have real incentives to share their knowledge, since most of them are at the top of their professional career. In addition, in the recent environment of constant layoffs, there is a fear for their jobs if they transfer their expertise to a computer solution. The recommended resolution is to define significant incentives in combination with high recognition of the value of the shared knowledge and a clear commitment to job security from the management.

Avoiding the Effects of the Peter Principle

In collecting experts' knowledge, we need to consider the bureaucratic influence of big hierarchical organizations on the quality of the shared information. In some sense, their influence is an effect of the famous Peter principle[1] on the nature of

---

[1] In a hierarchy, every employee tends to rise to his level of incompetence.

expertise with increasing hierarchical levels. According to the Peter principle, the technical expertise is located at the lowest hierarchical levels. It gradually evaporates in the higher levels and is replaced by administrative work and document generation. Unfortunately, in selecting domain experts as sources for knowledge acquisition, the effects of the Peter principle have not always been taken into account. As a result, some SMEs have been chosen based on their high hierarchical position and not on their technical brilliance. The damaging consequences of the recommendations given by such Peter-principle-generated pseudo-experts are very dangerous for the credibility of the resulting AI-based Data Science solutions. The real domain SMEs immediately find the knowledge gaps and probably do not use the models. Challenging the "knowledge" defined by their bosses is often not a good career move. The final absurd result is that instead of enhancing top expertise, such bureaucratically generated AI-based systems tend to propagate incompetence.

7.2.2   Problem-Related Documents

Another key source for knowledge acquisition is the available documentation related to the defined problem. This includes process schemes, manuals, and and documents about the business structure. The expectation is that data scientists will use these sources to gain a generic understanding of the main principles of operation of the targeted business units. It is not necessary, however, to go deep into the technical details. This is the role of the SMEs.

Process Diagrams

An important set of documents for understanding the principles of operation of a unit related to the problem is the process schemes. An example of a process diagram is shown in Fig. 7.1. This flowsheet represents part of the Tennessee Eastman Process, which is a well-known benchmark for exploring process control and fault detection problems.[2] The high-level process diagrams include only the main equipment in the unit, such as strippers, separators, and pumps, as shown in Fig. 7.1. This allows data scientists to add context to the data by understanding the important information, the material flows in the process, and the physical meaning of the process operations. The knowledge extracted is of critical importance in all subsequent steps of the Data Science workflow. It helps data scientists in the interpretation of patterns and features, and in the selection of models that satisfy the process physics. The decisions they make are beyond statistics and include process knowledge. As a result, they can exclude unfeasible solutions and models that cannot be explained by the process operations.

---

[2]J. Downs and E. Vogel, A plant-wide industrial-process control problem, *Computers & Chemical Engineering*, **17**, pp. 245–255, 1993.

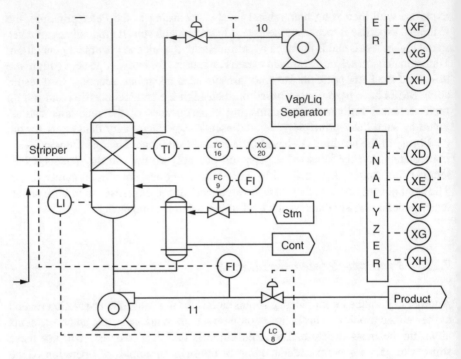

Fig. 7.1 A section of the Tennessee Eastman Process flowsheet

Process Documentation

Each organization keeps its knowledge in different documents, such as operation manuals, employee training materials, safety rules, and accounting practices. Most of them are accessible in electronic formats on the organization's Intranet. It is good practice that data scientists become familiar with a selected set of domain problem documents, recommended by the SMEs.

Examples of such documents for the emissions estimation project are the operation manuals and safety instructions for the unit, and the environmental agency's rules. Operation manuals are needed to extract knowledge about the key process control loops modes of operation, and parameter settings. Some information about alarms and action steps in the case of an emissions release can be extracted from the safety instructions as well. The sequence of the Relative Accuracy Test Audit, which is critical for the approval of emissions estimation model, is described in the corresponding state environmental commission document.

Business Structure

Knowledge of the organizational structure of the business is of critical importance for project organization and long-term success. An example of such a structure, related to the emissions estimation project, is shown in Fig. 7.2.

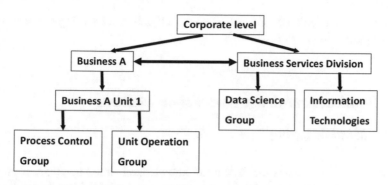

Fig. 7.2 An example of an organizational structure of a business

The organization has several hierarchical levels. At the top is the corporate level, in which the high-level management operations are concentrated. One level below is the main business operation level. This includes the supervision of all business units across a specific geography. On this level is the Business Services Division, which delivers specialized assistance to all of the business. Two groups that are involved in the emissions estimation project (Data Science and IT) are part of this large organization.

On the business side, each business unit includes specialized groups. The groups that are involved in the emissions estimation project are the Process Control and Unit Operation groups.

Of big importance for the success of the project is that the project leader has knowledge about the decision-making flows across the hierarchy and includes all needed individuals in the communication loop.

7.2.3 Publicly Available References

The fast development of AI-based technologies and the growth in their business applications requires a substep of exploring and using the available domain knowledge outside the organization. This step of estimating the current state of the art of a given problem is a must in academic and industrial R&D organizations but is still ignored by most business Data Science groups.

Fortunately, there are gazillions of references, such as papers, patents, reports, presentations, and social media responses, available on the Internet. Collecting and analyzing the essence of this external knowledge can protect the project team from "reinventing the wheel" mode and spawn many ideas for potential solutions. In the case of the emissions estimation project, several key journal papers gave a comprehensive summary of the current state of the art to the team. Selected patents confirmed the relevance of inferential sensors in emissions estimation. Some recent reports and presentations in the area of genetic programming and symbolic regression gave additional information about the potential for using this technology for the targeted inferential sensor.

7.3   Problem Knowledge Acquisition Methods

7.3.1   Mind Mapping

One of the best techniques for structuring knowledge is mind mapping. It involves writing down a central idea and thinking up new and related ideas that radiate out from the center. By focusing on key ideas written down in your own words, and then looking for branches out from them and connections between the ideas, you can map knowledge in a manner that will help in understanding and visualizing new information. An example of a mind map for the structure of this chapter is in shown in Fig. 7.3.

The creation of the map begins in the center with a block representing the key topic or theme of the map. In this specific case, this is the structure of the current chapter. The next elements in the mind map are the main branches that radiate from the central block. These branches signify the major subject areas. In this specific case, these are the key sections of the chapter. Because our brains can typically hold

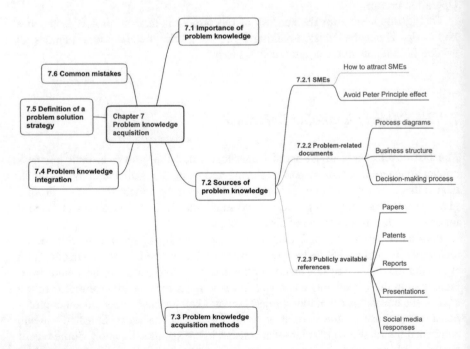

Fig. 7.3   A mind map of the current chapter

about seven pieces of information in our short-term memory, the recommended number of main branches per mind map is between five and nine.[3] If there is an order of information in the mind map, it is read clockwise, beginning with the main branch at the one o'clock position ("7.1" in Fig. 7.3). Each main branch can include a hierarchy of subbranches that refine the corresponding main topic. In Fig. 7.3, the subbranches include only two levels down, which are the substeps of the corresponding key steps of accesing the sources of problem knowledge (shown for Sect. 7.2). To avoid creating complex mind maps and to improve their visual impact, it is recommended that short names and branches be used, with no more than three hierarchical levels.

One of the advantages of mind mapping is that it is possible to look at the system at different hierarchical levels by collapsing or expanding it to a suitable level. For example, by selecting the level-one view of the mind map in Fig. 7.3, you can see only the main branches, which represent the key steps of the work process. There are other options for capturing knowledge in mind maps, such as adding notes, timings, priorities, relationships, resources, and so on. Readers who want to increase their proficiency in developing mind maps can find more details in a book by Buzan.[4]

7.3.2   Brainstorming Sessions

One of the key advantages of mind mapping is being able to represent expert knowledge shared during brainstorming sessions, when substantial information about the system is captured and classified in real time. This requires good facilitating skills, such as listening to the experts, making decisions about the number of details to be included in the mind map, reaching consensus on the proposed key words, and so on. To help the participants, it is strongly recommended that a mind map on a similar topic be shown at the beginning of the session. Ideally, the facilitator can prepare a prototype of the mind map with an idea of the initial structure illustrated with the main branches. An example of a generic template mind map for brainstorming business problem knowledge acquisition is shown in Fig. 7.4.

A known issue in brainstorming, especially with large teams of SMEs, is the lack of focus. As a result, extracting and structuring specific problem domain knowledge is inefficient. Often several brainstorming sessions are required until the necessary information is acquired and distilled. A potential solution is to navigate the brainstorming process via mind maps that impose a well-defined structure on the knowledge gaps that are expected to be filled.

The mind map shown in Fig. 7.4 is a generic template for the initial brainstorming session of the team. The key branches, defined by experience, show the main issues

---

[3]J. Nast, *Idea Mapping*, Wiley, 2006.

[4]T. Buzan, *The Mind Map Book*, third edition, BBC Active, 2003.

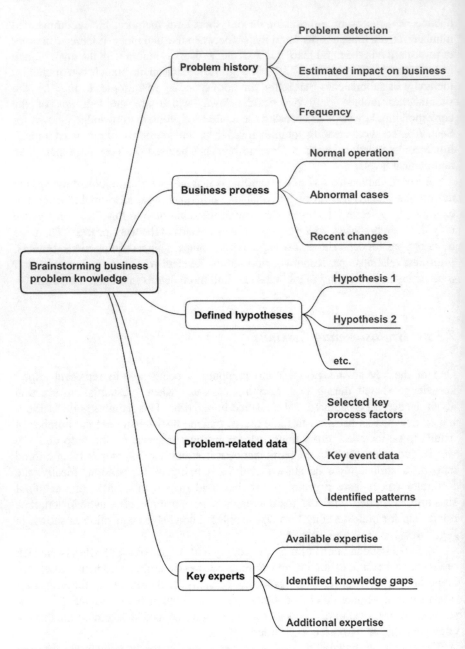

Fig. 7.4 A mind map template for brainstorming problem knowledge

of interest in filling gaps in problem knowledge, such as the history of the problem, details of the business process, definition of hypotheses, available data sources, and selecting the key SMEs. The next level of detail for each main branch is captured in the second level, of subbranches. For example, the business process main branch includes three subbranches: defining what normal operation means, identifying abnormal cases or criteria, and documenting recent changes in operating the business. The other subbranches are organized into similar important topics, related to the key branches.

Representing the structure of knowledge gaps in a mind map similar to the one shown in Fig. 7.4 helps all project stakeholders focus on the key issues. It provides a condensed view of the knowledge and reduces the risk of potential miscommunication. It pushes the experts to organize their knowledge in a structured way. In addition, typical template mind maps can be developed for more generic use and leveraged in future projects.

7.3.3 External Knowledge Acquisition

An inevitable step after extracting the available experience inside the business is to broaden the knowledge about the problem using external sources. Searching the Internet is the favorite technique, with many options—papers, software, patents, reports, presentations, etc. The issue is not finding many references but selecting the most informative ones and analyzing them. Aligning internal and external knowledge is also not trivial. Business domain SMEs should be actively involved in this process. They are the best indicators of the specific advantages of using external knowledge.

7.3.4 Knowledge Acquisition Skills

In order for knowledge acquisition to be successful, data scientists interviewing the domain experts have to have the following interpersonal skills:

- *Good and versatile communication skills* to interpret the whole variety of messages from the experts (verbal, written, even body language).
- *Tact and diplomacy* to avoid alienating the domain experts and facilitating possible conflicts between them.
- *Empathy and patience* to establish a reliable team spirit with the experts. The data scientist needs to encourage without being patronizing, argue without appearing self-opinionated, and ask for clarification without appearing critical.
- *Persistence* to keep faith in the project despite gaps and inconsistencies in knowledge.
- *Self-confidence* to lead the knowledge acquisition process. The data scientist needs to behave as the "conductor" of the domain expert "choir."

7.4 Problem Knowledge Integration

The success of problem knowledge acquisition depends on its effective integration into the Data Science workflow. The definition of success is increasing of the efficiency of the work process by reducing the overall time and effort of generating the targeted solution. The contributing factors are: high-quality data sets due to proper data preparation, fast variable selection based on identification of the correct process variables by the SMEs, and suitable model selection based on the domain experts' interpretation, to name a few.

Some of the key techniques for integrating problem knowledge are discussed briefly below.

7.4.1 Define Recommended Assumptions

Of special importance for executing an effective Data Science project is to clarify in advance the key assumptions about the available data, the modeling techniques, and how the derived solutions will be used. Accumulated problem domain knowledge is used to define the following assumptions:

- *Assumptions about process operation.* Define the tolerated ranges of process variables used for modeling (see Sect. 7.4.2 for details).
- *Assumptions about organizational changes of the business.* Define the expected changes in the current work process of the business when the solution is applied. In our example of the emissions estimation project, adding and tuning a control loop that uses the calculated estimate for product rate control is assumed. It is recommended also that some rules based on laboratory-measured emissions should be updated after the real-time estimate is available.
- *Assumptions about required training.* Define the team and the length of the desired training if the solution is applied. In the emissions estimation project, the assumption is that all control engineers and operators will be trained for at least 8 hours (the time suggested by the data scientists and domain SMEs).
- *Assumptions about maintenance and support.* Define the services that are expected to be needed to support the deployed solutions for 3 years, provided by an internal organization in Information Systems.

7.4.2 Define Normal/Abnormal Operating Conditions

Many business problems require a reliable differentiation between normal and abnormal operation. Without it, it is very difficult to handle outliers, extract patterns from the data, and build good models. The SMEs knowledge is critical in defining these conditions using key statistics from the data. Of special importance is domain experts' knowledge in the case of processes with multiple operating regimes and fast-changing operating conditions.

For the emissions estimation project, the SMEs defined normal operations with a recommended range of unit production rate. They also have documented 16 periods with significant alarms and 12 changes in the control loops related to production rate control.

7.4.3   Suggest Selection of Initial Variables

One of the most important deliverables from knowledge acquisition is a recommended starting list of process variables, economic factors, or customer features. Without this input from SMEs, statistical variable selection could be very inefficient. It is also critical to include all possible potential inputs to avoid significant redevelopment efforts.

For the emissions estimation project, 27 process variables were recommended as potential factors influencing the emissions.

Another important piece of information, given by problem domain SMEs is the suggested sampling time of the data. In the case of specialized data collection, as in the example with the original emissions data, taken using lab samples, details of the related procedure are needed. These include clarifying the timing of the grab sample, the duration of laboratory analysis, and entering the result into the process-monitoring system.

For the emissions estimation project, the SMEs recommended hourly averaged process data. This was synchronized with the hours when the lab samples were taken, which was usually done every 8 hours.

7.4.4   Define Qualitative Performance Metric

In addition to the quantitative performance metric given in the problem definition, important qualitative criteria for success of the project can be suggested by the SMEs. Examples of such measures are ease of use of the deployed solution, credibility of the developed model, and acceptability by the final users. It is recommended that data scientists understand very well from the domain SMEs the specific meaning of these qualitative criteria.

The expectation of "ease of use" of the deployed emissions estimator was defined by the experts as the ability to be integrated as a tag into the existing process information system and to be used by process operators in a way not different from any other tags in the system. The definition of "credibility of the developed model" for the same project was that the models should be interpretable and behave according to existing process knowledge and physics. According to the domain SMEs, the qualitative criteria for "acceptability by final users" meant a consistent use of the estimated emissions for closed-loop control by the process operators and control engineers. The process begins with training and continues with statistical performance tracking and analysis on a regular basis. Acceptability depend on model performance, i.e., the accuracy of emissions estimations in this example.

7.5   Definition of a Problem Solution Strategy

It is assumed that at this step of the Data Science workflow, the problem space has been sufficiently explored, and the project team is ready to switch to the solution space. We do not recommend beginning these efforts before clarifying the business problem and collecting the existing knowledge. Information gained in these steps significantly reduces the search space of possible solutions and allows a more effective definition of the modeling strategy and more effective project planning.

Some important components of the definition of the solution strategy are discussed below.

7.5.1   Define Solution Hypotheses

The first step in building a solution strategy is formulating the solution hypotheses. This step is based on the objectives of the problem and analysis of the available internal knowledge and public references. In the emissions estimation project, the key solution hypotheses for accomplishing the defined problem objectives were:

- Hypothesis 1. Emissions can be estimated based on nonlinear relationships with selected process variables.
- Hypothesis 2. Poduction rate is one of the key variables influencing emissions.
- Hypothesis 3. The developed models must have an average modeling error of less than 15% to pass the regulatory test.
- Hypothesis 4. The developed models will be explicit equations directly deployable into the existing process control system.

The defined hypotheses were supported by collected publicly available references. Many use cases for emissions estimation were found, as well as patents granted. The project team decided that there was sufficient evidence to move ahead with a specific plan to deliver appropriate solutions.

7.5.2   Define a List of Potential Solutions

The first step in this direction is selecting the type of solution. For the emissions estimation project, this type is prediction. The next step is the choice of modeling methods (see Table 4.1 for the recommended list). The requirement for models with an explicit functional form limited the selection to two AI-based techniques (in addition to the linear statistics benchmark model)—decision trees and evolutionary computation (genetic programming).

Unfortunately, most of the use cases for emissions estimation were based on neural networks that were black-box models, unacceptable to the project team. The chosen solutions were still in an exploratory phase with a short application record. However, some references[5] showed the big potential of genetic programming applications and the team decided to take the risk and use both genetic programming and decision trees as appropriate methods to develop the emissions estimator.

7.5.3 Define Issues and Limitations of Suggested Solutions

Understanding all details of the limitations of the selected methods is a recommended action item in developing the solution strategy. In addition to the generic issues discussed in Chap. 4, information related to problems in the specific application, obtained from collected references, should be considered. In the case of the emissions estimation project, the following generic and application-related issues were identified:

- The performance of developed data-driven models degrades beyond the range of the training data.
- The deployed models need parameter readjustment after significant changes in the process operating conditions.
- Developing genetic programming models requires training of data scientists.

7.5.4 Define Needed Infrastructure

The last step in defining the solution strategy is estimating the needed infrastructure in terms of hardware, software, and people. In the case of the emissions project, no specialized hardware was required for either model development or deployment. An additional software package for genetic programming was needed, however, to complement the available modeling tools. The data scientists would need to take classes in the method and software. Due to the expected frequent readjustment of the model parameters, more detailed model maintenance and support would need to be discussed and proposed.

---

[5] A. Kordon, L. Chiang, Z. Stefanov, and I. Castillo, Consider robust inferential sensors, *Chemical Processing*, September 2014.

7.6   Common Mistakes

7.6.1   Focusing on Data and Ignoring Problem Knowledge

Often Data Science projects begin with this scenario. Driven by the mindset that everything related to solution of the problem is hidden in the data, project development starts directly with data collection and analysis. Another dangerous pattern is context-free data preparation based on statistical metrics only (very often done automatically by the tools used). This may lead to losing data records with significant information content, incomplete data analysis, and developing models with reduced robustness.

7.6.2   SMEs Are Not Involved

The most obvious case of ignoring problem knowledge is not using domain experts at all. The negative consequences of this short-sighted approach to project organization influence each step in the Data Science workflow. Without selection by experts, the data collection could be either incomplete or too big. Data preparation will have the same issues as discussed in Sect. 7.6.1. The insight from data analysis could be incomplete without interpretation by the SMEs. Selecting developed models without exploration and interpretation by experts increases the risk of poor performance and early death of the models after deployment.

7.6.3   Not Validating SMEs Knowledge

This is a generic issue in using expert knowledge. The best way to verify some of the knowledge is by comparison with external references. Another way is by checking specific facts against collected data. It is recommended to discuss carefully the disagreement in the captured knowledge with the SMEs without questioning their expertise.

7.6.4   Reinventing the Wheel

An inevitable effect of ignoring external knowledge is the probability of spending time and effort on rediscovering publicly available solutions. Of special importance is awareness of known patents with similar solutions of the problems, to prevent potential legal issues.

7.7   Suggested Reading

- T. Buzan, *The Mind Map Book*, third edition, BBC Active, 2003.
- J. Nast, *Idea Mapping*, Wiley, 2006.
- D. Pyle, *Business Modeling and Data Mining*, Elsevier Science, 2003.

7.8   Questions

Question 1
What are the key benefits of using domain knowledge for solving business problems?

Question 2
What are the consequences of not using domain knowledge in data preparation?

Question 3
How does one find and attract problem domain experts?

Question 4
What process documentation is useful for knowledge extraction?

Question 5
What are the advantages of using mind maps in brainstorming sessions?

Chapter 8
Data Preparation

> *Torture the data, and it will confess to anything.*
> *Ronald Coase*

Data preparation includes data collection and integration, visual data exploration, and data preprocessing by imputing missing data, handling outliers, data transformation, and data balancing. Some of these actions are not trivial and require domain knowledge. Very often they need several iterations until data of acceptable quality is achieved.

Data quality is a decisive factor for the success of data analysis and model building. That makes data preparation the key productivity element in the Data Science workflow. If it is performed inappropriately, it can significantly reduce model development efficiency. In fact, data preparation solves the following dilemma: models or Garbage-In-Garbage-Out. Either we have low-quality data that is useless for solving the defined business problem and can be qualitatively characterized as garbage, or, in the case of data of acceptable quality, we can develop a variety of modeling solutions.

With its high probability of bad results and nasty scenarios, data preparation can be defined as a kingdom of Murphy's laws. A list of the effects of Murphy's law on data that presents these scenarios is given in Sect. 17.6.

The first objective of this chapter is to clarify the issues of data collection, such as identifying internal and external data sources, defining metadata, integrating the data, and performing a data sufficiency check. The second is to focus on several techniques for visual data exploration, such as identifying strange patterns, analyzing data distributions, and extracting insight from univariate, bivariate, and multivariate plots. The third is to discuss key data preprocessing capabilities such as handling missing data and outliers, important data transformations, and balancing the data.

© Springer Nature Switzerland AG 2020
A. K. Kordon, *Applying Data Science*,
https://doi.org/10.1007/978-3-030-36375-8_8

8.1   Data Collection

Data is collected in many different ways. The lifecycle of usable data usually involves capture, preprocessing, storage, retrieval, postprocessing, analysis, visualization, modeling, and so on.

Once captured, data is usually referred to as being structured, semi-structured, or unstructured. These distinctions are important because they are directly related to the type of database technologies and storage required, the software and methods by which the data is queried and processed, and the complexity of dealing with the data.

Structured data refers to data that is stored as a structure in a relational database or spreadsheet. Often it is easily searchable using SQL (structured query language), since the structure of the data is known. A record of manufacturing process variables is a good example. Each process variable has a data collection time, variable name, description, units, frequency of data collection, etc.

Unstructured data is data that is not defined by any schema, model, or structure, and is not organized in a specific way. A typical case is most social media data.

Semistructured data is a combination of the two. It is basically unstructured data that also has structured data appended to it. For example, smartphone photos in addition to the picture (unstructured data) have the date and time the photo was taken, the image size, etc. That is the structured part.

A popular approach to data collection is ETL (Extract, Transform, and Load). These three conceptual steps are how most data pipelines are designed and structured:

- Extract. This is the step where raw data waits for upstream data sources to land (e.g., an upstream source could be a machine- or user-generated log, a copy of a relational database, an external data set, etc.).
- Transform. This is the most important ETL job, where business rules and actions such as filtering, grouping, and aggregation to translate raw data into analysis-ready data sets are applied. This step requires a good understanding of the business and domain knowledge acquisition.
- Load. Finally, the processed data is loaded and transported to its final destination. Often, this data set can either be consumed directly by the final users or be treated as yet another upstream dependency on another ETL job, forming a so-called data lineage.

A short summary of other methods for data collection, such as Hadoop, is given in Chap. 12.

8.1.1   Data Sources

Most of the data sources are within the organization related to the defined business problem. They can be spread across different departments and geographies, however, and be based on various data collection platforms. Usually the domain SMEs have sufficient knowledge about the availability of data scources and the possible ways to access them.

Recently, there has been a growing trend toward using external data for improved data analysis and model building. Typical cases are macroeconomic and microeconomic data and data from social networks. A summary of both internal and external data sources is given below.

Internal Data Sources

In most cases the origin of the targeted problem-related data is inside the business. Examples are process-monitoring historians in manufacturing, and data warehouses in businesses. It is common practice that the internal data is extracted during model development by a person from the business who is familiar with the specific sources. Usually the final tables containing the targeted variables are delivered as Excel spreadsheets. However, these are aggregated data collected by scripts from the transaction tables in the corresponding corporate relational databases.

External Data Sources

One of the advantages of the Internet is the availability of many services offering almost any data of interest for building business-related models. Not all of these services, however, are equal in quality and in the number of available economic indicators. Some of the best-known sources are discussed below.

IHS Markit

The most well-known external source with almost universal collection of economic and financial data is IHS Markit (http://www.ihsmarkit.com.) It includes the main categories of global economic data, global financial data, and industry and sector data for over 200 countries and more than 170 industries. Some of this data includes forecasts for a defined forecast horizon. Access to the data needs a subscription.

FRED

The Federal Reserve Bank of St. Louis Economic Data (FRED) service offers more than 500,000 sets of american and international economic historical data for free. No forecasts are offered, however.

Indexes

Special types of external data is economic indexes. In the same way as the most famous stock market indexes, such as the Dow Jones and S&P 500, represent the state of the stock market in the United States, economic indexes capture the state of a national economy. Among the numerous available economic indexes, we will focus our attention on the Chicago Fed National Activity Index (CFNAI), which is the de facto normalized economic index for the state of the US economy. One of the advantages of this index is that it can be used as a leading economic indicator of a recession.

The Chicago Fed National Activity Index

The CFNAI is a monthly index based on a weighted average of 85 monthly indicators of US national economic activity. The 85 economic indicators that are

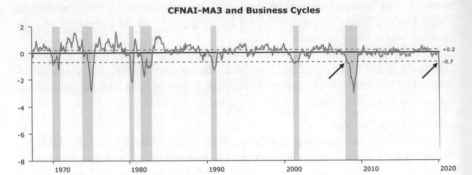

Fig. 8.1 CFNAI index and periods of recessions in the US economy. https://www.chicagofed.org/
publications/cfnai/index. Accessed on June 20, 2020

included in the CFNAI are taken from four broad categories of economic sectors:
(1) production and income; (2) employment, unemployment, and hours; (3) personal
consumption and housing; and (4) sales, orders, and inventories. It is a normalized
index, i.e., it has an average value of zero and a standard deviation of one. Since
economic activity tends toward trend growth rate over time, a positive index reading
corresponds to growth above trend and a negative index reading corresponds to
growth below trend.

An example of how the CFNAI relates to the recessions recently seen in the US
economy is shown in Fig. 8.1. Shading indicates official periods of recession, as
identified by the National Bureau of Economic Research. Following a period of
economic expansion, an increasing likelihood of a recession has historically been
associated with a CFNAI-MA3 ("MA3" means a moving average of the values from
the last 3 months) value below −0.70. Conversely, following a period of economic
contraction, an increasing likelihood of an expansion has historically been associated
with a CFNAI-MA3 value above −0.70 and a significant likelihood of an expansion
has historically been associated with a CFNAI-MA3 value above +0.20.

As shown by the first arrow in Fig. 8.1, the CFNAI rang the bell about the
upcoming great recession in January–February 2008 several months before many
businesses were heavily hit. It rang the bell for the upcoming recession due to
coronavirus 19 in February 2020, as shown by the second arrow.

Metadata Definition
The key procedure in data definition is clarifying the structure and content of the
data. This data about the data is called metadata and provides generic information
about key aspects of the data, such as:

- time and date of creation;
- data source;
- data description;
- purpose of the data

Table 8.1 Metadata of emissions estimation data

Variable name	Variable description	Units	PI tag	Frequency	Type
Var1	Plant production rate	Tons/day	FY317014C.DACA. PV	1 min	Manipulated
Var2	Material1 dosage	Lb/ton	AI317007.DACA. PV	10 min	Measured
Var3	Tower tempera- ture 1	Deg F	AI317432.DACA. PV	1 min	Measured
Var4	Tower pressure 1	Psig	WI317261A.DACA. PV	1 min	Measured
Var5	Tower inlet flow 1	Gpm	WI317261B.DACA. PV	1 min	Manipulated

- data type.

Above all, metadata is data. It is of critical importance for understanding the context of the data. An example of metadata for the emissions estimation project is shown in Table 8.1. This includes the key structural information about the variables used from the process historian, such as the variable name and description, the physical units, the tag in the system, the frequency of data collection, and type (is it only measured or can it be manipulated in control loops?)

On modeling sites, the metadata includes the types of variables available in the selected modeling tools. The most used data types are as follows:

- Numeric or interval type. Numeric variables.
- Categorical or nominal type. Character variables.
- Ordinal type. Categorical variables with clearly ordered categories.
- Binary type. Categorical variables with only two distinct classes.

Data Integration

In many real-world problems, integrating and synchronizing the different pieces of internal and external data is a challenging task. Two of the most important tech- niques for data integration are alignment of business-related data with the corresponding business structure and alignment with time in the case of time series data.

Alignment with Business Structure

The first question, after identifying the potential economic drivers, is their alignment with a specific business structure or economic category. This allows one to identify the specific data source on the Internet. In the case of the United States, Canada, and Mexico one can use an industry classification system called the North American Industry Classification System (NAICS). [1]

---

[1]http://www.census.gov/eos/www/naics/

Table 8.2 Food manufacturing NAICS code hierarchy

NAICS	Description
31–33	Manufacturing
311	Food manufacturing
3114	Fruit and vegetable preserving and specialty food manufacturing
31,141	Frozen food manufacturing
311,411	Frozen fruit, juice, and vegetable manufacturing
311,412	Frozen specialty food manufacturing
31,142	Fruit and vegetable canning, pickling, and drying
311,421	Fruit and vegetable canning
311,422	Specialty canning
311,423	Dried and dehydrated food manufacturing

Table 8.3 NAICS codes of some identified economic drivers for business forecasting

Economic driver	NAICS	Description
Trash bags	326,111	Trash bags, plastics film, single wall or multiwall, manufacturing
Liquid laundry detergent	333,312	Commercial laundry, drycleaning, and pressing machine manufacturing
Shampoo	81,211	Hair, nail, and skin care services
Bricks	327,331	Bricks, concrete, manufacturing
Cheese	311,511	Cheese, cottage, manufacturing
Napkins	322,121	Napkins, table, made in paper mills

NAICS is a six-digit hierarchical numeric classification system, with the first digit or two designating the broad industry sector and subsequent digits each reflecting more specific categories. For example, the first sector with the lowest first digit is agriculture, followed by mining, and then construction; manufacturing begins with the digit 3; wholesale trade, retail trade, and transportation are industry sectors in the middle of the list; codes beginning with 6, 7, or 8 are for services, and those beginning with 9 are for public administration. An example of the NAICS code hierarchy in the case of the food manufacturing sector is shown in Table 8.2.

An example of the aligninment of some identified economic drivers taken from a business forecasting project is given in Table 8.3.

An example of how to link a NAICS code to available historical data is shown in Fig. 8.2 for the case of shampoo production in the United States. Usually the economic indicator is the producer price index (PPI) of the industry sector where the driver of interest belongs. In the case of shampoo, the available PPI is for toilet preparation manufacturing, part of which is shampoo manufacturing. The starting time of the index is July 2007, when the index is 100%. The dynamics of the index until April 2018 is shown in Fig. 8.2.

Alignment with Time

Alignment of the internal and external data with a common time stamp is critical for the analysis of time series. The first obvious step in integrating the data from the

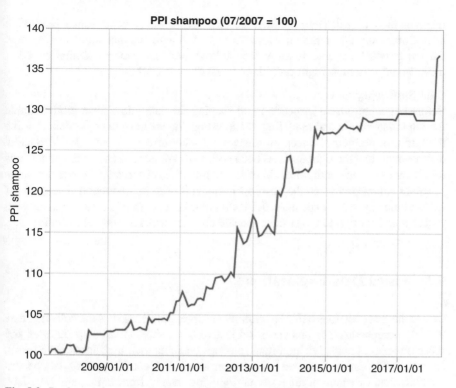

Fig. 8.2   Producer price pndex for shampoo (07/2007–04/2018)

different sources is to align it relative to the specific time window and frequency of the target (dependent) variable. This process requires expanding or contracting some of the data.

Often, not all of the data is for the same interval. Some might be annual, other data quarterly, and yet other data monthly. Thus, it all has to be brought into a form with the same interval. Sometimes, the time series data has to be converted from a finer interval to a coarser interval (monthly to quarterly). This is called contracting. Other times, an annual interval has to be converted to a quarterly interval. This is called expanding.

Converting time series data that is at a finer interval, such as monthly or weekly, can be simply a matter of summing or averaging data to get a coarser interval. Time series data such as demand or sales should be summed. Data such as indices should be averaged.

Sometimes, data for the independent variables is available only at a more aggregated level than the target variable. This means that the data has to be converted to a more frequent interval, such as going from annual to quarterly. Knowledge of the nature of the time series being expanded from one interval to another is definitely useful. The simplest approach is to divide by 4 (four quarters in a year), but this is generally not too informative. Expanding using some form of spline interpolation might be more useful. This approach would accommodate a trend from 1 year to the next. Combining the spline interpolation approach with seasonal adjustments when the independent variable is known to follow seasonal patterns similar to other independent variables might provide the best of both worlds.[2]

Data Sufficiency Check

Before beginning visual inspection and shaping the data into a form more suitable for extracting insight and modeling, the question "do we have the right data to solve the business problem?" needs an answer. Just because we have a lot of data, it doesn't mean that the right data has been collected. We need to ensure that the data is representative of the entire domain of interest–that the observations cover the range of values anticipated when the developed model is used in production. The perils of extrapolation have to be considered. Ideally, training data should include values with ranges broad enough to cover the expected changes when the model is deployed.

8.2   Visual Data Exploration

One of the first questions before beginning data preprocessing is: is the collected data visually acceptable? The answer is subjective but a short visual inspection of key process variables may give important information about their usefulness. As a result, some variables can be removed before preprocessing. A combination of data pattern identification, distribution analysis, and gaining insight from univariate, bivariate, and multivariate plots can help in the visual assessment of data quality. Several examples from manufacturing process data are given below.

8.2.1   Strange Data Patterns

The first task in visual data exploration is to look for patterns that reduce data quality, such as large deviations, constant values, or a high noise level. The degree of quality may decline in different ways, as is shown for three selected process variables in Fig. 8.3. This figure includes three plots of hourly averaged data from the same manufacturing process.

The first variable (Var1) has a curve with a relatively low noise level. The only exception is the big drop shown by an arrow. This significant deviation from the current trend was probably caused by a short process upset and will be treated as an outlier, which will probably be removed. This single outlier will not have a significant negative impact on the rest of the data, which can be reliably used in data preprocessing, analysis, and model development.

---

[2]T. Rey, A. Kordon, and C. Wells, *Applied Data Mining in Forecasting*, SAS Institute, 2012.

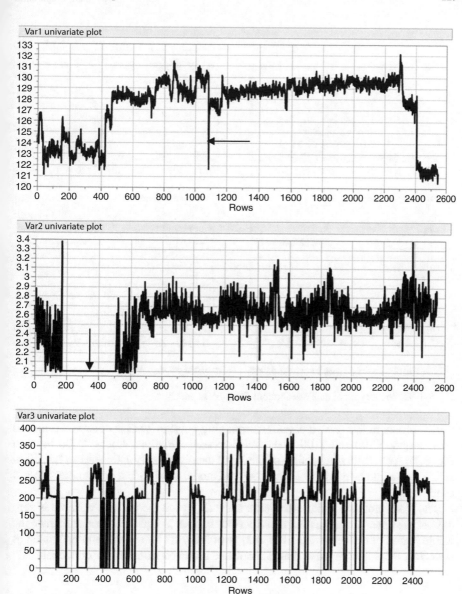

Fig. 8.3 Different patterns in the data for process variables

A more dangerous pattern is identified by an arrow in the Var2 plot. For almost 180 hours, the recorded value is a constant number. The hypothesis that the sensor was broken during this period of time was confirmed later by the SMEs, i.e., this part of the data cannot be used for analysis. In addition, for some reason, the new sensor shows large fluctuations until sample 700, which also cannot be used in the analysis. One option is to focus on the data after sample 700 and not use almost 25% of the

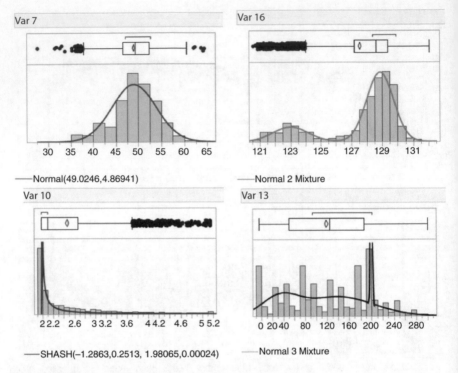

Fig. 8.4 Different types of data distributions for process variables

records. Even in this case, a pattern of changing noise in Var2 is observed (reduced noise around 1100 and 2100 rows and increased variations around 2400 rows in Fig. 8.3). As a result, the quality of the Var2 data is visually estimated as questionable, i.e., it could be used in data analysis and model building but only under selected conditions.

The Var3 plot shows the worst-case scenario, when the process variable is zero 45% of the time. Unfortunately, these periods happened repeatedly and influence the whole period of collected data. The quality of the Var3 data is visually estimated as poor, i.e., it is recommended that it should be removed from further data analysis.

8.2.2 Data Distributions

Exploring data distributions is another valuable source of preliminary insight about data quality. The first important piece of information is the type of distribution. Examples of different types of distribution are shown in Fig. 8.4.

The types of the distributions for all cases in Fig. 8.4 were identified by the modeling tool used.[3] Var7 is representative of the most common situation, of normally distributed data. Var16 is a case of a dual normal distribution, which in manufacturing indicates two operating regimes. Many process variables have a tailed distribution type, as is the case with Var10. Often these variables are transformed to make the distribution closer to normal. Sometimes it is difficult to fit a distribution type to the distribution pattern of the data, as in the example of Var13. This is another indicator of potential issues in analyzing the data.

The other benefit of exploring data distributions is identifying outliers. This will be discussed in Sect. 8.3.2.

8.2.3   Univariate Plots

Another way of extracting insight from collected data is by comparing univariate plots of the targeted and independent variables. Of special value is when the comparison is done with process variables sorted by the dependent variable. An example of such univariate plot, related to the emissions estimation project, with eight independent variables ($x1$–$x8$) sorted by the target variable (y, the emissions), is shown in Fig. 8.5.

A comparison between the trends of some independent variables, such as $x1$ and $x7$, and the trend of the emissions shows a lack of similarity. This is an indication of potential low correlation and a high probability of rejection of those variables. The opposite observation can be made, of a high similarity between the trends, for $x6$ and the emissions. Possible links between $x3$ and $x4$ and the emissions can be defined as well. All three could be good candidates for potential selection of independent variables.

8.2.4   Bivariate Plots

Bivariate plots give important information about potential relationships between variables of interest. They also define another set of outliers, identified by patterns outside the norm. The norm is usually the 95% confidence ellipse. An example of a bivariate plot of three independent variables from the emissions estimation project is shown in Fig. 8.6.

The insight from the plot can be summarized as follows. (1) Variables $x6$ and $x8$ are correlated to some extent but not highly, i.e., multicollinearity is not expected. (2) $x7$ is not correlated with either $x6$ or $x8$; and (3) there are several outliers outside the 95% confidence ellipse that need to be handled.

---

[3] SAS JMP 14.

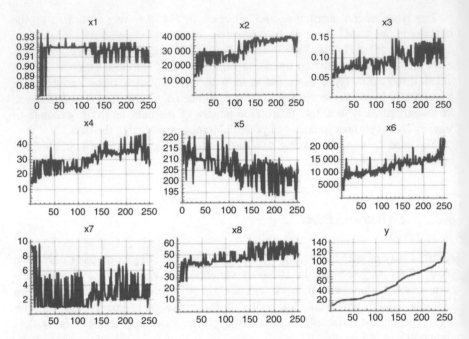

Fig. 8.5 Univariate plots of process variables sorted by target variable (emissions in this case)

Bivariate plots are not practical for data sets with a large number of variables, however. One potential option for visual exploration in this case is multivariate plots.

8.2.5   Multivariate Plots

As we'll discuss in Chap. 9, principal component analysis (PCA) is the most used method for multivariate exploration. The key result is a reduction of the dimensionality from the high number of original variables (dozens to thousands) to a low number of linear combinations of them, called principal components (usually less than 10). The most used plot is called the score plot, which is a scatter plot of the first two principal components. Examples of such plots from two different manufacturing processes are shown in Figs. 8.7 and 8.8.

The score plot, shown in Fig. 8.7, is based on the first two principal components of 89 original process variables. The first principal component explains 45.7% of the process variability, while the second principal component captures 14.4% of the process variability. This is a 2D view of the state of the process, based on 60% variability of the 89 original process variables, where each dot represents the state of the process at a corresponding period of time. Figure 8.7 shows that the manufacturing process operated most of the time in steady conditions with small variations

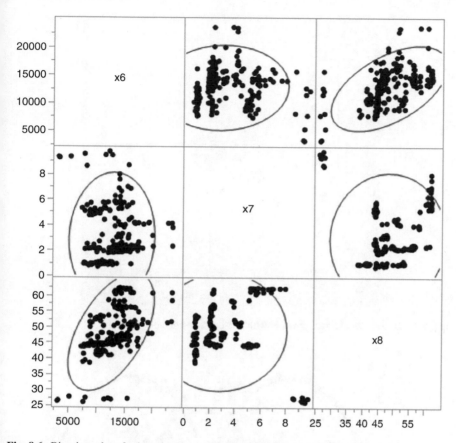

Fig. 8.6 Bivariate plot of selected process variables related to emissions

(represented by the gray cloud at the center of the plot). This is a clear example of process conditions with one operating regime. An advantage of this mode is that the insight and model development will be defined by this operating condition. The deviations from this mode, shown by black dots in Fig. 8.7, can be treated as outliers and their fate will be decided by discussions with domain SMEs.

In the case shown in Fig. 8.8, the number of original process variables is 186 and the first two principal components represent 59.8% of the process variability. The two operating regimes defined, represented by two clusters in the score plot, may require different modeling solutions for each operating regime or a robust model that satisfies the performance criteria for both situations. This is a more challenging case for developing and applying data-driven solutions than the single operating regime shown in Fig. 8.7.

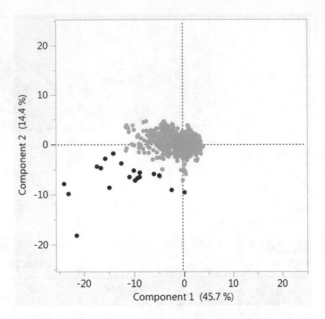

Fig. 8.7 Multivariate plot for a process with one operating regime

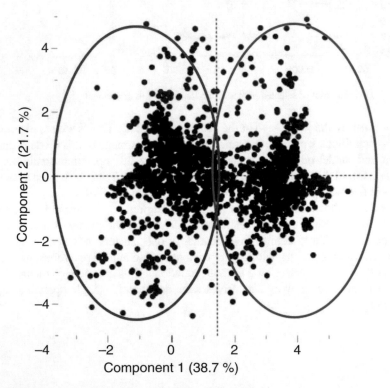

Fig. 8.8 Multivariate plot for a process with two operating regimes

8.3   Data Preprocessing

Data preprocessing is not trivial and is more an art than a science. It includes several steps that help to increase the informational content of the collected raw data. It is an iterative process, with several iterations until acceptable data quality is achieved. These steps are shown in Fig. 8.9 and discussed in this section.

8.3.1   Handling Missing Data

A missing value is a data point that has not been stored or gathered due to a faulty sensor or sampling process, cost restrictions, or limitations in the acquisition process. The treatment of missing values is difficult and has significant consequences in data analysis and modeling. Inappropriately handling the missing values can easily lead to poor knowledge being extracted, as well as wrong conclusions.

The main dilemma in handling missing data is: removal or imputation? The answer depends on the percentage of missing values in the data set, the variables affected by missing values, whether that missing data is a part of the dependent or independent variables, etc. Treatment of missing data becomes important, since the

Fig. 8.9   Key steps in data preprocessing

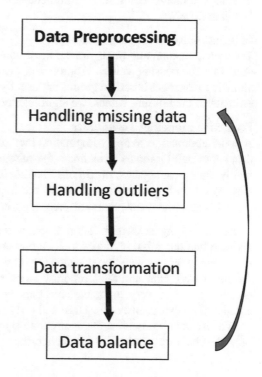

insights from the data or the performance of developed predictive model could be impacted if the missing values are not appropriately handled.

Removal

The first option is usually to discard those records that may contain a missing value. However, this approach is rarely beneficial, as eliminating records may introduce bias into the model development process, and important information can be discarded. There are two ways to implement the removal: (1) listwise and (2) pairwise.

1. Listwise. In this case, rows containing missing variables are deleted. It is recommended that the fraction of deleted rows does not exceed 30%.
2. Pairwise. In this case, only the missing observations are ignored and analysis is done on the variables present.

Both of the above methods suffer from loss of information. Listwise deletion suffers the maximum information loss compared with pairwise removal. But the problem with pairwise deletion is that even though it makes the data available, one cannot compare analyses because the data samples are different every time.

Imputation

Imputation is an attempt to keep the missing records by using replacement from available data. The two most popular techniques are averaging and building predictive models. In both cases, it is recommended to limit the fraction of imputed data to less than 20%.

Imputation by Averaging Techniques

The mean, median, and mode are the most popular averaging techniques which are used to infer missing values. Approaches ranging from a global average for the variable to averages based on groups are usually applied. Though we can get a quick estimate of the missing values, we significantly reduce the variation in the data set.

Predictive Modeling Techniques

A better approach is to apply imputation methods that use the other variables in the data set to build a model to estimate the missing data. The most used methods are multivariate least squares, maximum likelihood, and singular value decomposition. However, it is important to understand the imputation algorithms before applying them, because the results from different algorithms may be significantly different.

An interesting case is decision trees. A decision tree can be trained using a variable that has missing values as its target and all the other variables in the data set as inputs. In this way, the decision tree can learn acceptable replacement values for the missing values in the temporary target variable. This approach requires one decision tree for every input variable that has missing values, so it can become computationally expensive for large and dirty training sets.

We also have to consider the anonymous opinion that imputation is often thought of as the statistical equivalent of witchcraft.

8.3.2   Handling Outliers

Defining Outliers

Outliers can be defined as data points that violate the general pattern of smooth or otherwise regular variation seen in the data sequence.[4] A clear example of an outlier is the sudden drop in Var1 shown in Fig. 8.3. There is no rigid mathematical definition of what constitutes an outlier. Some statistical rules are available, but determining whether or not an observation is an outlier is ultimately a subjective exercise based on detection methods and domain knowledge.

There are several approaches to detecting outliers, which are classified into the following groups:[5]

- *Extreme value analysis.* This is the most basic form of outlier detection and only valid for univariate data. It is assumed that values which are too large or too small relative to the majority of the data are outliers. Examples of such statistical methods are the Z-test and Student's t-test. These are good heuristics for initial analysis of data but they cannot detect multivariate outliers. They can be used as final steps in interpreting the outputs of other outlier detection methods.
- *Probabilistic and statistical models.* These models assume specific distributions for the data. Then, using expectation-maximization methods they estimate the parameters of the model. Finally, they calculate the probability of membership of each data point in the calculated distribution. The points with low a probability of membership are marked as outliers.
- *Linear multivariate models.* These methods model the data into lower- dimensional subspaces with the use of linear correlations. Then the distance of each data point to a plane that fits the subspace is calculated. This distance is used to find outliers. PCA is an example of a linear model for anomaly detection. It uses an error metric, called T^2, that defines an upper confidence limit for outlier identification.
- *Proximity-based models.* The idea of these methods is to model outliers as points which are isolated from the rest of the observations. Cluster analysis, density-based analysis, and nearest-neighborhood methods are the main approaches of this kind.

It has to taken into account that some data analysis algorithms are more sensitive than others when dealing with outliers. Supervised machine learning algorithms that use a squared-loss function to determine the parameters that best fit the training data are heavily influenced by outliers. For example, gradient- boosting algorithms add large weights to observations that are considered to be hard cases, i.e., potential outliers. Some clustering algorithms, such as *K*-means clustering, can be quite

---

[4]R. Pearson, Mining Imperfect Data: Dealing with Contamination and Incomplete Records, SIAM, 2006.
[5]C. Aggarwal, *Outlier Analysis*, Springer, 2013.

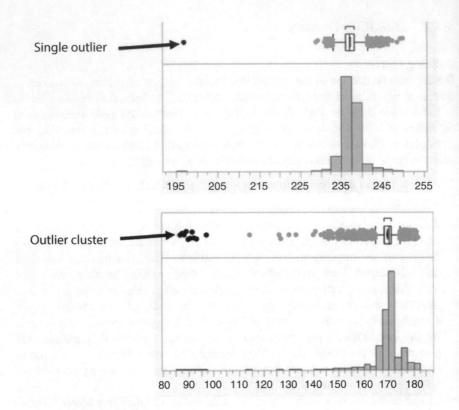

Fig. 8.10 Identification of outliers from univariate data distributions

sensitive to outliers because they try to minimize the sum of squared distances from cluster member points to the cluster means; a large deviation caused by an outlier receives a lot of weight.

Outlier Removal

Fortunately, outliers can usually be detected by some simple initial data exploration using univariate, bivariate, and multivariate plots. An example of identification of outliers from univariate data distributions is shown in Fig. 8.10.

First, we need to determine whether an outlying value is simply an invalid or erroneous entry that can be disregarded. If we have determined that an outlier provides no valuable information, it is acceptable to simply filter it out. This is the case with the single outlier shown in Fig. 8.10, where a single data point is far outside the normal distribution of the rest of the data. Handling the cluster of outlier, shown in Fig. 8.10, is not so trivial, though. In this case, a cluster of data points is far away from the majority of the data. On the one hand, removing the cluster will significantly improve the data distribution and potential for the model building. On the other hand, however, that cluster may include data for a potential new operating regime. The decision about its removal should be based on consulting the SMEs.

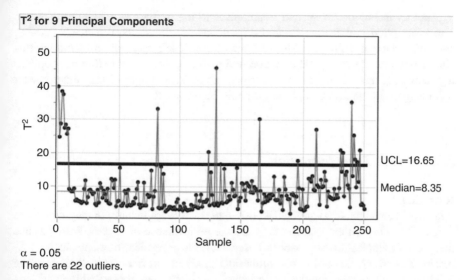

Fig. 8.11 Identification of multivariate outliers by PCA

A faster way to identify outliers is by multivariate methods, such as by using T^2 statistics derived from PCA. An example of such a determination is shown in Fig. 8.11.

Each dot in the plot is a value of the T^2 statistic, which is an indication of how close the data point is to what is statistically defined as the normal state. The important metric is the upper confidence limit (UCL), which defines the threshold between normal and abnormal behavior for a selected confidence level. Records with T^2 values above the threshold are identified as multivariate outliers. In the case shown in Fig. 8.11, 22 outliers were identified based on a UCL of 16.65 and a confidence limit of 95%. Handling these outliers can be done in different ways, however. The individual outliers with very high T^2 (for example >30) can be removed, but the fate of the set of outliers in the first 10 samples should be discussed with domain experts.

Special Treatment of Outliers

An acceptable reason for the removal of an outlier is when it provides no valuable information. However, if it has been determined that outliers might represent some real but rare relationship or if the information from the other features in those observations is too valuable to discard, then they can be used for special treatment by selected modeling techniques.

One such technique, used in time series modeling, is defining outliers as events. Some of the most common events involve a level shift up or down, a change in the trend, or a single observation point, also called a pulse. The events are integrated into the original data set by adding dummy variables with values of zero (no-related to the specific event) and one (related to the specific event).

Another technique for special treatment is to use outliers as a novelty detector. The assumption is that the deviated data points are the seeds of a new operating condition that can define a cluster of a new form. If this assumption is correct, this cluster will attract subsequent data and will gradually become the dominant operating regime. If this assumption is incorrect, the relative weight of this small cluster will decrease in the future, and then it could be removed.

8.3.3   Data Transformation

Rescaling

Real-world data includes variables whose values vary significantly in magnitude and range. These big differences can degrade the performance of several Data Science methods. Examples are approaches based on a distance measurement (such as K-means clustering), those that use numerical gradient information in their solution (such as neural networks and support vector machines), and those that depend on a measure of the variance (such as principal component analysis). The objective functions used in many of the relevant algorithms can be dominated by the variables that have a large variance relative to other variables, preventing the model from being able to learn the relationship with the other variables.

Two methods for rescaling data are well known—normalization and standardization. Normalization scales all numeric variables into a range of [0,1]. One possible formula is $x_{normalized} = \frac{x_{original} - x_{min}}{x_{max} - x_{min}}$.

where x_{min} and x_{max} are the minimum and maximum values, respectively, of the selected variable, $x_{original}$ is the value of the variable in the original numbers, and $x_{normalized}$ is its normalized value in the range [0,1].

The other option for rescaling is standardization of the data set. This transforms the original data to have zero mean and unit variance, for example by using the equation is $x_{stdrt} = \frac{x_{original} - \mu}{\sigma}$.

where $x_{original}$ is the value of the variable in the original numbers, μ is either the mean or the median of the selected variable in the original numbers, σ is the corresponding standard deviation also in the original numbers, and x_{stdrt} is the standardized variable.

Both of these techniques have their drawbacks. Normalization is very sensitive to outliers. If they are present in the data, normalizing the data will certainly scale the "normal" data to a very small interval. On the other hand, when standardization is used the new, standardized data is not bounded (unlike the case of normalization).

Improved Distributions by Power Transforms

The power transform family of functions are typically used to create monotonic data transformations. Their main significance is that they help in stabilizing the variance, keeping the data close to a normal distribution and making the data independent of the mean based on its distribution.

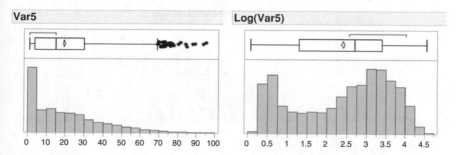

Fig. 8.12 Original tailed distribution onverted to dual normal distribution by log transform

The most popular transform in this family is the log transform. It can be represented mathematically as $y = \log_b(x)$.

This can then be translated into $b^y = x$.

The natural logarithm uses $b = e$, where $e = 2.71828$ is popularly known as Euler's number.

Log transforms are useful when applied to skewed distributions as they tend to expand the values which fall in the range of lower magnitudes and tend to compress or reduce the values which fall in the range of higher magnitudes. This tends to make the skewed distribution as normal-like as possible. An example of the benefits of a log transform is shown in Fig. 8.12.

As it is shown in Fig. 8.12, the original tailed distribution is transformed into a dual normal distribution that gives a better chance of discovering features and relationships.

Time Series Transformations

- Stationarity

A stationary time series is one whose statistical properties such as the mean, variance, and autocorrelation are all constant over time. One of the key reasons for trying to make a time series stationary is to be able to obtain meaningful sample statistics such as the mean, variance, and correlation with other variables. For example, if the series has a consistently increasing trend over time, the sample mean and variance will grow with the size of the sample, and they will always underestimate the mean and variance in future periods. If the mean and variance of a time series are not well defined, then neither are its correlations with other variables, which are the basis of data analysis and model building.

Unfortunately, most real-world economic time series are far from stationary, and they exhibit trends, cycles, seasonality, and other nonstationary behavior. An example of such a time series is shown in Fig. 8.13.

Several data transformations, such as differencing, detrending, and deseasonalizing need to be applied until stationarity is achieved. They are discussed briefly below.

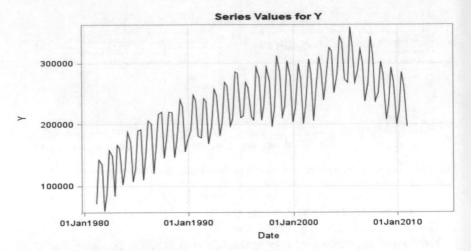

Fig. 8.13 Original real-world economic time series $Y(t)$

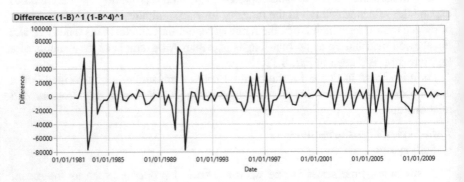

Fig. 8.14 Seasonal and first differences of the original time series shown in Fig. 8.13

Differencing

The first difference of a time series is the series of changes from one period to the next. For our example in Fig. 8.13, the time series $Y(t)$ is a quarterly time series. If $Y(t)$ denotes the value of the time series Y at period t, then the first difference of Y at period t is equal to $Y(t)–Y(t–1)$.

In the case of seasonal data (as is the case in Fig. 8.13), it is recommended that nonstationarity be reduced by using the seasonal difference (that is, the difference between an observation and the corresponding observation a year ago). For example, the first seasonal difference for the quarterly data $Y(t)$ with an annual period is equal to $Y(t)–Y(t–4)$. A plot of seasonal and first differences applied to the original data set $Y(t)$ is shown in Fig. 8.14.

When both seasonal and first differences are applied, it is recommended that the seasonal difference be done first because there is a chance that the resulting time

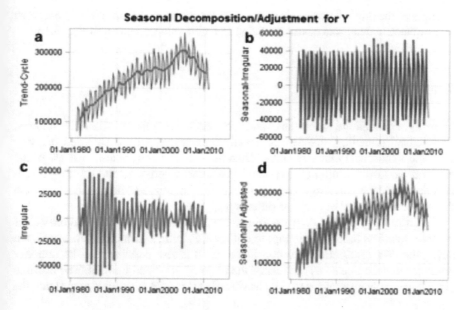

Fig. 8.15 Trend and seasonal decomposition of the original time series shown in Fig. 8.13

series will be stationary, and there will then be no need for a first difference. If stationarity cannot be accomplished by a first difference, it is possible to apply a second-order regular and seasonal difference to the data.

Detrending

A trend in a time series is a slow, gradual change in a property of the series over the whole interval under consideration. Unfortunately, identifying a trend in a time series is subjective because a trend cannot be unequivocally distinguished from low-frequency fluctuations. What looks like a trend in a short- segment of a time series often proves to be a low-frequency fluctuation (perhaps part of a cycle) in a longer series.

Detrending is the statistical transformation of removing a trend from a time-series. Many alternative methods are available for detrending. A simple linear trend in the mean can be removed by subtracting a least-squares-fit straight line. More complicated trends might require different procedures, such as curve fitting, digital filtering, and piecewise polynomials.

The trend decomposition of the original time series $Y(t)$ is shown in Fig. 8.15a.

Deseasonalizing

Deseasonalizing is the statistical or mathematical operation of removing seasonal patterns from a time series. Usually, the removal of the seasonal patterns is combined with decomposition of the time series into cyclical and irregular components. The major distinction between a seasonal and a cyclical pattern is that the former has a constant length and appears at regular intervals, and the latter varies in length. In our example of the original economic time series $Y(t)$, an annual seasonality is observed in the quarterly data. The deseasonalizing transformation is shown in Fig. 8.15b.

8.3.4   Data Balance

In many applications, such as fraud detection, analysis of machine failures, and medical screening, the data is heavily unbalanced, with a crowded majority class and a miniscule minority class. Often the minority class is less than 1% of the data and we face "needle in a haystack" problems, where machine learning classifiers are used to sort through huge populations of negative (uninteresting) cases to find the small number of positive (interesting) cases.

Unfortunately, conventional algorithms are often biased toward the majority class because their loss functions attempt to optimize quantities such as the error rate, not taking the data distribution into consideration. It is also possible that the minority examples may be treated as outliers of the majority class and ignored. The learning algorithm simply generates a trivial classifier that classifies every data record into the majority class.

The recommended solution is to resample the training set and create a new data set where both classes are balanced.

Resampling Training Data

Balance can be accomplished by two approaches—undersampling and ovesampling. Undersampling randomly downsamples the majority class. Ovesampling randomly replicates minority instances to increase their population. Some details are given below.

Undersampling

Undersampling balances the data set by reducing the size of the majority class. This method is used when the quantity of data is sufficient. By keeping all samples in the minority class and randomly selecting an equal number of samples in the majority class, a balanced new data set can be retrieved for further analysis and modeling. The principle is illustrated in Fig. 8.16.

Oversampling

In contrast, oversampling is used when the quantity of data is insufficient. Balance is accomplished by increasing the number of minority samples. Rather than getting rid of majority samples, new minority samples are generated by using repetition, bootstrapping or other methods. The principle of oversampling is illustrated in Fig. 8.17.

Existing experience from implementing data balancing shows that there is no absolute advantage of one resampling method over the other. Application of these two methods depends on the business application and the dataset itself. A combination of over and under ampling is often recommended for improving the results as well.

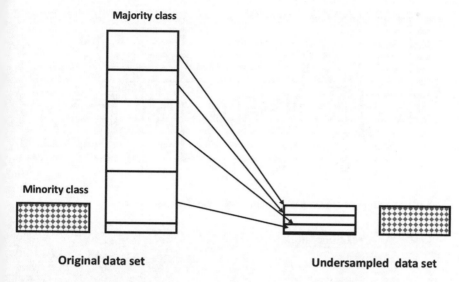

Fig. 8.16 Balancing data by undersampling

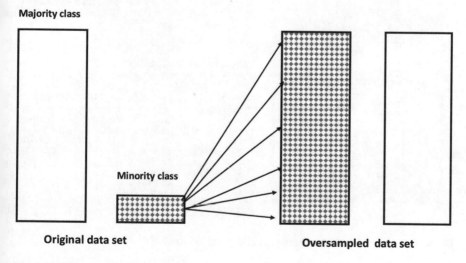

Fig. 8.17 Balancing data by oversampling

Training/Test Data Balance

Another step in data preparation that is very important for successful data analysis and modeling is achieving the proper balance between training and test data. The expectation for both manufacturing and business data is that the test data will be selected from the most recent data. Usually the balance between the most recent data and the rest of available data history is visualized by comparing the distributions of training and test data. An example of emissions data is shown in Figs. 8.18 and 8.19.

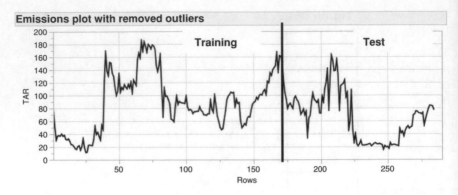

Fig. 8.18   An example of training and test data selection (for emissions data)

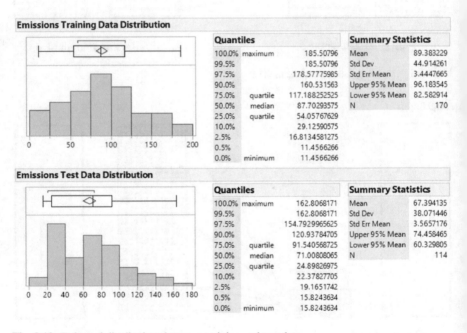

Fig. 8.19   Balanced distributions between training and test data

Figure 8.18 shows a plot of the target variable and the periods selected for the training (the initial 170 records of hourly averaged measurements) and test (the last 114 records of hourly averaged measurements) emissions data. The quality of the balance is best demonstrated by the ranges of the two distributions, shown in Fig. 8.19. Both the low range (11.4 for training vs. 15.8 for test) and the high range (162.8 for training and 185.5 for test) are close, which confirms the good balance between the data sets. In many cases, however, the balance is not trivial and some data rearrangement is needed.

8.3.5   *Data Quality Assessment*

The final step in data preparation is estimating the data quality after applying all the procedures discussed. The key question to be answered is: Is the data ready for productive data analysis and model development? Often the answer is not clear in black or white but has several shades of gray. The most generic division of data quality is into three broad categories: good, acceptable, and poor. It is not trivial to define firm quantitative criteria for this ranking, because these categories are strongly problem dependent. An attempt to describe qualitative measures for data quality of these three categories is given below:

- *Good-quality data.* The available internal and external data has a sufficient size, defined by the numbers of variables and records, to allow effective generation of solutions to the problem. The collected data contains a small amount of missing data and outliers. All of this have been successfully handled by removal or imputation. The final size of the data set after data preparation has not been reduced by more than 10%. Several well-balanced training and test data sets are available for effective model development. The recommended decision is to fully accept the prepared data and move on to the next steps of the Data Science workflow.
- *Acceptable quality data.* Most of the needed internal and external data has been collected but the size of the available data, defined by the numbers of variables and records, is insufficient to allow full-scale generation of solutions to the problem. Some data is costly and a decision to purchase some such data can be made dependending on the data analysis. The collected data contains a moderate amount, less than 20%, of missing data and outliers. Most of these have been successfully handled by removal or imputation. The final size of the data set after data preparation has been reduced to 70–80% of the original size. Balancing training and test data is difficult and the generated files are barely balanced. It is recommended to prepare several versions of such data sets for model development with different fractions of training and test data. The recommended decision is to cautiously accept the data and to be very careful in the data analysis and model development steps. One option for improving the data quality is not to select the poor-quality variables during variable selection.
- *Poor-or low-quality data.* There are big gaps between the needed and collected internal and external data due to sensor failures, long shutdowns, or a short history. The size of the available data, defined by the numbers of variables and records, is insufficient to generate solutions to the problem based on data-driven methods. The collected data contains a significant amount, greater than 30%, of missing data, outliers, and data with strange patterns. It is impossible to handle these by removal or imputation because the final size of the data set after data preparation would be be reduced to less than 50% of the original size. This is a clear example of the Garbage-In type of data. The recommended decision is to reject the data and discuss with the project team the potential options to the data quality, such as sensor replacement, improvement of the data collection

infrastructure, purchasing of needed external data, and longer data collection. A decisive message that continuing with this Garbage-In data will lead to a "spectacular" Garbage-Out fiasco should be sent to all stakeholders.

8.4   Common Mistakes

8.4.1   GIGO 2.0

As we have discussed several times in the book, one of the biggest dangers in applied Data Science is the almost religious belief that garbage in the data can be compensated for by the magic of advanced data analysis methods, especially those based on AI. We defined this mode of operation as Garbage-In-Gold-Out or GIGO 2.0, and gave more details in Chap. 1.

8.4.2   Problem Solving with Insufficient Data

Another typical mistake is to try to deliver a solution to the problem with insufficient data in terms of number of variables and with a limited number of records. One of the causes of this mode of operation is a lack of knowledge about the required variables and the available records. A good knowledge acquisition step should give an answer to this key question. An inevitable result of using such an informationally incomplete data set is deriving an imperfect data analysis and models with reduced performance. The issue of insufficient data can be resolved by adding new variables and collecting or purchasing more data records.

8.4.3   Problem Solving with Low-Quality Data

The most common mistake, typical of inexperienced data scientists, is to try to solve the problem at any cost even with low-quality data. They don't have the courage to say a clear NO and to stop a waste of time and effort after identifying the data quality as poor. They are afraid that killing the project at an early phase, such as data preparation, will be accepted very badly by the project sponsors and their reputation will suffer. The strategy preferred by them is to continue with data analysis and model development and only then to deliver the bad news. The argument is that everything possible has been done but, due to low-quality data, the solutions derived have poor performance. They believe that in this case the negative message is less damaging. The problem is that a lot of time and effort has been lost in fruitless work.

8.4.4   Low-Quality Data Preparation

A known habit of less experienced data scientists is ignoring the importance of data preparation. They are obsessed with the significance of modeling and prefer to reach that step of the workflow with minimal effort. The result is low-quality imputation, ineffective handling of outliers, and unbalanced training and test data sets. Often the impact of this fast-track data preparation on the quality of model generation is disastrous.

8.5   Suggested Reading

- C. Aggarwal, *Outlier Analysis*, Springer, 2013.
- R. Pearson, *Mining Imperfect Data: Dealing with Contamination and Incomplete Records*, SIAM, 2006.
- D. Pyle, *Data Preparation for Data Mining*, Morgan Kaufmann, 1999.

8.6   Questions

Question 1
What are the key internal data sources in your organization?

Question 2
Why does economic data need to be aligned with the business structure?

Question 3
What are the benefits of visual data exploration?

Question 4
Can outliers be removed automatically?

Question 5
How can data be balanced?

Chapter 9
Data Analysis

Data is not information. Information is not knowledge. Knowledge is not understanding. Understanding is not wisdom.

Clifford Stoll

As we discussed in Chap. 2, translating data into value requires several transformations. The first one, which is the focus of this chapter, is extracting insight from data. According to the Meriam-Webster dictionary, "insight is the act or result of apprehending the inner nature of things or of seeing intuitively." Obtaining insight from prepared data by data analysis includes, but is not limited to, obtaining a multivariate view of the process, understanding the key drivers, discovery of unknown patterns and relationships, and generation of data for model development. Most of this data analysis is based on applying statistics and machine learning but an important factor is also effective data visualization. Various plots to represent the data in the most informative way are available in almost any analytics software. However, the key factor for successful extraction of insight from data analysis is the imagination and problem domain knowledge of data scientists. They should identify the discoveries by thinking outside the box and communicate them clearly to all stakeholders. Since insight is strongly problem-dependent, it is recommended that the SMEs should be part of this process.

A critical condition for beginning data analysis is that the prepared data are of at least acceptable quality. Violating this rule is a clear recipe for wrong conclusions from the analysis and a modeling disaster in the next steps of the workflow.

The first objective of this chapter is to clarify the expected gains in translating data into insight. The second is describing the benefits of multivariate data analysis while the third is to discuss the selection of the most important variables. The fourth objective of the chapter is to present the methods for extracting relevant features from the data. The fifth is to emphasize the importance of data visualization, and the last objective is to give guidance on how to deliver a data story based on the analysis.

© Springer Nature Switzerland AG 2020
A. K. Kordon, *Applying Data Science*,
https://doi.org/10.1007/978-3-030-36375-8_9

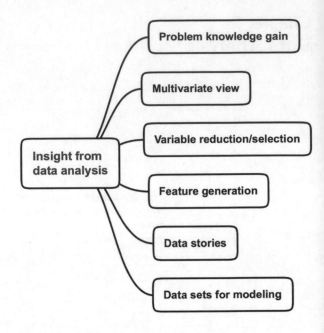

9.1   Translation of Data into Insight

Data scientists can extract different types of insight from analyzed data. The most
valuable insight is related to hypothesis testing, looking at the problem from a
multivariate point of view, understanding the most influential factors in the problem,
and extracting and defining patterns in the data. For some projects, the insight gained
is the final deliverable. A popular way to communicate the results from the data
analysis in this case is a data story, which requires a special skillset from data
scientist. For most AI-based Data Science projects, however, the insight captured
from data analysis is used in the final preparation of the data set for model development. These key issues of translating data into insight are shown in Fig. 9.1 and
discussed briefly below.

9.1.1   Problem Knowledge Gain from Data Analysis

It is expected that insight obtained from data analysis will add new knowledge about
the business problem. The best-case scenario is when new patterns and relationships
are discovered. Another important knowledge gain is from hypothesis testing. Most
of the initially defined hypotheses are validated with the available data in the data
analysis step. In our emissions estimation project example, the first two hypotheses
defined (see Sect. 7.5.1) were proved during data analysis (see Sect 9.3.2 for details
of Hypothesis 1 and Sect. 9.3.3 for Hypothesis 2).

An overall feasibility test for the targeted solution is another important insight from analyzing the data. For example, a lack of correlation between the independent and target variables raises serious questions about the statistical basis for building data-driven models. The opposite, finding highly correlated independent inputs to the dependent variable and stable operating conditions for the targeted solution, increase significantly the chance of developing good models.

All these problem knowledge gains contribute to improving the credibility of the suggested solution. Proved hypotheses and solution feasibility as well as selected important variables, and identified features, demonstrate that important constituents of the targeted problem solution are already available before the critical model development step.

9.1.2   Insight from Multivariate View

It is recommended to begin the translation process from data into insight by looking at the big picture, considering all factors. Identifying generic patterns derived from all related variables is a better strategy for analyzing the problem than drawing conclusions from individual factors. It is always preferable to look at the forest first before focusing on the trees.

The multivariate view of the problem allows one to investigate the impact of all process variables on process behavior. One important insight from multivariate analysis is related to defining operating conditions and their associated features. The insight from the multivariate view is also beneficial in fault detection and diagnosis problems. In this case a multivariate definition of normal and abnormal modes of operation is critical for a reliable solution of the problem. More details are given in Sect. 9.2.

9.1.3   Insight from Understanding Key Drivers

Often, as a result of a data deluge or insufficient problem knowledge, the initial number of independent variables for the targeted solution is bigger than necessary. Finding the most influential drivers and reducing the number of variables is a key expectation of data analysis. However, variable selection from big data with thousands of inputs is a challenging task. Using only linear methods is insufficient, because the variables may have a low linear correlation with the target variable but could be highly related in a nonlinear fashion, individually or in features together with other variables. Applying several linear and nonlinear methods for variable selection, derived from different scientific principles, is highly recommended. Inputs, selected by several methods have higher statistical credibility. The final test for acceptance, however, is done by the SMEs. They will challenge any selected variable that does not match process behavior.

The key principles and methods for variable selection are discussed in Sect. 9.3.

9.1.4   Insight from Discovered Features and Patterns

Feature extraction and pattern discovery from data are another key contributor to insight from data analysis.[1] Examples of patterns in time series data, such as trends, seasonality, and cyclicality, have been shown in Sect. 8.3.3. Other examples of patterns are the clusters and segments discovered in the data, and newly defined operating regimes and diagnostic signatures.

The best-known example of features based on linear transformations is the principal components derived by principal component analysis. Numerous simple nonlinear transforms also provide a popular way to manually generate features. Some examples have been given in Sect. 8.3.3. One of the advantages of genetic programming is the automatic generation of diverse nonlinear transforms that can be selected as features. More details of this topic are given in Sect. 9.4.

9.1.5   Insight from Data Analysis as the Final Problem Solution

For some projects, data analysis is the final delivery step. This is the case for descriptive analytics solutions focusing on the question: What has happened? A typical example is the analysis of the root-cause of manufacturing problems, such as equipment breakdowns, unit shutdowns, and poor process control. The expected technical deliverables are a list of the key variables and features that caused the problem, a demonstration of the impact of selected features on observed problems, an understanding of the root cause by means of physical considerations, and recommended action items for process changes to reduce or eliminate the problem.

A popular recent approach to communicating the extracted insight when data analysis is the final step of the project step is data stories. Some recommendations on how data scientists can prepare a data story are given in Sect. 9.6.

9.1.6   Insight for Final Data Preparation for Modeling

For most projects, however, the insight gained in data analysis is used to begin the critical data-modeling step with better data and the process knowledge gain. As a

---

[1]We make a distinction between input variables and features. A feature consists of a composite of transformed input variables.

Fig. 9.2 Visualization of
the first two principal
components

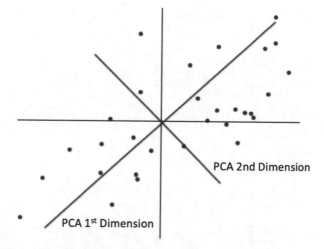

PCA 2nd Dimension

PCA 1st Dimension

result of multivariate analysis, the modeling strategy could be updated based on the
observed operating regimes. The insight extracted from variable selection and
generated features could be used for arranging the final data sets for modeling.
These final data sets have lower dimensionality and are focused on the right
operating conditions.

9.2   Multivariate Data Analysis

9.2.1   Principal Component Analysis

The most popular method for multivariate data analysis is PCA. It is based on the
idea of constructing an independent linear combination of the original variables in
which the coefficients capture the maximum amount of variability in the input data
itself. The linear combination with the largest variance in the data is called the first
principal component (PC1). The second principal component is orthogonal to the
first principal component, and it captures the largest variation in the rest of the data.
A visualization of the first two principal components of a data set is shown in
Fig. 9.2.

The next principal components are organized in a similar fashion. As a result, the
system can be described by a low-dimensional set of principal components that
capture the lion's share of variability in the data. One can compute as many principal
components as the number of independent variables, which can be further analyzed
and retained on the basis of the variability explained by them. Selecting the proper
number of principal components is one of the critical decisions in multivariate
analysis. The statistical decision is based on linear algebra calculations (eigenvalues)
and is presented in a special plot, called a scree plot. An example of such a plot for

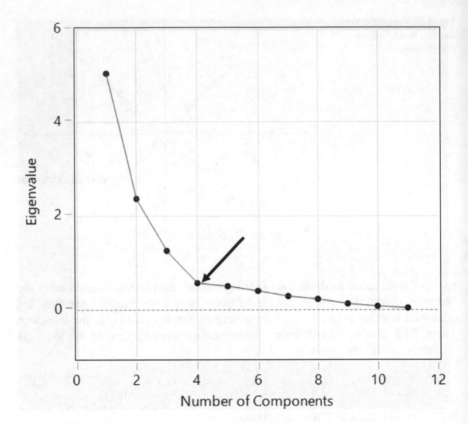

Fig. 9.3 Scree plot for selecting the number of principal components

emissions estimation data set is shown in Fig. 9.3. The number of principal components is selected based on the slope of the decrease of the eigenvalues. The
recommended selection is in the area of the knee of the slope. This is shown by an
arrow in Fig. 9.3, where the chosen number of PCs is four. Usually the number of
principal components is much smaller than the number of original variables. In the
emissions estimation data set, the number of original inputs was 11.

It has to be made clear that PCA is not a variable selection but a dimensionality
reduction method. Each principal component includes a linear combination of all
original variables with different weights (scores), i.e., the number of variables has
not been reduced. An example of three principal components for the emissions
estimation data set is shown in Fig. 9.4, where the number m of original inputs is
11 and their scores are given in Table 9.1.

The benefit of PCA as a dimensionality reduction method is that it delivers a 2D
or 3D view of the high-dimensional space of the many original variables. This
low-dimensional view is a source for extracting insight from multivariate patterns.
Some examples are given below.

Fig. 9.4 Example of the structure of a principal components calculation

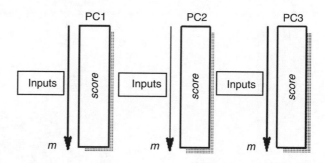

Table 9.1 Scores for the first three PCs of emissions estimation data

Variable	PC1	PC2	PC3
Var1	4.01	27.07	1.08
Var2	16.77	1.58	1.38
Var4	17.42	0.91	0.00
Var5	6.55	15.25	3.54
Var6	7.97	13.53	1.65
Var7	9.77	2.39	6.52
Var8	12.21	1.93	1.07
Var9	6.28	9.47	24.84
Var10	3.38	23.95	0.12
Var11	8.79	0.50	34.19

9.2.2 Multivariate Patterns

Among the different multivariate patterns that can be found, based on the first two or three principal components, we'll focus on those that may deliver insight into defining operating conditions, root cause analysis, grouping the original variables, and multivariate diagnostics.

Defining Operating Conditions

The most used multivariate view of a process is the score plot obtained from the first two principal components. The patterns and structures that can be recognized in it give insight to SMEs and data scientists about the state of the process. The shape of the pattern and its intensity can be used to define various operating conditions. A well-known type of insight is defining operating regimes. An example of a process with one operating regime is shown in Fig. 8.7, and one with two operating regimes is shown in Fig. 8.8.

In addition to defining operating regimes, score plots are used to get insight into the nuances of operating conditions. An example for the process explored in the emissions estimation project is shown in Fig. 9.5.

The first two principal components, based on the 11 original variables, explain 67.4% of the process variability and are a solid statistical basis for representing the whole process. Most of the data points in Fig. 9.5, represented by gray dots, are

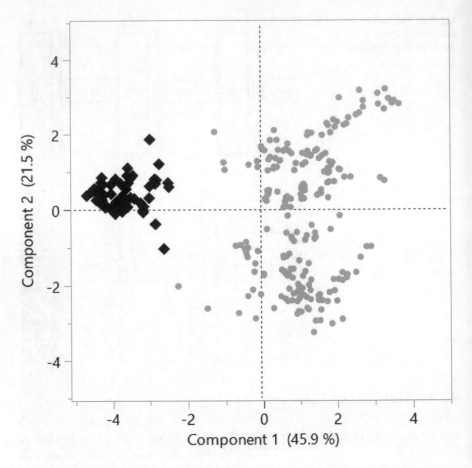

Fig. 9.5 Score plot for emissions estimation data

spread over the right side of the plot and define the normal operating condition. However, a different cluster of data, represented by black diamonds on the left side of the plot, needs a different interpretation. It is not populated enough to be defined as a second operating regime (as is the case in Fig. 8.8), nor is it diluted (as is the case in Fig. 8.7). An investigation of the behavior of two independent variables (Var4 and Var7) and the emissions showed that the process conditions in this cluster were not concentrated in one time period. The univariate plots of the selected variables are shown in Fig. 9.6, where the periods during the cluster-defined conditions are shown by black dots.

As shown in Fig. 9.6, some patterns in the original variables support the different operating conditions defined by the multivariate cluster. For example, both Var4 and the emissions operate with very low values. In the opposite direction is the behavior of Var7, which operates with very high values. Analyzing these patterns may help

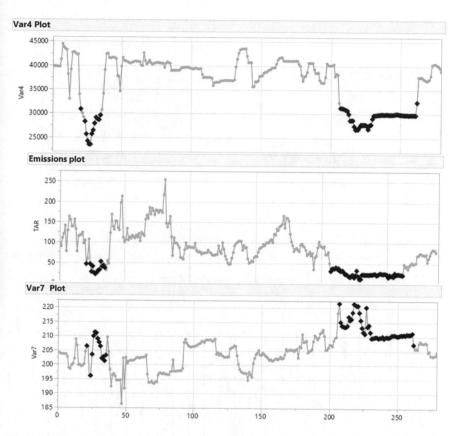

Fig. 9.6 Univariate plots of emissions and two independent variables. The black dots are part of the cluster

SMEs and data scientists to better understand the drivers of this change in operating condition and to improve process performance.

Root Cause Analysis

Often a deep dive into defined multivariate operating conditions can lead to identifying the root cause of a problem. Usually the problem is driven by several interrelated factors. Identifying them from univariate analysis is extremely challenging. We'll illustrate the power of multivariate root cause analysis with a frequent problem in manufacturing processes.

The problem relates to the manufacturing process moving from normal operation with optimal productivity to other, less efficient modes. An example of a process with 28 original variables with two distinct operating regimes, defined by two multivariate clusters, normal operation (cluster 1) and low production (cluster 2), is shown in a score plot in Fig. 9.7, where the two axes are based on the first two principal components.

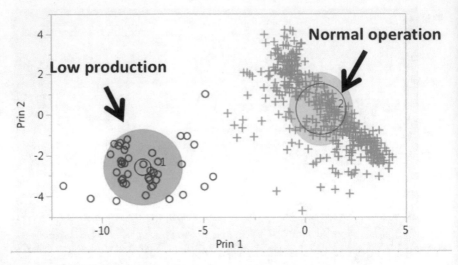

Fig. 9.7 Normal operation and low-production operation regimes

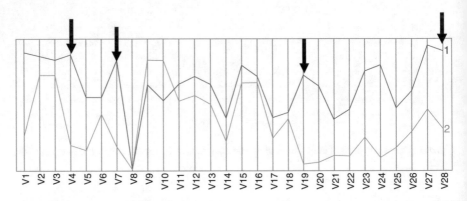

Fig. 9.8 Cluster means of variables

The negative financial impact of the low-production mode is significant and the objective of the study is to find the most influential process variables that drive the move. The basis of the analysis is the difference in the contributions of the original variables to the two regimes. A very simple way to estimate this is by comparing the cluster means of the variables. The cluster means of all 28 variables, V1 to V28, are shown in Fig. 9.8.

As is shown in Fig. 9.8, several original variables, indicated by arrows, have big differences between their cluster means. Their individual behavior is quite different as well. Examples of plots for both operating conditions for Var 4 and Var 7 are shown in Figs. 9.9 and 9.10, respectively.

Both variables operate at significantly lower values in the second operating regime. V7 is in a steady mode of around 40% of the normal-operating-regime

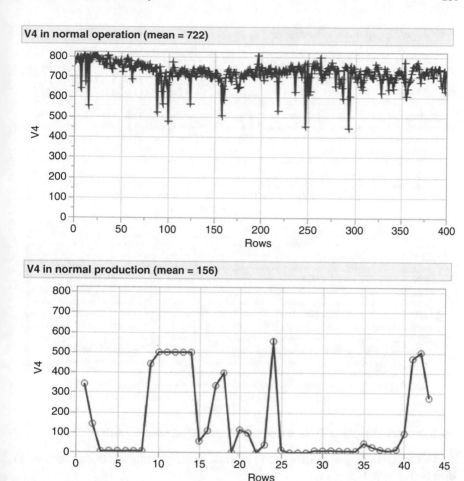

Fig. 9.9 Univariate plot of V4 in normal and low-production operating regimes

value, while V4 has big swings. It is even close to zero for almost half of the time. These two variables illustrate different components of the puzzle. It is of special importance to root cause analysis to combine the pieces and find a physical interpretation of this strange behavior.

Grouping of Original Variables
Another insight from multivariate analysis is exploring how the original variables are related to each other. Some questions about the existence of clusters of variables and the sign of any potential correlation between variables can be answered by a loading plot of the first two principal components. A loading plot is a plot of the relationship between the original variables and the principal components. It is used for interpreting relationships among the original variables.

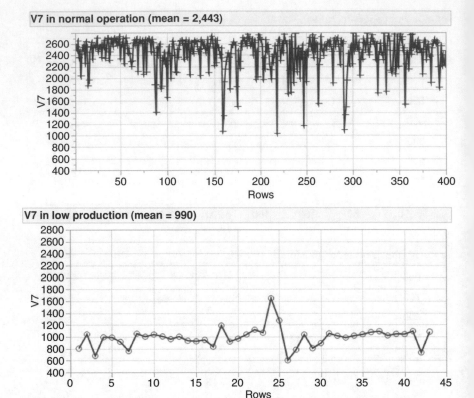

Fig. 9.10 Univariate plot of V7 in normal and low-production operating regimes

An example of a grouping of the 11 original variables in the emissions estimation data is shown in Fig. 9.11.

Analyzing the patterns in Fig. 9.11 shows that half of the variables (Var2, Var11, Var4, Var8, Var12, and Var9) are in a group with similar correlations. Var5 and Var6 form a small group with the same sign of correlation as the previous group. Var1, Var7, and Var10 are not associated with any other variables and have an opposite sign of correlation to the rest of the inputs.

Multivariate Diagnostic

The multivariate diagnostic is based on a special metric, called the T^2 statistics, calculated from selected number of principal components. It defines the normal operating condition and suggests an upper confidence limit as a threshold between normal and abnormal behavior. It is mostly used in the detection of multivariate outliers. An example was given in Sect. 8.3.2.

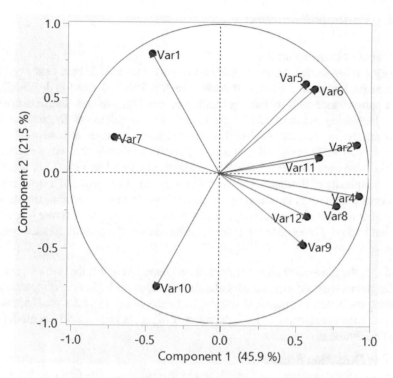

Fig. 9.11 Grouping of original variables for emissions data

9.3   Variable Selection

Selecting the most important variables related to the target variable is one of the most important tasks of data analysis. Usually it is done in two steps. First, a data dimensionality reduction is accomplished by removing those input variables that are highly correlated among themselves. This can be done by exploring multicollinearity, applying PCA or clustering the input variables. Second, the reduced number of variables are ranked by several linear and nonlinear methods for their relevance to the target variable. Only those, that have been consistently ranked high in most of the methods, are selected for model development in the final third step.

Why is variable selection so important?

Removing a redundant variable helps to improve accuracy. Similarly, inclusion of a relevant variable in the model has a positive effect on model performance. Too many variables may result in overfitting, which means that the model is not able to generalize. Other benefits of variable selection are facilitating visualization and understanding of the data, reducing training and utilization times, and defying the curse of dimensionality to improve prediction performance.

9.3.1   Variable Reduction

The Curse of Dimensionality
The key challenge that arises in analyzing many variables is best described by a phenomenon known as the *curse of dimensionality*. This is defined as degradation of model performance with increasing number of input dimensions (explanatory variables). Increasing the number of inputs increases the volume of the feature space exponentially; in turn, this higher-dimensional space requires exponentially more data points to sufficiently fill that space in order to ensure that combinations of variable values are accounted for. The effect is illustrated in Fig. 9.12, where the increased sparsity of the data is shown with a very simple example of a small number of observations in one, two, and three dimensions. In one dimension, the observations cover the domain sufficiently, in two dimensions, they still cover it relatively well, but in three dimensions the sparsity of the data in the overall domain becomes obvious.

The higher the dimensionality of the space, the more likely it is that we will not have the data points we need to make a local estimate. Also, the higher the dimensionality, the more we have to extrapolate instead of interpolating to make predictions. For these reasons, it is essential to limit the number of inputs in a data driven model to an absolute minimum.

Variable Reduction Sequence
Dimensionality reduction is accomplished in several steps. The first step is based on domain knowledge and common sense. It includes a broad selection of potential factors influencing the target variable. The average range for manufacturing problems is between 300 and 500 variables.

The next step in the reduction process is removing those inputs with low data quality during data preparation. On average, 20–25% of variables are in this category. The third step is getting rid of those variables that are highly correlated among themselves. This option is discussed in detail next.

Fig. 9.12 Data sparsity in increased dimensions

9.3.2   Handling Multicollinearity

Analyzing correlations among variables is a must in data analysis. The most well-known measure of correlation is the Pearson product-moment correlation, which quantifies the strength and direction of the relationship between two variables. The calculated correlation coefficient is in the range from -1.0 to $+1.0$. The closer the coefficient is to $+1.0$ or $--1.0$, the greater the strength of the linear relationship. A zero value of the correlation coefficient indicates no linear relationship.

The first key insight from correlation analysis is about the hypothesis that a relationship based on high correlations between available variables and the target variable can be derived. The answer is given by the values of the correlation coefficients of the inputs with respect to the target variable. As a practical rule of thumb, if the maximum absolute correlation coefficient is less than 0.3, the chance of deriving a useful data-driven solution with the data set being explored is minimal, i.e., the hypothesis has not been proved by the available data. In this case it is strongly recommended to communicate this critical negative result and suggest looking for other data sources or problems.

An example of such a correlation analysis, related to the emissions estimation project, is the correlation ranking of the inputs related to emissions. The ranking based on the absolute correlation coefficient is shown in Fig. 9.13, where the input variables are named Var1, Var2, Var3, and Var12 and the target variable is named TAR. The absolute correlation values are on the x-axis.

As shown in Fig. 9.13, several variables have an absolute correlation greater than 0.7, which is a solid foundation for developing relationships with the emissions, i.e., Hypothesis 1 in the project, defined in Sect. 7.5.1, has been supported by the data.

The second key insight from correlation analysis is concerened with data multicollinearity. A correlation matrix among all variables is used for this purpose, and an example of such a matrix for the emissions estimation data is shown in the table in Fig. 9.14.

This is a symmetric matrix with ones in the diagonal representing the autocorrelation of each variable and the cross-correlation coefficients between the variables in the nondiagonal cells. Our focus is on analyzing the correlation coefficients of the two-pair combinations of independent variables only. It is assumed that a high correlation coefficient between two independent variables implies redundancy, indicating the possibility that they are measuring or calculating the same data point. In such a scenario, only one of these inputs is needed and the rest have to be removed. The threshold correlation coefficient for defining multicollinearity is problem-dependent but an acceptable generic value is about 0.95.

In the example in Fig. 9.14, all cells above the threshold are highlighted. In this particular data set, only two input variables (Var3 and Var4) are highly correlated. After discussion with the SMEs, the recommendation was to remove Var3 and keep Var4. With more complex data sets, when many inputs show multicollinearity, selecting the representative variable is not trivial and requires experts'opinion.

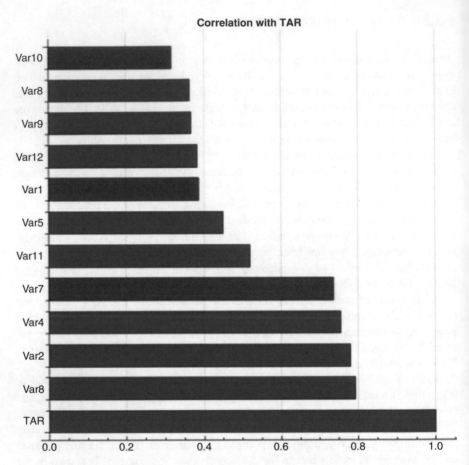

Fig. 9.13 Ranking of absolute correlation of input variables with emissions

In addition to the correlation coefficient, there are other criteria to detect multicollinearity and remove variables. One of the most used metrics is the variance inflation factor. This provides information on how large the standard error of a predictor is compared with what it would be if the variables were uncorrelated with the other predictor variables in the model. It is calculated for each input variable, and those with high values are removed. A problem-dependent threshold value is calculated as well.

However, there is no clear rule of thumb for variable reduction based on high collinearity. Different practitioners use different ways of handling this issue, and the chances of success of the different methods depend on the severity of the collinearity and the business problem at hand.

Variable	TAR	Var1	Var2	Var3	Var4	Var5	Var6	Var7	Var8	Var9	Var10	Var11	Var12
TAR	1.00	-0.39	0.78	0.74	0.74	0.45	0.35	-0.74	0.79	0.37	-0.32	0.51	0.38
Var1	-0.39	1.00	-0.24	-0.54	-0.50	0.12	0.10	0.47	-0.47	-0.55	-0.43	-0.17	-0.55
Var2	0.78	-0.24	1.00	0.79	0.80	0.59	0.67	-0.53	0.72	0.45	-0.46	0.71	0.39
Var3	0.74	-0.54	0.79	1.00	1.00	0.36	0.45	-0.68	0.72	0.61	-0.26	0.60	0.63
Var4	0.74	-0.50	0.80	1.00	1.00	0.38	0.48	-0.67	0.72	0.61	-0.31	0.61	0.62
Var5	0.45	0.12	0.59	0.36	0.38	1.00	0.69	-0.34	0.33	-0.04	-0.56	0.31	0.21
Var6	0.35	0.10	0.67	0.45	0.48	0.69	1.00	-0.24	0.27	0.04	-0.53	0.29	0.27
Var7	-0.74	0.47	-0.53	-0.68	-0.67	-0.34	-0.24	1.00	-0.60	-0.32	0.12	-0.26	-0.48
Var8	0.79	-0.47	0.72	0.72	0.72	0.33	0.27	-0.60	1.00	0.42	-0.16	0.39	0.48
Var9	0.37	-0.55	0.45	0.61	0.61	-0.04	0.04	-0.32	0.42	1.00	0.07	0.61	0.19
Var10	-0.32	-0.43	-0.46	-0.26	-0.31	-0.56	-0.53	0.12	-0.16	0.07	1.00	-0.39	-0.11
Var11	0.51	-0.17	0.71	0.60	0.61	0.31	0.29	-0.26	0.39	0.61	-0.39	1.00	0.06
Var12	0.38	-0.55	0.39	0.63	0.62	0.21	0.27	-0.48	0.48	0.19	-0.11	0.06	1.00

Fig. 9.14 Correlation matrix of emissions estimation variables

9.3.3   Linear Variable Selection

Variable selection is differentiated from variable reduction in that it considers the relationship between the target and the exploratory variables. The objective of variable selection methods is to reduce the input set to a subset that includes only inputs relevant to the target variable. Unfortunately, the criteria for relevancy do not guarantee optimal selection of this subset (that is, the selected subset is generally suboptimal). A by-product of variable selection is variable ranking, indicating the relative importance of the input variables to the target variable.

It is assumed that variable selection begins after all attempts of variable reduction have been exhausted. Out of the linear variable selection methods, we will focus on two of the most popular approaches—stepwise regression and the importance of variables in projection ranking based on partial least-squares models.

Variable Selection Based on Stepwise Regression
The most used methods for variable selection are based on multivariate regression modeling techniques. The key techniques are forward selection, backward elimination, and stepwise regression.[2]

- *Forward selection.* This method begins with the assumption that there are no inputs into the model. The procedure begins by fitting an intercept-based model, and then the number of inputs into the model is increased one itput at a time. Only those inputs related to the target variable above a predefined level of statistical significance are added to the model. In general, once an input variable has been

---

[2]T. Rey, A. Kordon, and C. Wells, *Applied Data Mining in Forecasting*, SAS Institute, 2012.

integrated into the model, it is never removed. The forward selection process stops when all remaining inputs are statistically insignificant.

- *Backward elimination.* This method begins with the assumption that all possible inputs are part of the model. The procedure begins with removing the input with the least statistical significance relative to the others. A detail of importance is that the level of statistical significance (called the stay level) might differ from the level for early entry into the model in forward selection. The backward selection process stops when all remaining inputs are statistically insignificant relative to the significance threshold, defined by the early stay level.
- *Stepwise regression.* This method is a modified version of the forward selection method. In each step of the stepwise method, the statistical significance of all previously entered inputs is reassessed. In this process, it is possible that an input added in an earlier step might become statistically insignificant, and it is then dropped from the model. This procedure requires two significance levels: an entry significance level to allow an input to be kept in the model, and a stay significance level to allow the algorithm to remove inputs already selected in the regression model. The stepwise regression process stops when all remaining inputs are statistically insignificant relative to the stay significance level. One of the advantages of this method is that the final model is selected based on a larger set of potential model evaluations than in the other two methods.

The key issue in stepwise regression is that different stepwise methods generate different variable selections from the same set of data. On top of that, none of the methods guarantees that the selected set of variables is optimal. It is recommended in the statistical literature that the backward elimination algorithm be applied because it is less adversely affected by the correlative structure of the regressors than forward selection.[3]

The variable selection for the most important inputs related to the emissions estimation project, based on both forward and backward stepwise regression, is shown in Fig. 9.15.

In this case, both forward and backward stepwise selection are consistent and have removed two inputs: Var5 and Var11.

Variable Selection Based on Partial Least Squares

Partial least squares is a multivariate method for building predictive models based on principal components. It can be used for variable selection in two ways. The first way is by ranking the original variables based on their regression coefficients in the predicted target equation. The second way is by using a special metric, the variable importance in the projection (VIP), which represents the relative importance of each original input to the principal components. The rule is that if an input has a relatively small coefficient (in absolute value) and the VIP value is below some threshold, then it is removed.

---

[3]Montgomery, E. Peck, and G. Vining, *Introduction to Linear Regression Analysis.* fourth edition, Wiley, 2006.

Entered	Parameter	Estimate	nDF	SS	"F Ratio"	"Prob>F"
✓	Intercept	2006.29365	1	0	0.000	1
✓	Var1	-1521.5763	1	6400.048	19.667	1.33e-5
✓	Var2	0.14288486	1	25430.55	78.147	1.1e-16
✓	Var4	0.00277734	1	6701.572	20.594	8.46e-6
☐	Var5	0	1	1416.868	4.407	0.03669
✓	Var6	-2.0894863	1	15929.39	48.951	2e-11
✓	Var7	-2.2385926	1	23073.37	70.904	2e-15
✓	Var8	0.00286204	1	9458.702	29.066	1.5e-7
✓	Var9	-8.4912567	1	15323.01	47.087	4.4e-11
✓	Var10	-19.319553	1	7183.758	22.076	4.13e-6
☐	Var11	0	1	323.6223	0.994	0.31952
✓	Var12	-1.1797253	1	13776.39	42.335	3.6e-10

Fig. 9.15 Variables selected for emissions estimation by forward and backward stepwise regression

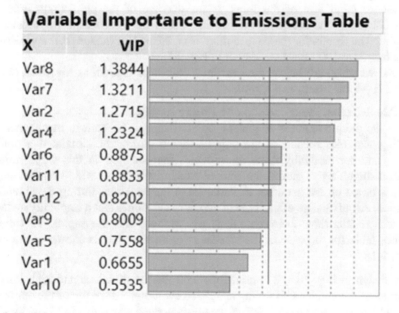

Fig. 9.16 Selected variables for emissions based on VIP metric

In the case of the emissions estimation data, the VIP variable ranking is shown in Fig. 9.16.

Three variables are below the calculated threshold (shown as a vertical line in Fig. 9.16)—Var5, Var1, and Var10.

9.3.4   *Nonlinear Variable Selection*

While the linear variable selection methods are based on linear correlations of independent variables with the target variable, the nonlinear methods are based on diverse principles. We'll discuss two nonlinear methods—variable ranking based on decision trees/random forest and the presence of variables in the nonlinear equations derived by genetic programming.

Variable Selection Based on Random Forest Method
Decision trees are based on following a sequence of rules that forms a series of partitions that divide the target values into a small number of homogenous groups that formulate a tree-like structure. Each split is performed on the basis of the values of one of the input variables that best partitions the target values. By examining which input variables are used to split the nodes near the top of the tree, one can quickly determine the most important variables. The importance of variables can be obtained by analyzing all of the splits generated by each variable and the selection of surrogate splitters.

A random forest consists of a large number of decision trees, for example 1000. The overall prediction of the forest is the average of the predictions from the individual trees. Variable selection is also based on the average importance of variables. The results for the contribution of variables to a random forest emissions model based on 1000 trees are shown in Fig. 9.17.

Three variables are below the calculated threshold (shown as a vertical line in Fig. 9.17)—Var1, Var5, and Var10.

Variable Selection Based on Genetic Programming
One of the unique features of genetic programming is its built-in mechanism for selecting variables related to the target variable during the simulated evolution process and for gradually ignoring variables that are not. In this way, variable selection driven by evolutionary computation principles is used. The ranking of inputs is based on the presence of variables in the equations during the simulated evolution. An advantage of this type of variable selection is that it implicitly includes nonlinear relationships between the target variable and the inputs. A variable presence table for the case of the emissions estimation data is shown in the table in Fig. 9.18.[4]

As shown in Fig. 9.18, 324 equations with unique structures and high- quality performance were generated during simulated evolution. At the top of the ranking is Var7, which was selected in 100% of the GP-generated equations, followed by Var2 with a corresponding value of 99.4%, and Var8 with 80.9%. Unfortunately, there is no theoretically derived method to define a threshold for variable selection using GP based on statistical significance. One practical solution is to define this threshold as a

---

[4]All results of data analysis and model development related to genetic programming in this book are based on the software package DataModeler of Evolved Analytics (www.evolved-analytics.com)

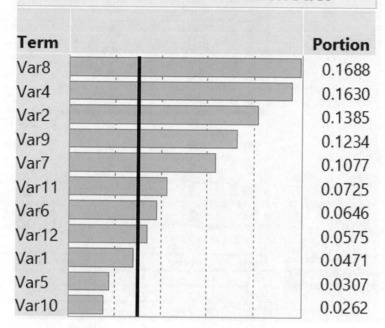

Fig. 9.17 Selected variables for emissions based on contributions of variables in random forest model

5% presence limit. This assumes that 95% of any linear or nonlinear equationsgenerated by GP will include the selected set of inputs. The selected variables below this threshold (shown as a horizontal line in Fig. 9.18) are Var11, Var9, Var12, Var12, and Var1.

Final Variable Selection

It is recommended that the inputs in the final variable selection should be chosen only if supported by several methods. The bare minimum is selection by at least one linear and one nonlinear method. The results for the emissions estimation data obtained from all methods applied are given in Table 9.2.

Four explanatory variables, Var2, Var4, Var6, and Var8, were selected by all methods. Three other variables, Var7, Var9, and Vae12 were highly ranked by three of the methods. The final selection is based on these seven variables. Fortunately among them is the production rate (Var4) and this is a proof of Hypothesis 2, that this critical variable is linked to the emissions.

Fig. 9.18 Variable
presence table for emissions
models, generated by
genetic programming

	♯ Models	% of Models	Meaning
1	324	100.0	var7
2	322	99.4	var2
3	262	80.9	var8
4	257	79.3	var6
5	58	17.9	var10
6	49	15.1	var4
7	8	2.5	var11
8	5	1.5	var9
9	4	1.2	var12
10	1	0.3	var5
11	1	0.3	var1

Table 9.2 Variable selection
for emissions-related process
variables by different methods

Inputs	Stepwise	PLS	Random forest	GP
Var1	x			
Var2	x	x	x	x
Var4	x	x	x	x
Var5				
Var6	x	x	x	x
Var7	x		x	x
Var8	x	x	x	x
Var9	x	x	x	
Var10	x			x
Var11		x	x	
Var12	x	x	x	

9.4   Feature Extraction

Feature extraction is the process of transforming the existing inputs into a lower-
dimensional space, typically generating new features that are composites of the
existing variables. There are a number of techniques that reduce dimensionality
through such a transformation process. The result is, simpler models, shorter training
times, improved generalization, and a greater ability to visualize the feature space.

We'll focus on two main approaches to extracting features—feature engineering
performed by data scientists, and automatically generated features.

9.4.1   Feature Engineering

Feature engineering involves finding connections between variables and packaging them into a new single variable, called a feature. Feature engineering is a vital component of the modeling process, and it is the toughest to automate. It takes domain expertise and a lot of exploratory analysis of the data to engineer features. Some generic forms of potential features are given below:

1. Simple transformations of the original variable x: $1/x$, $\sqrt{x}$, log (x), e^x, e^{-x}, etc.
2. Differencing (see Sect. 8.3.3).
3. Moving average (using averaged data over a selected time window to reduce noise).
4. Interactive terms (adding terms, such as products of variables).
5. Scaling/normalization (see Sect. 8.3.3).
6. Time series decomposition (see Sect. 8.3.3).
7. Prewhitening time series data (applying a filter that can transform the input of a time series to white noise).
8. Special transforms, such as fast fourier transform, noise filters, and wavelet transforms.

Data scientists explore some of these options and select the appropriate features based on the resulting information gain. In the case of big data, however, feature engineering can be a long and inefficient process. A better alternative is that the features are generated automatically.

9.4.2   Automatically Generated Features

Several statistical and AI-based methods generate features automatically. A typical example is PCA, where each principal component is a feature based on a linear combination of the original variables. Another example is support vector machines, which create features dependent on the selected feature generator, called the kernel (see the example in Sect. 3.3.3). Recently, the most powerful automatic feature generator has been deep learning. Examples of a set of features extracted from the different layers of the deep learning network were shown in Fig. 1.6. Unfortunately, most automatically generated features, especially those created by deep learning, are very difficult to interpret and are expected to be blindly accepted by the data scientist.

One of the few exceptions (in addition to PCA) is genetic programming. A side effect of the severe fight for high fitness of the competing equations in simulated evolution is the generation of simple transforms that survive the competition. Some of them appear as building blocks in many equations and, as such, can be defined, analyzed, and selected as features. Selecting the proper transforms is not trivial, however, and is limited to simple equations with a limited number of original variables. Validating the physical dimensionality of the transform is highly recommended. SMEs do not accept features with incorrect physical dimensions.[5]

An example of selected simple transforms of input variables with high fitness related to emissions and generated by genetic programming is given below:

- Transform 1: $1/Var1.^2$
- Transform 2: Var4*Var8.
- Transform 3: Var2*Var8.
- Transform 4: $(Var1*Var8)/Var7^3$.

These transforms were selected according to the suggestions, given in the reference cited below. The transforms are simple and based on up to three variables. All of them have high fitnes and significant presence in the generated models. However, the dimensionality of Transform 3 does not have physical meaning, and this transform was not selected by the SMEs.

9.5 Data Visualization

The active use of data visualization techniques provides a powerful way of identifying important structures and patterns in the data very quickly. Visualization also provides the user with feedback from the data analysis that is easy to understand. It is an important way of communicating the insight obtained from data analysis to the team and the final users.

Data visualization is only successful to the degree that it encodes information in a manner that our eyes can discern and our brains can understand. One of the biggest challenges in data visualization is deciding which visual presentation should be used to best represent the information. To ensure that data visualizations have real business value, we need to understand how to graphically represent the data so that the message we are trying to convey is immediately apparent to the audience. The first step to accomplish this is the selection of the apropriate types of chart that are best suited to a specific category of analysis of a given data set. Fortunately, data scientists have a rich set of visualization options for chart selection. Many analytics tools automate this process and use intelligent autocharting to create the best possible visual solution for the given data set. This data exploration capability of the tools decides whether to display the data as a pie chart or a bar graph, or as a matrix or maps.

It is recommended, however, that data scientists have full control over the visualization process and use it for effective communication of the insight extracted from the data. In general good data visualization is more an art than a science and the

---

[5] Some guidance on selection of transforms is given in F. Castillo, A. Kordon, and C. Villa, Genetic programming transforms in linear regression situations, in *Genetic Programming Theory and Practice VIII*, R. Riolo, T. McConaghy, and E. Vladislavleva (eds.), Springer, pp. 175–193, 2011.

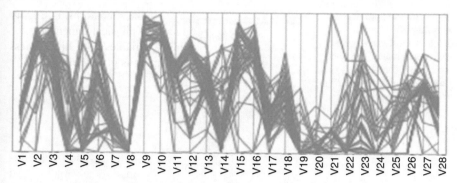

Fig. 9.19 A parallel coordinates plot of the 28 variables in the low-production operating regime

final success is up to the imagination of data scientists. There are generic rules, though, and it is good to understand and follow them.[6]

Among the many available types of chart, we recommend the use of the following three useful, but not so popular plots: parallel coordinates, chord, and contour plots.

9.5.1   Parallel Coordinates Plot

A parallel coordinates plot is a possible way of visualizing high-dimensional data and analyzing multivariate data when PCA is not used. An example of such a plot for the 28 variables in the low-production operating regime case discussed in Sect. 9.2.2 is shown in Fig. 9.19.

The plot gives an overall view of the different behavior of all variables analyzed. Some obvious examples of insight that can be extracted are:

- V9 and V10 operate in their high range of values, while V8 is operates in its low range.
- All three of these variables have low variability, while V6 and V23 operate with very high variability.
- V12 operates in two modes—a high-values mode most of the time and a distinct low-values mode.

---

[6]A recommended reference for developing a successful data visualization strategy is N. Yau, *Data Points: Visualization That Means Something*, Wiley, 2013.

Fig. 9.20 A chord diagram showing the interrelationships between emissions- related variables with linear correlation coefficients greater than or equal to 0.5

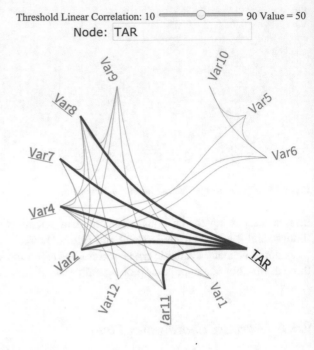

9.5.2   Chord Diagram

A chord diagram is a graphical method of displaying the interrelationships between data in a matrix. The data is arranged radially around a circle with the relationships between the data points typically drawn as arcs connecting the points. An example of such a plot, visualizing the relationships among the variables in the emissions project, is shown in Fig. 9.20.

The plot shows how variables of interest are related at a selected level of linear correlation (in this specific case this correlation coefficient is 0.5, or 50% in this particular software implementation). The inputs correlated with the emissions (TAR) are linked by thick arcs, while the cross-correlated inputs are linked by thin arcs.

9.5.3   Contour Plot

A contour plot represents a 3D surface by plotting n contour layers on a 2D surface. For a selected value of the constant n, lines are drawn that connect the 2D coordinates where the n values occur.

An example of a contour plot that illustrates the 2D patterns of how two of the inputs, Var2 and Var6, are related to the emissions (TAR) is shown in Fig. 9.21.

Five contour layers are defined for the emissions ($n = 5$), with the corresponding levels shown in Fig. 9.21. One insight from the plot is that the ranges of both

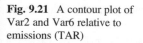

Fig. 9.21 A contour plot of Var2 and Var6 relative to emissions (TAR)

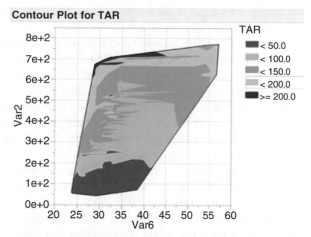

variables for low emissions levels, less than 100, are relatively well demarcated. For emissions between 150 and 200, however, the ranges of Var2 and Var6 are difficult to define, and they overlap with other ranges. The ranges for the highest emissions, greater than 200, where the focus of attention is, are defined better, at the top left corner of the plot.

9.6 Data-Analysis-Driven Storytelling

Insight from data analysis is the final deliverable for some business problems. A typical case is identifying important variables that lead to discovery of the root cause of a problem. The standard way to deliver the findings is by a document or presentation in a meeting to all stakeholders. Data scientists are well prepared to communicate the insight in this way.

There is a growing trend, however, toward delivering the findings in a more emotional and attractive way by using storytelling techniques. The advantages of this approach are many. Stories are memorable, connect with humans on an emotional level and inspire action. Effectively crafting and pitching captivating data-driven stories to project stakeholders in particular and to managers, potential customers, partners, and peers in general is a vital skill that data scientists need to learn.[7] Unfortunately, the needed skillset is outside the comfort zone of most data scientists.

To tell illuminating data stories we begin with an idea, taken from the problem definition. Then, we consider the target audience and of the objectives of the data story by answering the following questions: Why should they care about insight

---

[7]A recommended reference for developing successful storytelling techniques is: P. Smith, *Sell with a Story: How to Capture Attention, Build Trust, and Close the Sale*, Amacom, 2016.

extracting from data? What motivates them? Most importantly, what actions do we recommend to be taken next?

After clarifying the audience and the goals of the data story, effective ordering and placement of supporting visuals becomes much easier to do. By using a storyboard approach, we can combine narratives with visuals that guide the analysis to the desired conclusion. Some key recommendations are given below.

First, organize the story in a dramatic way with the context at the beginning, some challenge and conflict in the middle, and insight at the end.

Second, add an emotional component with data scientists' attitudes, histories, responses to findings, conflicts, etc.

Third, focus on challenges, issues, and doubts during data analysis.

Fourth, use the principle of showing, not telling. This captures the audience's attention and sends them a message of participation in the story.

Fifth, let the audience draw their own conclusions and offer their own recommendations. This is one of the objectives in storytelling. When we are done telling the story, we pause and let the audience react. If we are presenting data in the standard reporting fashion, we make our recommendation up front and tell the audience exactly what conclusions to draw. Storytelling is the opposite.

We'll give a short example of the structure of a data story based on the problem of low-productivity operation, discussed in Sect. 9.2.2.

The objective of the data story is to communicate insight for the purpose of root-cause analysis of the key factors in reduced productivity in a manufacturing unit. The targeted audience includes the unit management, process and control engineers, and selected process operators. The majority of the audience has no statistical or data science background. One possible structure of the data story is as follows:

- Begin the story with a unit manager's emotional narrative about his painful attempts to fix the issue of reduced productivity with his team and decision he made to contact data scientists to solve the problem.
- Show the key findings of the data analysis, explaining figures, such as Figs. 9.7– 9.10.
- Focus on conflicting interpretations of the behavior of variable V4 in the low-production mode.
- Discuss the different opinions of stakeholders about the findings.
- Ask the audience for their hypotheses about the root causes of moving from normal to low-production mode.
- Compare their answers with the root causes discovered by the data scientists.
- Discuss the next steps for fixing the problem in a dialog with the audience.

9.7 Common Mistakes

9.7.1 Data Analysis with Low-Quality Data

One of the biggest mistakes of inexperienced data scientists is making a compromise on data quality and performing the analysis with low-quality data. The negative consequences of this wrong decision are many. First, finding patterns and relationships will be significantly more difficult due to the high level of noise in the data. Second, variable selection and feature extraction can be wrong for the same reason. Third, the extracted insight, based on erroneous patterns, relationships, and variable selection, can be incorrect. As a result, the credibility of the data analysis is ruined.

There are two key types of low-quality data: (1) based on crappy raw data and (2) based on poor data preparation. It is recommended that data scientists return to data preparation until an acceptable data quality is achieved for the second category. For the first category, of crappy raw data, there is only one solution—stop and don't waste your time!

9.7.2 Ignoring Multivariate Analysis

Another common mistake in extracting insight from data is missing the enormous capabilities of multivariate analysis. One explanation for this behavior is that some data scientists are not familiar with the corresponding approaches. Another is that these capabilities are not part of some data analysis packages.

The insight obtained from data analysis without a multivariate view of the problem is incomplete. It is impossible to compensate for this with more detailed univariate or bivariate analysis. The higher the dimensionality of the problem (the number of variables), the higher the need for and value of multivariate analysis.

9.7.3 Low-Quality Variable Selection and Feature Extraction

Some inexperienced data scientists do not pay the necessary attention to variable reduction and selection, or to feature extraction. The most frequent case is skipping multicollinearity reduction, which could lead to very inefficient variable selection and modeling structures.

Another observed bad habit is limiting variable selection to one method only. If there are hundreds or more variables, the selection derived is not conclusive, especially if it is limited to stepwise regression. As is recommended in this book, important variables should be selected by at least one linear and one nonlinear method.

The third bad habit in this category is skipping feature extraction or exploring very limited options for feature engineering, mostly due to lack of experience.

9.7.4   Inappropriate Visualization

Mastering visualization requires a lot of learning, experience, and creativity. Not all data scientists are at the level needed to visualize insight extracted from data in an effective way. As a result, the findings from data analysis are communicated badly, with ambiguous or even confusing conclusions. Typical examples of bad visualization are a wrong plot choice, inappropriate scales, and overcrowded plots. Usually the lack of self-explanatory visualization is compensated with additional verbal or written explanations, which reduce more the efficiency of the insight message derived from the data analysis.

9.8   Suggested Reading

- M. Berthold and D. Hand, *Intelligent Data Analysis*, second edition, Springer, 2007.
- K. Healy, *Data Visualization: A Practical Introduction*, Princeton University Press, 2019.
- M. Kuhn and K. Johnson, *Applied Predictive Modeling*, second edition, Springer, 2016.
- D. Montgomery, E. Peck, and G. Vining, *Introduction to Linear Regression Analysis*. fourth edition, Wiley, 2006.
- B. Ratner, *Statistical and Machine-Learning Data Mining: Techniques for Better Predictive Modeling and Analysis of Big Data*, third edition, CRC Press, 2017.
- T. Rey, A. Kordon, and C. Wells, *Applied Data Mining in Forecasting*, SAS Institute, 2012.
- P. Smith, *Sell with a Story: How to Capture Attention, Build Trust, and Close the Sale*, Amacom, 2016.
- N. Yau, *Data Points: Visualization That Means Something*, Wiley, 2013.

9.9   Questions

Question 1
Is principal component analysis a data reduction or a data selection method?

Question 2
How is the number of principal components selected?

Question 3
What is the difference between variable reduction and selection?

Question 4
What is the difference between variable selection and feature extraction?

Question 5
Extract insight from the parallel coordinates plot in Fig. 9.19.

Chapter 10
Model Development

> *A model is like a political cartoon. It picks up the substantial part of a system and exaggerates it.*
>
> *John Holland*

The second transformation of data into value, which is the focus of this chapter, is generating a solution to a problem from available data and derived insight. In most cases, the expected solution is a data-driven mathematical model (or models). Developing high-fidelity models based on first principles or econometrics is outside the scope of this book. Usually the development of this type of model requires solid fundamental knowledge and the availability of top experts. As a result, the total cost of ownership is very high.

The first objective of this chapter is to discuss the nature of the translation of insight into actions via model development. The second is to define the key components of a good model while the third is to emphasize the importance of fighting overfitting of the data for robust model development. The fourth objective is to present and illustrate with examples the model generation and validation process. The fifth is to discuss the power of model ensembles.

10.1 Translation of Insight into Actions

Model development is the most critical technical step in the AI-based Data Science workflow. It is expected that the solution to the problem will be delivered and reliably validated. If accepted, it will be the necessary tool to complete the defined objectives of the technical project. This will also allow one to design the actions by which the derived model will begin to create value and accomplish the financial objective of the project. In the case of our emissions estimation project, an acceptable solution is a generated and validated non-black-box predictive model that satisfies the performance requirements. The predicted emissions per se, however, do not create value. An action, such as a decision about how to run a plant more efficiently translates the prediction into potential value.

© Springer Nature Switzerland AG 2020
A. K. Kordon, *Applying Data Science*,
https://doi.org/10.1007/978-3-030-36375-8_10

This raises the relevant question: Why do we need models? Can we directly design decisions based on the insight extracted from the data and existing problem knowledge? In fact, this was the expert systems approach in the late 1980s. The experience from it, however, was not very positive. The expert-defined rules didn't have the necessary fidelity to generate accurate predictions. Another issue is the static nature of the rules and their limited complexity. The biggest drawback, however, is the subjective nature of the dependencies used for the predictions. Objective validation of these expert-defined relationships is difficult to accomplish.

The recommended solution is to use math models in the decisions that create the final value. The advantages of mathematical models are the following:

- *Models represent objective relationships among the factors of interest.* According to Wikipedia, "a mathematical model is a description of a system using mathematical concepts and language." In the case of AI-based Data Science, the developed models are data-driven and the statistical foundation of the quantitative relationships they derive is correlation. They are objective but cannot claim causality.
- *Models can be validated.* In contrast to expert-defined decisions, models can be validated during model development and tested by various methods on entirely different data sets. This gives a more realistic metric for the performance of models after deployment.
- *Models can be generated by different methods and appropriately combined.* The trust in using relationships derived by various scientific principles (statistics, machine learning, decision trees, genetic programming, etc.) is higher if the performance is similar. Applying ensembles of models generated by either one method or different methods increases the confidence in the model predictions as well.
- *Some models have confidence limits.* Confidence limits are used to assess how reliable a prediction is. Wide confidence intervals mean that the derived model is poor (and it is worth investigating other models), or that the data is very noisy. An advantage of models with confidence limits is that corresponding decisions will be based not only on a prediction but also on recommended low and high limits. As a result, the action defined by the decision has additional flexibility and higher trust.

10.2   What Is a Good Model?

The technical quantitative definition of a good model is when it satisfies the performance criteria outlined in the project objectives. While this is sufficient in an academic environment, in real-world applications we need more features to describe the behavior of a good model. Most of them are qualitative and beyond the known statistical metrics. The key features are shown in Fig. 10.1 and described below.

Fig. 10.1 Key features of a good model

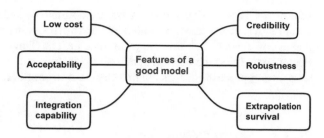

10.2.1   High Credibility

At the top of the list is the requirement that the derived solution is trustworthy under a wide range of operating conditions. Usually credibility of a model is based on its principles, performance, and transparency. First-principles-based models have built-in trust by the laws of nature. Statistical models, with their confidence metrics, are also accepted as credible by the majority of users with a statistical background. All other solutions, built on either empirical or expert knowledge methods, are treated as lacking a solid theoretical basis and must fight for their credibility with almost flawless performance. An additional factor that may increase a data-driven model's credibility is the level of its transparency to the user. At the one extreme are complex black boxes, such as neural networks or support vector machines, which are among the least trusted models. At the other extreme are simple symbolic regression equations, especially if they can be physically interpreted. Our experience shows that users have high confidence in this type of data-driven solution.

The key factor for credibility, however, is model performance according to the expected behavior. The erosion of credibility begins with questioning some predictions of the model and grows into raising doubts about its validity after a series of erroneous behaviors. The final phase of lost credibility is the silent death of the solution applied, which is gradually eliminated from use. Unfortunately, it is almost impossible to recover an application with lost credibility. Even more, an accidental credibility fiasco in a specific application may create a bad image for the whole technology and poison the soil for future implementation for a very, very long time.

10.2.2   High Robustness

Performance credibility depends on two key features of the models applied: (1) the ability to handle minor process changes and (2) the ability to avoid erratic behavior when moving to unknown extrapolation areas. Process changes driven by different operating regimes, equipment upgrades, or product demand fluctuations are more the rule than the exception in industry. It is unrealistic to expect that all the variety of process conditions will be captured by the training data and reflected in the empirical or fundamental models developed. One of the potential solutions to obtain increased robustness is to derive models with an optimal balance between accuracy and complexity, such as those generated by Pareto-front genetic programming. A current survey of several industrial applications in The Dow Chemical Company has demonstrated that the derived symbolic regression models show improved robustness during process changes relative to neural network-based models.

10.2.3   Extrapolation Survival

Robust models have limits on the range in which they operate with acceptable performance (within 10% outside the original range of model development), though. When these limits are crossed, the behavior in the extrapolation mode becomes critical for the fate of the application. It is expected that a potential solution can manage its performance degradation in a gradual fashion until it is at least 20% outside the training range. Switching from normal to erratic predictions due to limitation of the model within the range of the training data (as is usually the case with neural networks) can destroy the credibility of the application in minutes. A potential empirical solution to the problem is to select statistical or symbolic regression models with a low level of nonlinearity. Another technique that may help models to avoid an extrapolation disaster is to add a self-assessment capability using built-in performance indicators or by using combined predictors and their statistics as a confidence indicator of the model's performance.

10.2.4   Integration Capability

Since industry has already invested in developing and supporting infrastructure for existing work processes, the integration effort for any new technology becomes a critical issue. The solution applied must be integrated within the existing infrastructure with minimal effort. The best-case scenario is when one can use a well-established dialog with the user and the existing software for deployment and maintenance. It is assumed that different tools and software environments can be used during model development, but the final run-time solution has to be integrated into the known environment of the user. Very often, this envonment could be the ubiquitous Excel.

10.2.5   Broad Acceptability

A very important requirement for the success of real-world applications is support from all stakeholders. Several factors contribute to a model's acceptability. The first factor is the credibility of the modeling principles, which has already been discussed.

Another critical factor is making the development process user-friendly with minimal tuning parameters and specialized knowledge required. Evolutionary computation models are developed with a minimal number of assumptions, unlike, for example, first-principles models, which have many assumptions stemming from physical or statistical considerations. Another factor in favor of acceptance of a model is a known use case for a similar problem. In the real world, proven practical results speak louder than any purely theoretical arguments.

In many applications related to banks, insurance companies, and healthcare providers, interpretability of a model becomes a critical requirement for acceptance. Interpretability has been defined as "the ability to explain or to present in understandable terms to a human." [1] The complexity of the derived model is directly related to its interpretability. Generally, the more complex the model, the more difficult it is to interpret and explain. A possible way to assess interpretability is by analyzing the functional form of a model. The following list describes the functional forms of models and discusses their degree of interpretability in various use cases:

- *High interpretability—linear, monotonic functions.* Functions created by traditional regression algorithms are probably the most interpretable class of models. We refer to these models as "linear and monotonic," meaning that for a change in any given input variable (or sometimes a combination of variables or a function of an input variable), the output of the response function changes at a defined rate, in only one direction, and by a magnitude represented by a readily available coefficient. Monotonicity also enables intuitive and even automatic reasoning about predictions.
- *Medium interpretability—nonlinear, monotonic functions.* Although most response functions generated by AI-based methods are nonlinear, some can be constrained to be monotonic with respect to any given independent variable. Although there is no single coefficient that represents the change in the response function induced by a change in a single input variable, nonlinear and monotonic functions do always change in one direction as a single input variable changes. Nonlinear, monotonic response functions are therefore interpretable and potentially suitable for use in regulated applications.
- *Low interpretability—nonlinear, nonmonotonic functions.* Most AI-based Data Science algorithms create nonlinear, nonmonotonic response functions. This class of functions is the most difficult to interpret, as they can change in a positive or negative direction and at a varying rate for any change in an input variable. Typically, the only standard interpretability measures these functions provide are relative variable importance measures. A combination of several techniques, discussed in the book by Hall and Gill refered to above, should be used to interpret these extremely complex models.

---

[1] D. Hall and N. Gill, *An Introduction to Machine Learning Interpretability*, O'Reilly Media, 2018.

10.2.6 Low Total Cost of Ownership

The sum of the development, deployment, and expected maintenance costs of the real-world application must be competitive versus the other alternative solutions. In the case of the introduction of new technology, such as AI-based Data Science, the potentially higher development cost must be justified be the expected benefits, such as a competitive advantage or an expected solution leverage that will reduce the future cost.

10.3 Fighting Overfitting the Data

In the same way as obesity is a symptom of potential health problems, developing complex, "fat" models can lead to poor performance and potential model degradation. The fight for fit models with the right complexity is the central topic in AI-based Data Science model development. Among the many methods to accomplish this task we'll focus on the fundamental capabilities of statistical learning theory with its structural risk minimization principle and the popular regularization approach.

10.3.1 The Overfitting Issue

Overfitting is the use of models or procedures that include more terms than are necessary or have been generated by more complicated methods than are necessary. Typical examples of the first type of "fat" models are statistical models using high-order polynomials and neural networks with a large number of neurons in the hidden layer. An example of the second type of "fat" model is a very complex deep learning model when a much simpler neural network or random forest solution would be sufficient for solving the problem.

A "fat" model will add a level of complexity without any corresponding benefit in performance or, even worse, it will have poorer performance than the simpler model. Adding irrelevant inputs can make predictions worse because the coefficients fitted to them add random variation to the subsequent predictions.

Overfitting of models is widely recognized as a concern. It is less recognized, however, that overfitting is not an absolute but involves a comparison. A model overfits if it is more complex than another model that fits equally well. This means that recognizing overfitting involves not only the comparison of a simpler and a more complex model but also the issue of how you measure the fit of a model.[2]

---

[2]D. Hawkins, The problem of overfitting, *Journal of Chemical Information and Computer Sci*ince, **44**, 1–12, 2004.

10.3.2 Statistical Learning Theory

The key topic of statistical learning theory is defining and estimating the capacity of a machine to learn effectively from finite data. Excellent memory is not an asset when it comes to learning from limited data. A machine with too much capacity to learn is like a botanist with a photographic memory who, when presented with a new tree, concludes that it is not a tree because it has a different number of leaves from anything she has seen before; a machine with too little capacity to learn is like the botanist's lazy sister who declares that if it's green, it's a tree. Defining the right learning capacity of a model is similar to developing efficient cognitive activities by proper education.

Statistical learning theory addresses the issue of learning-machine capacity by defining a quantitative measure of complexity, called the Vapnik–Chervonenkis (VC) dimension. The theoretical derivation and interpretation of this measure require a high level of mathematical knowledge.[3] The important information from a practical point of view is that the VC dimension can be calculated for most of the known analytical functions. For learning machines that are linear in their parameters, the VC dimension is given by the number of weights, i.e., by the number of "free parameters". For learning machines that are nonlinear in the parameters, the calculation of the VC dimension is not trivial and must be done by simulations. A finite VC dimension of a model guarantees generalization capabilities beyond the range of the training data. An infinite VC dimension is a clear indicator that it is impossible to learn with the functions selected, i.e., the model cannot generalize. The nature of the VC dimension is further clarified by the structural risk minimization principle, which is the basis of statistical learning theory.

10.3.3 Structural Risk Minimization Principle

The structural risk minimization (SRM) principle defines a trade-off between the quality of the approximation to the given learning data and the generalization ability (the ability to predict beyond the range of the learning or training data) of the learning machine. The idea of the SRM principle is illustrated in Fig. 10.2, where the y-axis represents the error or the risk of prediction and the x-axis represents the complexity of the model.

In this particular case, the complexity of the model, defined by its VC dimension, is equal to the order of the polynomial, which ranges from h_1 to h_n. The approximation capability of the set of functions (fitting the data) is shown by the empirical risk (error) which declines with increased complexity (polynomial order). If the learning machine uses too high a complexity (for example, a polynomial of order

---

[3]Curious readers may look for details in the book by V. Vapnik, *The Nature of Statistical Learning Theory*, second edition, Springer, 2000.

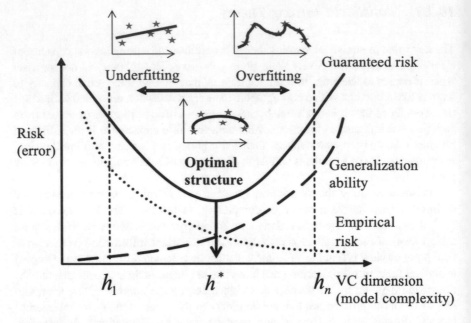

Fig. 10.2 The structural risk minimization principle

10), the learning ability may be good, but the generalization ability will not be. The learning machine will overfit the data to the right of the optimal complexity, with a VC dimension of h^*, in Fig. 10.2.

On the other hand, when the learning machine uses too little complexity (for example, a first-order polynomial), it may have good generalization ability, but not impressive learning ability. This underfitting of the learning machine corresponds to the region to the left of the optimal complexity. The optimal complexity of the learning machine corresponds to the set of approximating functions with lowest VC dimension and lowest training error (for example, a third-order polynomial).

There are two approaches to implementing the inductive SRM principle in learning machines:

1. Keep the generalization ability (which depends on the VC dimension) fixed and minimize the empirical risk.
2. Keep the empirical risk fixed and minimize the generalization ability.

Neural network algorithms implement the first approach, since the number of neurons in the hidden layer is defined a priori and therefore the complexity of the structure is kept fixed.

The second approach is implemented in the support vector machines method, where the empirical risk is either chosen to be zero or set to an a priori level and the complexity of the structure is optimized. In both cases, the SRM principle pushes toward an optimal model structure that should match the capacity of the learning machine with the complexity of training data.

Fig. 10.3 Key steps in
model generation

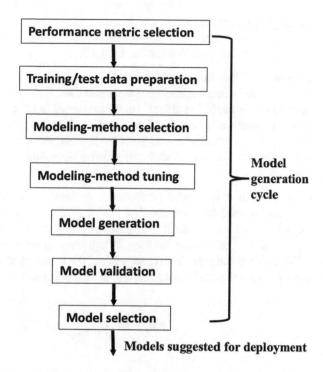

10.3.4   *Regularization*

The most popular method to fight overfitting of the data is regularization. This
automatically penalizes inputs that make the model too complex. Regularization is
useful if we wish to keep the selected inputs and we just need to penalize them but
not remove them completely. Regularization penalizes all inputs used, not a selected
subset.

A regularization parameter, called lambda (λ), controls how the inputs are
penalized. It is a hyperparameter with no fixed value. As its value increases, there
is higher penalization of the inputs. As a result, the model becomes simpler. As its
value decreases, there is lower penalization of the features and thus the complexity of
the model increases. Zero lambda means no removal of inputs at all.

10.4   Model Generation

10.4.1   *Model Generation Cycle*

Model generation requires several iterations until the appropriate solution is derived.
It includes several steps, shown in Fig. 10.3, in three consecutive phases:

(1) preparation, (2) generation, and (3) selection. The preparation steps are discussed in this section, model generation is covered in the next section, and model validation and selection are considered in Sect. 10.4.3.

Selection of Performance Metrics
The first step before beginning model generation is defining the criteria for technical success of the modeling efforts. As we discussed in Chap. 3, each solution type has specific performance metrics. For example, the most popular metric for forecasting models is their mean absolute percentage error. Another example as a widespread metric for classification models, the receiver operating characteristics plot, shown in Fig. 3.5.

In addition, several other performance metrics are available for each solution type and can be used for evaluation of model quality. A typical case is the two most popular performance metrics for prediction-type models—the coefficient of determination, or r^2 (discussed in Chap. 3) and the root mean square error, or RMSE.

The root mean square error represents the standard deviation of the differences between the predicted values and actual values according to the formula

$$RMSE = \sqrt{\frac{1}{n} \sum_{j=1}^{n} (y_j - \widehat{y}_j)^2} \qquad (10.1)$$

where y_i is the actual value and $\widehat{y}_j$ is the predicted value.

Why is it recommended to select both of these metrics for performance evaluation of predictive models? The advantage of r^2 is that it is a standard statistical measure of goodness of fit. However, some users with no statistical background may have difficulties in understanding it. The RMSE is more universal and is understood by most users. A combination of the two gives a better measure of model performance based on the goodness of fit and the physical nature of the prediction error.

Model Improvement Metric
In the case of prediction solution, it is suggested to begin with a linear regression model as a benchmark, and then train more complex models such as a neural net, random forest, or symbolic regression model generated by genetic programming. The hypothesis is that the more complex model will lead to a performance improvement, which can be estimated by the following performance metric:

$$MI = \left(\frac{R^2_{selected} - R^2_{benchmark}}{R^2_{benchmark}} \right) \times 100 \qquad (10.2)$$

where MI is model improvement expresses as a percantage, $R^2_{benchmark}$ is the goodness of fit of the linear benchmark model, and $R^2_{selected}$ is the goodness of fit of the final selected model.

This is the metric that quantifies potential the model improvement obtained by a different method relative to the benchmark.

Training/Validation/Test Data Preparation
In many cases the preprocessed data are divided into two data sets—a training set (to develop and tune the model) and a test set (to select the model and evaluate its performance on a different data set). This approach was discussed in Sect. 8.3.4.

Another option for organization of data for model development is to divide the data into three parts—training, validation, and test data sets:

- **Training set** The training set is typically 60% of the data. As was discussed in Sect. 8.3.4, this is the data used to train the model over the selected method structure (neural network, deep learning neural network, SVM, etc.)
- **Validation set** The validation set is also called the development set. This is typically 20% of the data. The key purpose of this data set is to test the quality of the trained model and to navigate model selection. A typical case is the selection of the number of neurons in the hidden layer of the neural network (one of the key structural parameters of this method) based on the performance on the validation data set. However, even though this set is not used for training, the fact that it has been used in model selection raises questions abouit using it for reporting the final accuracy of the model.
- **Test set** The accepted solution is to allocate a part of the data (typically 20% of the data) for reporting the accuracy of the final model.

Guidance on Selection of Methods
This issue has been discussed in Sect. 4.3.3.

Tuning of Modeling Methods
In order to generate high-performance models, each modeling method requires tuning of some selected parameters, called hyperparameters. Examples of such hyperparameters are the number of hidden units in a neural network and the maximum depth and number of leaves in a decision tree.

Hyperparameter tuning is an optimization task, and each proposed hyperparameter setting requires the model-training process to derive a model for the data set and evaluate results on the validation data set. After evaluating a number of hyperparameter settings, the hyperparameter tuner provides the setting that yields the best-performing model. The last step is to train a new model on the entire data set (which includes both the training and the validation data) with the best hyperparameter setting. Many software vendors offer options for automatic hyperparameter tuning, which is of great help in effective model development.

The key hyperparameters of the key AI-based Data Science methods are discussed briefly below.

Hyperparameters of Neural Networks
Neural networks with a multilayer perceptron architecture require selection of the following hyperparameters: the number of hidden layers, the number of neurons in

the hidden layers, the type of activation function, the stopping criterion (the number of epochs or steps when the training should be stopped), the momentum, the learning speed, and the type of weight initialization. In most of the available modeling packages, the majority of these hyperparameters have universal default values and the only tuning parameter is the number of neurons in the hidden layers.

Hyperparameters of Deep Learning Networks
Due to their high complexity, deep learning networks require many hyperparameters for tuning their structure and learning algorithms. The most important are the following:

- **Type and number of hidden layers and units**. The hidden layers are the layers between the input layer and the output layer. Different types of layers, such as a convolution layer, pooling layer, or LSTM layer, with their specific hyperparameters, can be selected.
- **Hidden-layer sizes.** The number and size of each hidden layer in the model.
- **Dropout.** This is a regularization technique to avoid overfitting. Generally, one should use a small dropout value for the neurons, with 20% providing a good starting value. A probability too low has minimal effect, and a value too high results in under-learning by the network.
- **Activation function.** The activation function to be used by the neurons in the hidden layers. The following functions are available:

  - Tanh: the hyperbolic tangent function (same as a scaled and shifted sigmoid).
  - Rectifier: the rectified linear unit function. This chooses the maximum of $(0, x)$ where x is the input value.
  - Maxout: This chooses the maximum coordinate of the input vector.
  - ExpRectifier: the exponential rectified linear unit function.

- **Loss function.** The loss (error) function to be minimized by the model. The absolute, quadratic, and Huber functions are applicable for regression or classification, while CrossEntropy is only applicable for classification.

SVM Hyperparameters
SVMs require several tuning parameters, such as: the type of kernel (linear, polynomial, radial basis function, sigmoid, etc.), the complexity parameter C, the type of optimizer (linear or quadratic), and the type of loss function (linear or quadratic).

Decision Tree Hyperparameters
The key hyperparameters for decision trees are given below:

- **Splitting criterion**. This selects the criterion by which attributes will be selected for splitting. For each of these criteria, the split value is optimized with regard to the chosen criterion. This criterion can have one of the following values: information gain, gain ratio, or GINI index (a measure of inequality between the distributions of label characteristics).

- **Maximum depth**. The depth of a tree varies depending upon the size and characteristics of the data set. This parameter is used to restrict the depth of the decision tree.
- **Minimum gain**. The gain of a node is calculated before splitting it. The node is split if its gain is greater than the minimum gain.

Random Forest Hyperparameters

As a more complex approach than decision trees, the random forest method has more hyperparameters, such as:

- **Number of trees in the forest**. The number of trees to be grown and then averaged.
- **Number of terms sampled per split**. The number of predictors to be considered as splitting candidates at each split. For each split, a new random sample of predictors is taken as the candidate set.
- **Bootstrap sample rate**. The proportion of observations to be sampled (with replacement) for growing each tree. A new random sample is generated for each tree.
- **Minimum splits per tree**. The minimum number of splits for each tree.
- **Maximum splits per tree**. The maximum number of splits for each tree.
- **Minimum-size split**. The minimum number of observations needed in a candidate split.

Evolutionary Computation Hyperparameters

From the many evolutionary computation methods, we give an example of the hyperparameters for genetic programming:

- **Terminal set**. This is usually the inputs used for model generation.
- **Functional set (selected functions used for genetic operations)**. The generic set includes the standard arithmetic operations +, −, *, and /, and the following mathematical functions: square root, logarithm, exponential, and power.
- **Genetic operator tuning parameters:**
  - Probability for random vs. guided crossover.
  - Probability for mutation of terminals.
  - Probability for mutation of functions.

- **Simulated evolution control parameters:**
  - Number of generations.
  - Number of simulated evolutions.
  - Population size.

Usually the parameters of the genetic operators are fixed for all practical applications. They are derived after many simulations and represent a good balance between the two key genetic operators: crossover and mutation. In order to address the stochastic nature of genetic programming, it is suggested to repeat the simulated evolution several times (usually 20 runs are recommended). The other two parameters that control the simulated evolution—population size and the number of generations—are problem-size-dependent.

Swarm Intelligence Hyperparameters

The hyperparameters of the two key types of swarm intelligence algorithms, ant colony optimization and particle swarm optimization, are entirely different. Some ACO hyperparameters, such as the number of ants, the stopping criterion, based on either accuracy or a prescribed number of iterations, or the number of repetitive runs, are generic. The rest of the tuning parameters, such as pheromone range and rate of evaporation are problem-specific.

Usually the selection of an ACO algorithm requires several simulation runs on a test case with fine adjustment of the tuning parameters until acceptable performance is achieved. Then the developed algorithm can be used in other similar applications.

A typical PSO setting includes the following hyperparameters:

- **Neighborhood structure**. A version of the algorithm with a global neithborhood structure is faster but might converge to a local optimum for some problems. The local version is a little bit slower but has a lower probability of being trapped in a local optimum. One can use the global version to get a quick result and then use the local version to refine the search.
- **Number of particles in the population**. The typical range is from 20 up to 50. Actually, for most problems 10 particles are sufficient to get good results. For some difficult or special problems, one can try 100 or 200 particles as well.
- **Dimension of particles**. This is determined by the problem to be optimized.
- **Range of particles**. This is also determined by the problem to be optimized; one can specify different ranges for different dimensions of particles.
- V_{max}. This determines the maximum change one particle can make during one iteration. Usually the range of V_{max} is related to the upper limit of X_{max}.
- **Learning factors**. The parameters c_1 and c_2 are usually equal to 2. However, other settings have also been used in different references. But usually c_1 equals c_2 and is in the range [0, 4].
- **Inertia weight** w. This is usually in the range from 0.4 to 1.4.
- **Maximum number of iterations**. This is problem-dependent; the typical range is 200 to 2000 iterations.

Due to the stochastic nature of PSO, it is recommended that the algorithm is run with at least 20 simulations for a given set of parameters. The final solution is selected after exhaustive tuning and validation under all possible conditions. It could then be applied as a run-time optimizer for similar problems.

10.4.2   Model Generation by Multiple Methods

Model generation begins with developing a relevant benchmark solution. In the case of prediction, the accepted benchmark is a linear model. The next step is to develop a

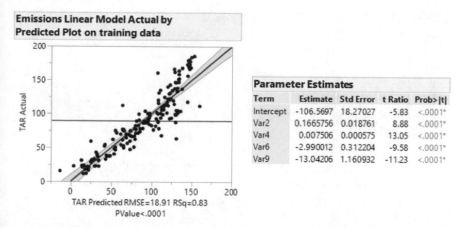

Fig. 10.4 Actual-by-predicted plot of training data and parameter estimates from a linear benchmark model for emissions estimation

sequence of models derived by the selected modeling methods until the desired solution is chosen. This potential solution is compared with the benchmark and is considered for deployment if it has better performance, measured as the model improvement, calculated by Eq. (10.2).

We'll illustrate this process with the generation of multiple models for the emissions estimation example. In all cases we use the training and test data sets discussed in Sect. 8.3.4 and shown in Fig. 8.18. The starting list of dependent (input) variables is based on the variable selection, described in Sect. 9.3, and includes seven variables: Var2, Var4, Var6, Var7, Var8, Var9, and Var12. Let's recall the key performance criterion for the developed models—their r^2 on the test data should be greater than 0.85. The problem definition also limits the type of modeling method to non-black box options, in this case linear statistical models, decision trees, and symbolic regression generated by genetic programming. However, we'll develop a neural net model to explore if a black-box model with higher performance can be derived as an alternative. The structure of the generated models derived from the training data is presented in this section. Model validation on test data and model selection are discussed in next section.

Generating a Benchmark Model

In the case of prediction, the benchmark model is an optimal linear regression model. For our emissions estimation problem, the performance of the derived benchmark model on the training data (shown by its key statistics and actual-by -predicted values plot) and its parameter estimates are shown in Fig. 10.4.

Unfortunately, the performance of this model on the training data, as shown in Fig. 10.4 (r^2 of 0.83), is below the required r^2 of 0.85. From the seven variables explored only four have been selected in the model (Var2, Var4, Var6, and Var9), shown in the parameter estimates table in Fig. 10.4. Their impact on the estimated

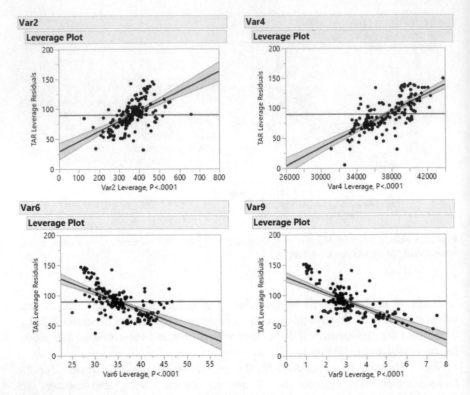

Fig. 10.5 Leverage plots of process variables used in the linear benchmark model for emissions estimation

emissions is illustrated by the leverage plots in Fig. 10.5. The direction of the effect of the input variables on the target variable (the estimated emissions) is in accordance with process knowledge (the emissions increase by increasing Var2 and Var4 and decreasing Var6 and Var9).

Generating Nonlinear Models

- Generating a Decision Tree Model

The seven starting process variables were used in generating a decision tree model, and a section of the tree of the selected model is shown in Fig. 10.6.

This model has much better performance on the training data (its r^2 is 0.91) than the benchmark linear model. However, it has a much more complex structure of five layers (only four layers of one of the key trees are shown in Fig. 10.6 for good visualization). As a result, the structure is difficult to interpret and its acceptance by the final users is questionable.

According to the importance of the variable for splitting the trees, the most influential variables for estimating the emissions are Var4, Var2, and Var7. The

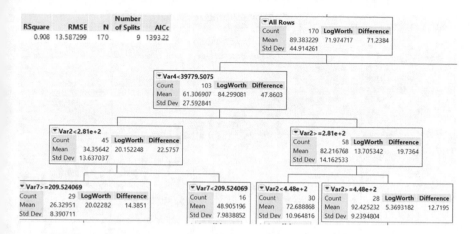

Fig. 10.6 A section of the decision tree model for emissions estimation

last variable was not selected as statistically significant by the linear model (see Fig. 10.4).

Generating Genetic Programming Models
In order to take account of the stochastic nature of genetic programming, 20 simulated evolutions were run for model generation. The basic mathematical functions were used as building blocks and the generated nonlinear models were linearized, i.e., the coefficients in each term in the derived equations have a regular statistical meaning.

From all generated models, a quality box of the best candidate models was selected, and the Pareto front of this selection is shown in Fig. 10.7.

Three models at the knee of the Pareto front with slightly different complexity and number of variables (4–6), shown by arrows in Fig. 10.7, were selected for exploration. Their r^2 on the training data, around 0.9, is above the critical threshold of 0.85 and their equations are given below. All of them are simple and interpretable and their expressions are given in the equations in Figs. 10.8, 10.9, and 10.10.

Neural Networks Models
Even when black boxes are not accepted by clients, as is the case with the emissions estimation project, developing alternative neural networks models is still recommended to exploring potential gain. Several neural networks models with one hidden layer and different numbers of neurons (3, 5, 7, and 10) were developed. All of them have r^2 on the training data above the critical threshold of 0.85, as is shown in Table 10.1 . The selected optimal structure, with four inputs (Var2, Var4, Var6, and Var7) and five hidden nodes, is shown in Fig. 10.11.

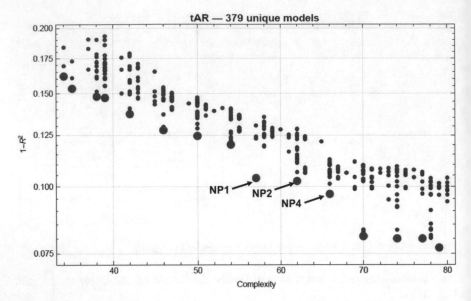

Fig. 10.7 Pareto front of generated models for emissions estimation. The selected models are shown by arrows

$$-22951.97 + 0.13\, \text{var}_2 + (1.49 \times 10^{-2})\, \text{var}_4 + \frac{15244.18}{\text{var}_6} -$$
$$47.65\, \text{var}_6 - (3.51 \times 10^{-4})\, \text{var}_4\; \text{var}_6 + \frac{2502196.30}{\text{var}_7} + 45.86\, \text{var}_7 + 0.35\, \text{var}_6\; \text{var}_7$$

Fig. 10.8 Equation corresponding to NP1

$$-21616.70 + 0.13\, \text{var}_2 - (8.98 \times 10^{-3})\text{var}_4 + \frac{0.39\, \text{var}_4}{\text{var}_6} - 56.94\, \text{var}_6 + \frac{2447961.20}{\text{var}_7} + 45.18\, \text{var}_7 + 0.33\, \text{var}_6 \text{var}_7$$

Fig. 10.9 Equation corresponding to NP2

$$-10203.68 + 0.17\, \text{var}_2 - (2.00 \times 10^{-2})\, \text{var}_4 + \frac{(1.11 \times 10^{-5})\, \text{var}_4^{\,2}}{\text{var}_6} +$$
$$9.86\, \text{var}_6 + \frac{1089078.40}{\text{var}_7} + 23.61\, \text{var}_7 - (6.95 \times 10^{-5})\, \text{var}_8 + \frac{38.13}{\text{var}_6}$$

Fig. 10.10 Equation corresponding to NP4

10.4.3   Model Validation and Selection

Model validation is the decisive factor for the final selection of the derived solution. Usually, quantitative validation is based on the performance metrics of the generated

Number of neurons	r^2 training	r^2 test
3	0.94	0.89
5	0.92	0.91
7	0.95	0.9
10	0.96	0.8

Table 10.1 Performance of neural network models on training and test data

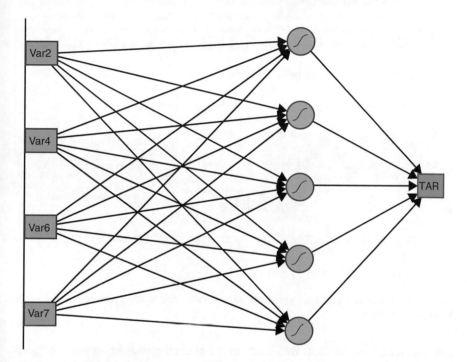

Fig. 10.11 Selected neural net model for emissions estimation

models on test data. In the case of small data sets with insufficient data for allocating separate validation/test data sets, the cross-validation method is used.

In addition to quantitative performance metrics, various qualitative validation criteria, such as residual stationarity, model interpretability and usability are considered. These key topics are discussed briefly in this section.

Cross-Validation

Cross-validation is based on a simple idea: instead of separating the available data set into training and test data, we can use all of the data and randomly divide it into training and test data sets. In this case the condition for using separate data for model generation and testing is fulfilled using all available data.

There are several types of cross-validation methods, for example LOOCV (leave-one-out cross-validation), the holdout method, and k-fold cross-validation. We'll

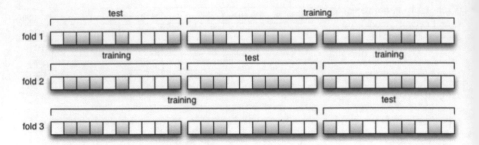

Fig. 10.12   An example of k-fold cross-validation

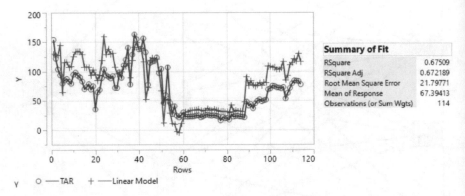

Fig. 10.13   Actual and predicted emissions on test data from a linear benchmark model for emissions estimation

give an example of the most frequently used k-fold cross-validation method. The k-fold method basically consists of the following steps:

1. Randomly split the data into k subsets, also called folds.
2. Fit the model to the training data (or k—1 folds).
3. Use the remaining part of the data as the test set to validate the model and calculate the test performance metrics.
4. Repeat the procedure k times.

An example of k-three fold cross-validation is shown in Fig. 10.12.

Model Validation on Test Data
The obvious first step is to start with the validation of the benchmark model, followed by validation of the other generated models and calculating the modeling gain relative to the benchmark. We'll illustrate this sequence with the validation of the emissions estimation models.

We begin with the linear benchmark model, which has a disappointing performance of r^2 of 0.68 on the test data, much below the required threshold of 0.85 (see the summary of fit in Fig. 10.13). The plot of the actual values and the model

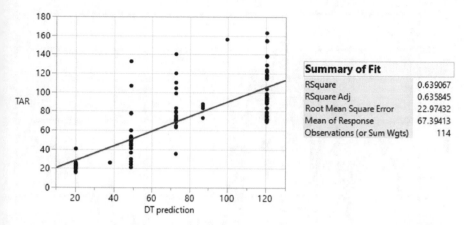

Fig. 10.14 Actual-by-predicted plot of test data and summary of fit from a decision tree model for emissions estimation

Expression	$-10203.68+0.17\,\mathrm{var}_2-(2.00 \times 10^{-2})\mathrm{var}_4 + \dfrac{(1.11\times10^{-5})\,\mathrm{var}_5^{\,2}}{\mathrm{var}_6} +$ $9.86\,\mathrm{var}_6 + \dfrac{1089078.40}{\mathrm{var}_7} + 23.61\,\mathrm{var}_7 - (6.95 \times 10^{-5})\,\mathrm{var}_8 + \dfrac{38.13}{\mathrm{var}_9}$
R–Squared	0.892051
Adjusted R–Squared	0.883826
Noise Power	0.0263405
Root–Mean–Square Error	12.5492

Fig. 10.15 Summary statistics for model NP4 on test data

prediction shown in the same figure demonstrates the difficulties of this model in terms of accurate emissions estimates, especially after sample 90.

The performance of the generated decision tree model on the test data is summarized in Fig. 10.14. The decision tree model has a disappointing performance, even worse than the linear benchmark, with an r^2 of 0.64 on the test data, much below the required threshold of 0.85.

Fortunately, the three nonlinear models generated by genetic programming have similar performance on the test data with an r^2 of 0.89, above the required threshold of 0.85. An example of some summary statistics of the performance of model NP4 on the test data is shown in the table in Fig. 10.15, and the good agreement between the predicted and observed emissions over the whole range of the test data is illustrated in Fig. 10.16. Of special importance is the good agreement when estimating high emissions greater than 120.

The performance of the alternative neural network models on the training and test data is given in Table 10.1. As can be seen in the table, the neural network structure with five hidden layers has the best performance on the test data, and that is why it was selected. Its good performance on the test data is illustrated in Fig. 10.17.

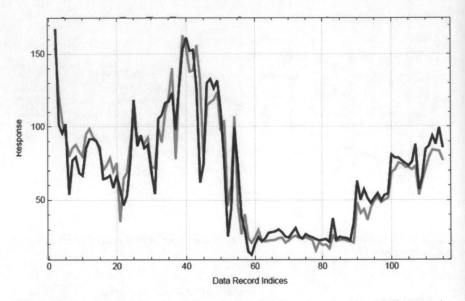

Fig. 10.16 Actual and predicted emissions for test data from the selected symbolic regression model NP4 for emissions estimation

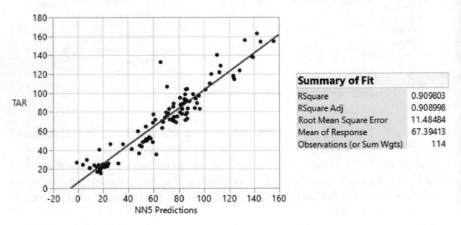

Fig. 10.17 Actual-by-predicted plot of test data and summary of fit from a potential neural net model for emissions estimation

However, the genetic programming nonlinear model NP4 has a similar performance and much less complexity, i.e., there is no significant performance improvement if an alternative black box-model is used.

Comparison of Model Performance on Test Data

The obvious first step is to compare the selected statistical performance metrics and to start with a comparison with the linear benchmark. In the case of our emissions estimation model, the generated nonlinear model NP4 has a much better r^2 than the

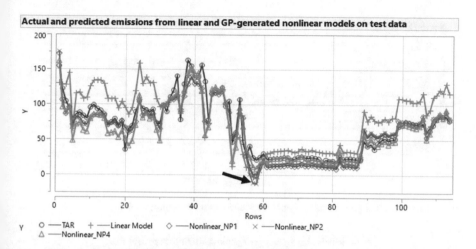

Fig. 10.18 Actual and predicted emissions for test data from the linear benchmark and the selected symbolic regression model NP4 for emissions estimation

linear benchmark (0.89 vs. 0.675). The performance gain, calculated by Eq. (10.2), is an impressive 31.85%!

A popular next step in this comparison is to use the actual emissions and those predicted by different models. A comparative plot is shown in Fig. 10.18, where the poor predictions of the linear model until sample 30 and after sample 90 are obvious. In addition, several predictions of negative emissions values, shown by an arrow, are unacceptable to the final user.

Another popular plot for comparison is a model residuals plot. Residuals refer to the difference between the actual value of a target variable and the predicted value of the target variable for every record in the test data. The residuals of a well-fitting model should be randomly distributed because good models will account for most phenomena in a data set, except for random errors. If strong patterns are visible in the plotted residuals, this is a symptom that there are problems with the model. Conversely, if the model produces randomly distributed residuals, this is a strong indication of a well-fitting, trustworthy model.

An example of a residual plot for a model with issues (the linear benchmark model) is shown in Fig. 10.19.

While the mean of the residuals is around zero, there is a pattern of growing deviation with predicted value (the so-called "funnel" effect). The arrow shows the negative emissions estimates.

An example of a relatively acceptable residual plot (based on the prediction of the nonlinear model NP4 prediction on the test data) is shown in Fig. 10.20. With the exception of an obvious outlier (shown by an arrow), the deviation of the residuals is within the limits of ±20.

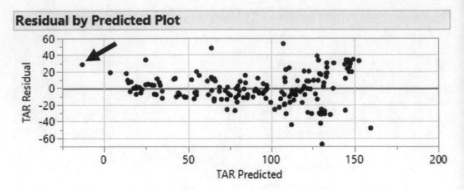

Fig. 10.19 Residual plot for linear benchmark emissions estimation model

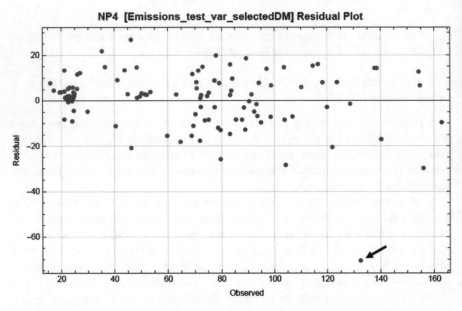

Fig. 10.20 Residual plots of test data for the selected symbolic regression model NP4 for emissions estimation

Model Selection

The final result from model comparison is model selection. In this book, we'll focus mostly on model selection based on statistical performance metrics. There are numerous other nonquantitative criteria, such as interpretability, usability, and acceptability, which can be used but are difficult to formalize.

In our emissions estimation example, out of the several generated models based on statistics, decision trees, genetic programming, and neural networks, the best option is a nonlinear model derived by genetic programing given in the table in

Fig. 10.15. Its performance on the test data is much better than that of the linear benchmark model and the alternative decision tree model. The only method that shows similar performance is neural networks, but as a black-box method it is not acceptable to the client.

10.5 Model Ensembles

Another option for model generation that has growing popularity is developing model ensembles. The key hypothesis is that by using many models in collaboration, combining their strengths and compensating for their weaknesses, the resulting model will generalize better for future data.

The key ensemble methods, namely bagging, boosting, and stacking, are discussed briefly below.

10.5.1 Bagging

Bagging operates by using a simple concept: build a number of models, observe the results of these models, and settle on the majority result. "Bagging" is short for "bootstrap aggregation," so named because it takes a number of samples from the data set, with each sample set being regarded as a bootstrap sample. The results of these bootstrap samples are then aggregated. Bagging algorithms are highly parallelizable.

The most popular bagging method is the random forest method. It combines the predictions of multiple decision trees that are trained on different samples by using random subsets of the variables for splitting.

In random forests, each tree in the ensemble is built from a sample drawn with replacement (i.e., a bootstrap sample) from the training set. In addition, instead of using all the inputs, a random subset of inputs is selected, further randomizing the tree. An example of the use of the random forest method for emissions estimation modeling is given below.

The specification of the model and its performance on the training data are given in the table in Fig. 10.21.

The performance on the training data is high ($r^2 \sim 0.89$) but its predictive capabilities on the test data, shown in Fig. 10.22 and measured to have an r^2 of 0.65, are below the required threshold for r^2 of 0.85.

An example of a random forest ensemble of four decision trees is shown in Fig. 10.23.

Each decision tree has a different structure and different process variables. The interpretation of the combined structure, however, is questionable.

Specifications					
Target Column:			TAR	Training Rows:	104
				Validation Rows:	66
Number of Trees in the Forest:			4	Test Rows:	0
Number of Terms Sampled per Split:			1	Number of Terms:	4
	RSquare	**RMSE**	**N**	Bootstrap Samples:	104
Training	0.893	14.418528	104	Minimum Splits per Tree:	10
Validation	0.892	15.078688	66	Minimum Size Split:	5

Fig. 10.21 Specification of random forest model

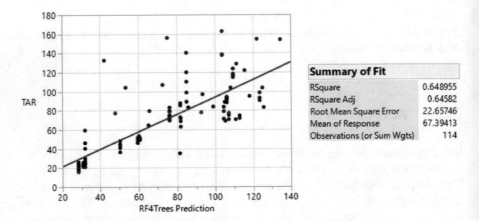

Summary of Fit	
RSquare	0.648955
RSquare Adj	0.64582
Root Mean Square Error	22.65746
Mean of Response	67.39413
Observations (or Sum Wgts)	114

Fig. 10.22 Actual-by-predicted plot of test data and summary of fit from a potential random forest model for emissions estimation

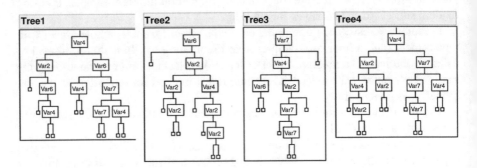

Fig. 10.23 Decision tree structure of a random forest model for emissions estimation

10.5.2   Boosting

The main principle of boosting is to fit a sequence of simple models that are only slightly better than random guessing, such as small decision trees, to weighted versions of the data. More weight is given to examples with higher error (misclassification).

A boosted tree consists of simple decision trees that are generated sequentially in layers. These layers consist of decision trees with a few splits in each tree. The boosting algorithm can use hundreds of layers in the ensemble.

The predictions are then combined through a weighted majority vote (classification) or a weighted sum (regression) to produce the final prediction. The principal difference between boosting and committee methods, such as bagging, is that the base models are trained in sequence on a weighted version of the data. An example of the use of boosting for emissions estimation modeling is given below.

The specification of the boosted tree model and its performance on the training data are given in the table in Fig. 10.24.

Similarly to bagging, the performance on the training data is high ($r^2 \sim 0.9$) but its predictive capabilities on the test data, shown in Fig. 10.25, are below the required threshold (r^2 is 0.69 with a required value of 0.85).

An example of a bootstrap ensemble of simple decision trees is shown in Fig. 10.26. This shows only the structure of three layers out of the total of 20 in the model. The generated decision trees are very simple, and some of them have the same structure.

10.5.3   Stacking

Stacking is a bit different from the other two ensemble-building techniques as it trains multiple individual models, as opposed to various incarnations of the same model. While bagging and boosting use numerous models built using various instances of the same machine learning algorithm (decision trees), stacking builds

Fig. 10.24 Specification of boosted tree model

Specifications

Target Column: TAR   Number of training rows: 122
Number of Layers: 20   Number of validation rows: 48
Splits per Tree: 3
Learning Rate: 0.1

Overall Statistics

	RSquare	RMSE	N
Training	0.896	14.542539	122
Validation	0.893	14.157041	48

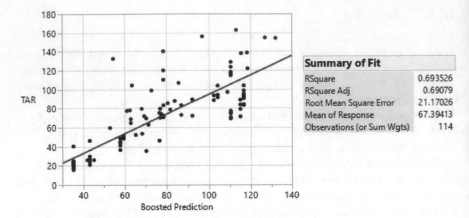

Fig. 10.25 Actual-by-predicted plot of test data and summary of fit from a potential boosting model for emissions estimation

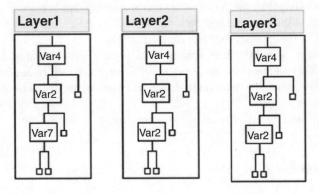

Fig. 10.26 Decision tree structure of selected layers of a potential boosting model for emissions estimation

its models using different modeling algorithms (linear regression, neural networks, genetic programming, or some other combination of models).

A combination algorithm is then trained to make the final predictions using the predictions of the other algorithms. This combiner can be any ensemble technique or even a custom combination. An example in the case of the emissions estimation project of a custom combination of the three accepted genetic programming models NP1, NP2, and NP4 is shown in Fig. 10.27.

The performance of the combined stacked model is slightly better than that the "best" individual model, NP4 (ensemble $r^2 = 0.91$ vs. NP4 $r^2 = 0.89$). However, the negative predictions (shown by an arrow in Fig. 10.27) of the ensemble are not acceptable to the user. The individual model NP4 does not generate negative predictions, as is shown in Fig. 10.14, and was selected for deployment.

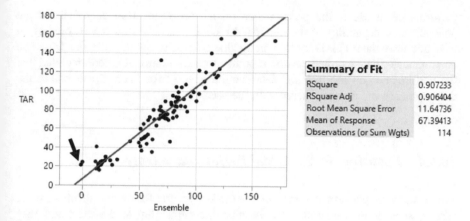

Fig. 10.27 Actual-by-predicted plot of test data and summary of fit from an ensemble of combined symbolic regression models for emissions estimation

10.6   Common Mistakes

10.6.1   Model Development without Variable Reduction/ Selection

Ignoring variable reduction and selection, discussed in detail in the previous chapter, leads to inefficient model development due to the larger-than-necessary number of inputs. As a result, the generated models have high complexity and are overparametrized. Usually their robustness is very low and they need frequent retuning.

10.6.2   Not Fighting Data Overfitting

A more common mistake is underestimating the importance of taking measures to prevent overfitting. The negative effect is similar to that of the previous mistake—developing models with evaporating credibility due to low robustness.

10.6.3   Communicating Optimistic Model Performance Based on Training Data

As a general rule, the performance on the training data is a metric of how well the developed model fits the data. It is an "internal" measure used during model generation and is an important criterion for model developers. However, the real performance metric is the performance on the test data. This demonstrates the generalization capability of the generated model, i.e., how well it will predict on new, unknown data. This is the true metric that is of interest to the user and should be communicated as the only measure of performance. It is strongly recommended that the performance on the training data should be kept for internal use within the development team and not shared with the user.

10.6.4   Focusing on Statistical Performance Only

While statistical performance metrics are the key criteria for model selection, other factors, such as interpretability of the model, acceptability to the user, and easy deployment have to be considered. Defining these qualitative criteria in advance of model development is strongly recommended.

10.6.5   Using One Modeling Method Only

The variety of modeling methods is one of the key advantages of AI-based Data Science. Not using this advantage is an inexcusable waste of opportunity.

10.7   Suggested Reading

- D. Abbott, *Applied Predictive Analytics: Principles and Techniques for the Professional Data Analyst*, Wiley, 2014.
- D. Hall and N. Gill, *An Introduction to Machine Learning Interpretability*, O'Reilly Media, 2018.
- M. Kuhn and K. Johnson, *Applied Predictive Modeling*, second edition, Springer, 2016.
- D. Montgomery, E. Peck, and G. Vining, *Introduction to Linear Regression Analysis*, fourth edition, Wiley, 2006.
- B. Ratner, *Statistical and Machine-Learning Data Mining: Techniques for Better Predictive Modeling and Analysis of Big Data*, third edition, CRC Press, 2017.
- S. Tuffery, *Data Mining and Statistics for Decision Making*, Wiley, 2011.
- Z. Zhou, *Ensemble Methods: Foundations and Algorithms*, CRC Press, 2012.

10.8   Questions

Question 1
What are the key features of a good model?

Question 2
What are the important steps in developing a good model?

Question 3
How can overfitting the data be avoided?

Question 4
How does one select the proper methods for model development?

Question 5
What are the advantages of model ensembles?

Chapter 11
The Model Deployment Life Cycle

> *A model is as strong as its deployment.*
>
> Olav Laudy

The focus of this chapter is on the last transformation of data into value—translating the developed models into a value-generation problem solution by effective model deployment. Model deployment is the process of moving a model from an offline development environment to an online system or a large, powerful, and secure database or server where it can be used simultaneously by many mission-critical processes. Another popular name for this step is "moving the models into production."

The model development and deployment environments are inherently different because the production environment is continuously running. Data is constantly coming in, being processed and computed into KPIs, and going through models that are retrained frequently. These systems, more often than not, are written in different languages than the model development environment.

Model deployment is the most underestimated and misunderstood step in the AI-based Data Science workflow. The predominant perception is that the project consists of building a single successful model, and after it has been coded into the production environment, the project is completed. Industrial reality is quite different, however. Most business applications require working with multiple models in production at any given time. In addition, setting up processes for scoring models on new data, auditing those scores, doing continuous model validation against new observations, and setting up procedures to handle replacement if models do not work as expected, are needed. In order to extract value and guarantee the success of the project, all those processes need to be carefully organized and executed.

The first objective of this chapter is to focus on the different ways of extracting value from the deployed solution. The second is to discuss the key methods of model exploration to help the user make the final decision about deploying the selected models while the third is to emphasize the importance of model-based decisions in extracting value from an AI-based Data Science project. The fourth objective is to describe briefly the model deployment options. The fifth is to present the

© Springer Nature Switzerland AG 2020
A. K. Kordon, *Applying Data Science*,
https://doi.org/10.1007/978-3-030-36375-8_11

performance-tracking possibilities, and the sixth is to discuss long-term model maintenance strategies.

11.1   Translating Actions into Value

It is expected that the developed modeling solution will solve the defined business problem. However, demonstrating it in a development environment shows only the potential for value creation. For example, the generated nonlinear model for emissions estimation NP4 can deliver estimates with an acceptable accuracy of r^2 of 0.89 (see Fig. 10.15). By developing this model, the technical objective of the project has been accomplished. But not the financial objectives.

This accurate prediction is needed for improved process control of the unit. The real mechanism of value generation is the actions of the control system after the model is moved into production. These actions use the accurate emissions prediction to operate the unit at higher rates while not violating the critical emissions limits. This translates into value from higher productivity and reduced risk of environmental violations.

There are many ways to translate actions, driven by developed models, into value. The most popular are discussed briefly below.

11.1.1   Improved Performance by Using Developed Models

The most popular approach to translating the actions of deployed models into value is by demonstrating consistently improved performance according to expectations. In order to accomplish this transition, however, several steps need to be accomplished. First, the developed models should be understood, explored, and accepted by the users. Second, the users must be properly trained to understand the nature and limitations of the deployed models and in how to respond to model deterioration patterns. Third, an effective model maintenance process should be put into action.

The key requirement for accomplishing a successful performance improvement by using developed models in production is to convince the users to implement them when necessary. Often, however, due to lack of incentives or to politics, this is a challenging task even after all recommended steps have been accomplished.

11.1.2   Documented Value Creation

A critical action item in the process of translating actions into value and accomplishing the financial objectives is quantifying the value created after model deployment. This requires data collection for a benchmark of created value before the solution has been moved into production. This data collection can begin at the start of the project and should be based on a reasonable period of time.

The basis of good documentation for this is a well-defined value-tracking mechanism. Including financial experts in this process is highly recommended. Often the value created from the applied solution is difficult to isolate from other factors that may contribute to improved performance. The potential contribution has to be discussed in advance and an agreement should be reached with all related stakeholders.

It is recommended to perform periodic value assessment and to compare the real value with that expected in the project definition. In the case of the emissions estimation project, the two sources of created value, (1) increased production rate and (2) reduced environmental violations, were assessed every quarter based on a defined methodology for assessing the contributions to value. The benchmark of value creation included 2 years of data before deploying the model.

11.1.3 Improved Decision-Making by Using Models

Another popular way to translate the actions obtained from deployed models into value is by using them in the decision-making process. This assumes that the developed model becomes part of decisions, made most of the time by the user. However, there are several applications, such as fraud detection, loan approval, and product recommendations where the decision is made automatically by the system.

The value creation mechanism in the case where the user defines new rules, based on the developed models requires clarification. It is assumed that the current decision-making process is not based on models. Improved decision-making requires by definition of a new rule (or set of rules) based on the developed models. It is also recommended that the advantages of the new rule are clearly defined and illustrated with examples. All potential users should be trained in the proper use of the new rule and agree to implement it when necessary. Only after finishing this process is the value tracking reliable.

A typical example of this type of value creation is the use of predictive models for price forecasting. An accepted forecast creates value only if a successful price negotiation is accomplished. The key role of the models in this process, in addition to expert knowledge about prices, is in defining an effective new rule based on the forecast, its confidence limits, and the observed trends.

11.1.4 Final Assessment of Project Value

It is expected that a well-crafted AI-based Data Science project will reach the stage of final assessment of the value of the project. At the end, the key question to be answered is: Was the project a success or a failure? In most cases the answer depends on an effective move into production and proper use of deployed models.

If the models are not used due to low robustness or lack of engagement, the occurrence of an implementation fiasco is clear and usually is evident in a short period of time after deployment. In this case, it is recommended to accept the failure, define the lessons learned, and analyze the mistakes made.

If the models are used properly and the expected value is created and documented, it is good practice to advertise this success. The ultimate achievement, however, is in leveraging the accumulated experience to solve other problems.

11.2   Model Exploration

Delivering models with the expected performance is the final step of model development, but it is only the start of the long model deployment life cycle process. The next intermediate phase, before moving the models into production, is model exploration. Its objective is to convince the users of the qualities of the derived solutions. It includes understanding the models, exploring their behavior in all possible situations, finding optimal parameters under different constraints, etc. The models are given into the hands of the users who are encouraged to disturb them and extract as much knowledge as possible via "what-if" scenarios. This practice has shown that this is the best way to convince the users about the qualities of the derived solutions and give them the freedom to either accept or reject the models for future deployment. Ignoring this step and pushing unconvinced users to move unexplored models into production is one of the biggest mistakes a data scientist can make.

Some of the most popular techniques used for model exploration are discussed briefly below.

11.2.1   Model Exploration Techniques

Prediction Profilers

The most popular tools for playing "what-if" scenarios with the developed models are 2D prediction profilers. Profiling is an approach to visualizing response surfaces by seeing what would happen if just one or two factors were changed at a time. An example of a profiler for one of the developed nonlinear emissions estimation models, NP1 (see Sect. 10.4.2), based on four process variables, is shown in Fig. 11.1.

This profiler includes several sections linked to the full range of change of each process variable. The key data in each section is the model prediction of the response variable (the emissions). A clear advantage of profilers is that the nature of the dependence between the corresponding input and the target variable is visualized. For example, the profiler in Fig. 11.1 shows that (1) emissions increase with an

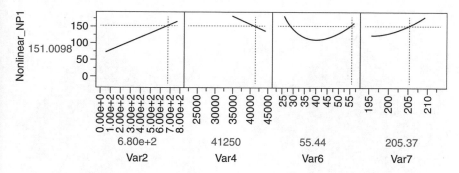

Fig. 11.1 An example of a prediction profiler for the nonlinear model NP1 for emissions estimation

increase in Var2 and (2) the relationships between Var6 and Var7 and the emissions are nonlinear. For each input variable, the value above the input name is its current value. Input values can be changed with sliders (dragging the vertical dotted lines in Fig. 11.1). The horizontal dotted line shows the current predicted value of the Y variable for the current values of the X variables.

The importance of an input can be assessed to some extent by the steepness of the prediction trace. If the model has curvature terms (such as nonlinear terms), then the traces may be curved (as is the case with Var6 and Var7).

Response Surface Plots

A surface plot is a three-dimensional plot with one or more dependent variables represented by a smooth surface. Surfaces can be defined by a mathematical equation, or through a set of points defining a polygonal surface. A surface can be displayed as a smooth surface or as a mesh. An example of three response surface plots of combinations of Var2, Var4, and Var6 for the nonlinear model NP1 for emissions estimation are shown in Fig. 11.2.

In the case of nonlinear behavior, response surface plots give a better visual representation of the nature of the relationships among the inputs of the model and potentially dangerous areas. An example of such a possible problem is the sharp increase in the emissions estimate in the high range of Var4 and the low range of Var6, shown in the bottom plot in Fig. 11.2.

Contour Plots

A contour profiler shows response contours for two inputs at a time. Other inputs can be set to specific values to show how this affects the contours. An example of three contour plots of combinations of Var2, Var4, and Var6 for the nonlinear model NP1 for emissions estimation are shown in Fig. 11.3.

Contour plots give a different view of the dependences between the inputs and the predicted variable. They are especially useful in defining areas of concern. Examples of such areas, on the plots with high emissions above 200 (the thick line drawn) are shown in the bottom right corner of each contour plot in Fig. 11.3.

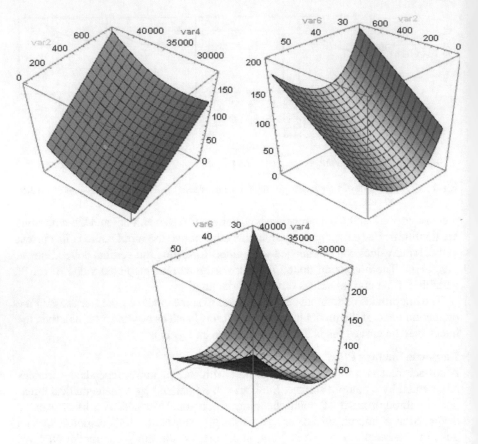

Fig. 11.2 Examples of response surface plots for the nonlinear model NP1 for emissions estimation

Sensitivity Analysis

Sensitivity analysis investigates whether model behavior and predictions remain stable when the data is intentionally perturbed or other changes are simulated in the data. The key concern is whether the developed model can make drastically different predictions with minor changes in the values of input variables. Sensitivity analysis enhances understanding because it shows a model's likely behavior and predictions in important situations, and how they may change over time.

Sensitivity analysis increases trust if the models agree with experts' domain knowledge and expectations when interesting situations are simulated, or as data changes over time. An example of a plot of input sensitivities for one of the developed nonlinear estimation models, NP1, based on four process variables, is shown in Fig. 11.14.

Sensitivities are presented as a sensitivity indicator, shown by a triangle whose height and direction correspond to the value of the partial derivative of the profiled

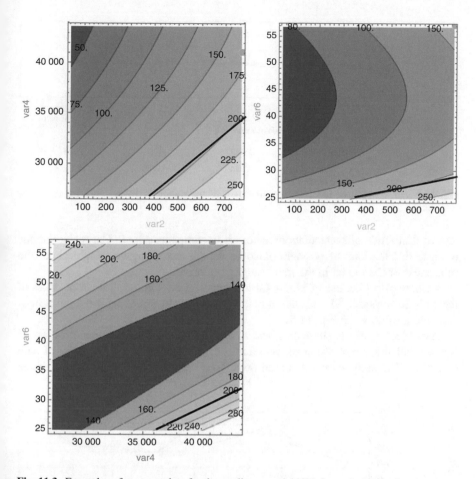

Fig. 11.3 Examples of contour plots for the nonlinear model NP1 for emissions estimation

function (emissions) at its current value. As can be seen in Fig. 11.4, Var2 and Var6 are the most sensitive process variables with respect to emissions.

Monte Carlo Simulations

A very important exploration task is testing the robustness of developed models under specific ranges of operating conditions. Such simulations enable one to discover the distribution of model predictions as a function of random variation in the model inputs and noise.

The most popular method in this category is Monte Carlo simulation. It performs risk analysis by building models of possible results by substituting a range of values—a probability distribution—for any input that has inherent uncertainty. It then calculates results over and over, each time using a different set of random values from these probability functions. Depending upon the number of uncertainties and the ranges specified for them, a Monte Carlo simulation can involve thousands or

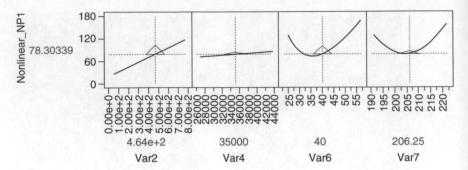

Fig. 11.4 An example of inputs sensitivities plot of the nonlinear model NP1 for emissions estimation

tens of thousands of recalculations before it is complete. Monte Carlo simulations produce distributions of possible outcome values that can give insight about the robustness of the model in the input ranges explored.

An example of the use of Monte Carlo simulations in exploring the robustness of the emissions model NP1 is shown below. The initial settings of the input ranges explored are shown in Fig. 11.5.

The objective of the simulation was to explore how robust the predictive model was around the target value of emissions of 150, which is on the edge of safe operation. The simulation was based on a fixed value of Var2 and random values

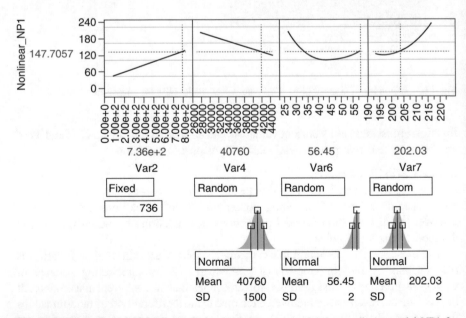

Fig. 11.5 Example of initial Monte Carlo simulation settings for the nonlinear model NP1 for emissions estimation

of Var4, Var6, and Var7 from a normal distribution, with means and standard deviations given in Fig. 11.5. The distribution of predicted emissions after 10,000 simulations is shown in Fig. 11.6.

Unfortunately, the robustness of the model in this broad range of input changes is not satisfactory, because 2.5% of the emissions are too high (>180) according to the 97.5% quantile in the table in Fig. 11.6.

A new set of simulations were run, and the final settings with narrower ranges are given in Fig. 11.7.

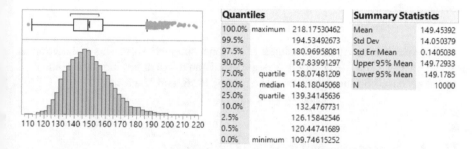

Quantiles			Summary Statistics	
100.0%	maximum	218.17530462	Mean	149.45392
99.5%		194.53492673	Std Dev	14.050379
97.5%		180.96958081	Std Err Mean	0.1405038
90.0%		167.83991297	Upper 95% Mean	149.72933
75.0%	quartile	158.07481209	Lower 95% Mean	149.1785
50.0%	median	148.18045068	N	10000
25.0%	quartile	139.34145636		
10.0%		132.4767731		
2.5%		126.15842546		
0.5%		120.44741689		
0.0%	minimum	109.74615252		

Fig. 11.6 Example of initial Monte Carlo simulation of predictions of the nonlinear model NP1 for emissions estimation

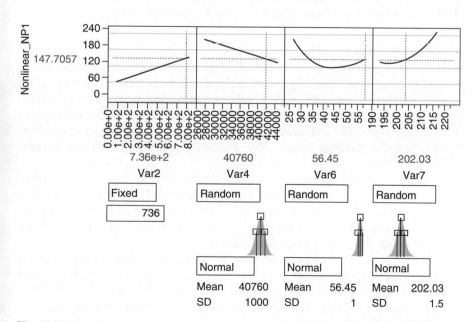

Fig. 11.7 Example of final Monte Carlo simulation settings for the nonlinear model NP1 for emissions estimation

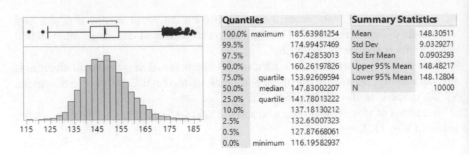

Quantiles			Summary Statistics	
100.0%	maximum	185.63981254	Mean	148.30511
99.5%		174.99457469	Std Dev	9.0329271
97.5%		167.42853013	Std Err Mean	0.0903293
90.0%		160.26197826	Upper 95% Mean	148.48217
75.0%	quartile	153.92609594	Lower 95% Mean	148.12804
50.0%	median	147.83002207	N	10000
25.0%	quartile	141.78013222		
10.0%		137.18130212		
2.5%		132.65007323		
0.5%		127.87668061		
0.0%	minimum	116.19582937		

Fig. 11.8 Example of final Monte Carlo simulation of predictions of the nonlinear model NP1 for emissions estimation

The distribution of the predicted emissions obtained from this simulation is shown in Fig. 11.8. It is within the expected range for allowing safe control of emissions based on the tighter ranges of Var4, Var6, and Var7.

11.2.2   Acceptance by Users for Deployment

After exploring the developed models as much as possible, the final users with the guidance of data scientists, should check the readiness of the models before making the final decision to put them into production. An example of a checklist is shown below.

Model Deployment Checklist
- *Does the performance of the developed model satisfy the project objectives?* This applies to all requirements given in the problem definition. In the case of the emissions estimation example, this, includes a prediction accuracy with $r^2 > 0.85$, non-black-box models, and easy integration of the developed model into the existing process information and control system (see Sect. 2.4). All three developed potential nonlinear models (NP1, NP2, and NP4) satisfy these requirements (see Sect. 10.4).

    In the case of unsatisfactory performance of the developed models, an analysis of what is achievable with the available data and the methods used is recommended. Potential alternative solutions can be suggested and discussed with the users. A decision to either continue the project with more realistic objectives or stop the effort should be made.
- *Is the behavior of the model according to expectations?* The exploration of the models has to agree with the domain experts' knowledge. An example of model exploration for one of the models for emissions estimation (model NP1) has been shown in this section. This demonstrates that this model has the expected behavior in all tools explored.

If issues arise during model exploration, several actions such as more comprehensive training of users in exploration methods, more detailed exploration, and addressing the observed concerns, are taken until an agreement to accept the models is reached. A new model development cycle is possible in the course of handling these issues.

- *Do all users understand and accept the model?* In the specific case of the emissions estimation models, all potential users were given the chance to explore the models and give their comments. All of them, without exception, accepted the models.

  In the case of a lack of understanding and acceptance of the models, the source of this attitude should be identified. It should not be technical at this point (it is assumed that the technical issues have been resolved in the previous step.) The solution could be more comprehensive training and better explanation of the benefits of the deployed solutions.

- *Is it clear how value creation will be documented?* The answer to this question includes (1) an agreement about how the value will be measured and by whom, and (2) a completed benchmark value estimate of the current situation before the solution is deployed. Unfortunately, in most cases, projects are moved into production without developing this necessary infrastructure for value tracking. As a result, the critical value creation step is either missing or without a solid financial basis, and thus could be questioned.

  In the case of the emissions estimation project, the value creation tracking was well defined by financial experts as increased unit productivity above the average values in the benchmark period and decreased environmental penalties. This period was well documented by data from the last 2 years before implementing the model.

- *Are there any issues in moving the model into production?* A negative answer to this question assumes that there is full agreement of all stakeholders and management support to deploy the selected solutions. If not, the specific issues should be discussed until an appropriate agreement is reached.

Decision about Deployment

Positive answers to all question in the checklist are a good basis for making the final decision to move developed models into production and initiate the corresponding model deployment steps.

11.3   Model-Based Decision-Making

Defining the specifics of how the developed models will be used in production is the most critical and unstructured part in transferring data into value. In most of the cases it requires using the models in the decision-making process by either defining new or updating old business rules. Usually these decisions and rules are expressed by the users but data scientists have to be entirely involved in this process, which is not trivial. Some guidelines how to accomplish this task are given below.

11.3.1   Why do we Need Model-Based Decisions?

One of the key reasons for using model-based decisions is that only recently have we had an opportunity to obtain broad access to high-volume data from a variety of sources and analyze it with very sophisticated algorithms. The hypothesis is that these two factors, when combined, dramatically increase the probability of defining something like "truth" and do so in a time window that is actionable. This allows businesses to make more correct decisions (based on "truth") and have a competitive advantage over those using heuristic-based decisions (based on gut feeling).

Some data supports this hypothesis. According to a recent survey, companies in the top third of their industry in the use of data-driven model-based decision-making are, on average, 5% more productive and 6% more profitable than their competitors.[1]

Let's begin with analyzing the key limitations of decision-making without models.

Limitations of Heuristic-Based Decisions

It is assumed that the heuristic-based rules represent the current best practices of the business. They include a combination of logic schemes, defined by experts, and related data representing important historical patterns and recent KPIs. While this set of subjective decisions represents the best collective knowledge about specific business problems, it has the following inherent limitations:

- *Broad range of uncertainty*. It is difficult to define quantitative ranges of suggested actions in a decision without validated models on a solid statistical basis. In general, heuristic-based ranges are uncertain and difficult to test. This increases the chance of incorrect actions outside well-known business conditions.
- *Qualitative metrics about the future*. Without predictive or forecasting models, decisions about future actions are not based on numerical values but on qualitative recommendations. For example, a heuristic-based rule for price negotiation usually looks like "It is recommended to purchase x amount of raw material y at an expected small price increase, according to a current report from source z." The decision-maker has quantitative information only about the past. The information about the future is ambiguous and pushes the user to make difficult quantitative decisions with a high level of uncertainty (for example, what is the quantitative meaning of "small price increase"?) All available information for decision-making in this case is visualized in Fig. 11.9. As a result, all specific decisions made by the expert are based on quantitative data from the past and qualitative information about the future.
- *No multivariate analysis*. With very few exceptions, heuristic-based decisions are limited to a few important variables about which experts have collective knowledge. The critical insight from multivariate analysis (described in Chap. 9) is absent. This is a serious gap in making decisions based on all relevant process variables and their total impact on the problem.

---

[1] A. McAfee and E. Brynjolfsson, *Big Data: The Management Revolution*, HBR, October 2012.

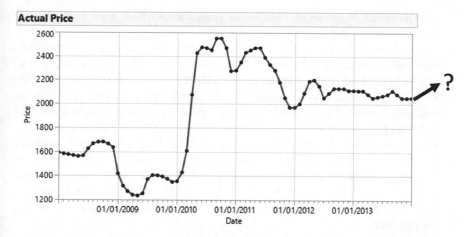

Fig. 11.9 Example of available information in heuristic-based decision about price negotiation

- *Difficult rule validation*. The performance of the defined business rules depends on the availability and professional level of domain experts. The "bag of tricks" they offer is their subjective heuristic model of the specific problem within their area of expertise. Often it is difficult to justify the defined rules of thumb either by physical interpretation or by statistical analysis of the available data. The subjective nature of represented human intelligence raises fundamental questions about how adequate the recommended actions are to the objective nature of the problem to which they are applied.
- *Vulnerable to politics*. Politics is an inevitable factor of business life, especially when decision-making is involved. Unfortunately, heuristic-based decisions are influenced by the current and past political environment and business policies. As a result, some politically "correct" business rules could be technically inferior.

Advantages of Model-Based Decisions
- *Objective nature of decisions*. As was discussed in Chap. 10, mathematical models represent objective relationships among the factors influencing a business problem. Using their predictions, with their confidence limits, in the decision-making process increases the unbiased nature of the process. Most business rules are still defined subjectively by experts. The quantitative metrics in them, however, are based on objective relationships.
- *Decisions based on multivariate factors*. Many AI-based Data Science-generated decisions use predictions from multivariate approaches or models that include many factors influencing the business problem. These decisions are more adequate to the complexity of real-world problems.
- *More accurate decisions*. Model-based decisions transform estimates and forecasts into much more precise, accurate predictions that remove enormous chunks of risk from the decision-making process. This radical increase in accuracy translates into higher degrees of confidence throughout an organization, which

in turn drives fundamental changes in behavior and high trust in the overall decision-making process.

- *Quantitative estimates about the future.* A big advantage of predictive and especially forecasting models is their numerical most probable forecast, with its low and high confidence limits over a specific forecasting horizon. It allows one to specify the "shape of the future" in quantitative terms and narrows down the uncertainty in making the decision. An example of price forecasting will be given in Sect. 11.3.2.
- *Improved existing business rules.* The most popular approach in using developed models in decision-making is to enhance existing business rules with more accurate predictions. This is the best way to integrate existing problem knowledge, based on previous experience, with more objective quantitative information, based on discovered complex relationships between the factors influencing the problem.

11.3.2   Defining Model-Based Decisions

The key steps in defining model-based decisions are (1) selecting candidate decisions related to developed models, (2) integrating these models within the decisions, and (3) communicating the chosen decisions to the users. A short description of each step is given below.

Selecting Candidate Decisions

Not all of the available business rules are suitable for defining model-based decisions related to a problem to be solved. In particular, appropriate decisions need to be repeatable, nontrivial, and measurable.[2]

- *Repeatable decisions.* If a decision is not repeatable, there is no value in using it as a business rule. To be considered repeatable, a candidate decision must pass four tests:

  The decision is made at defined times.
  Each time the decision is made, the same information is available, considered, and analyzed.
  The set of possible actions remains consistent between decisions.
  The way in which the success of these actions is measured in terms of business outcomes remains consistent.

- *Nontrivial decisions.* It is assumed that the selected decision has some degree of complexity, i.e., it is not an elementary rule. All candidates for model-based

---

[2]This subsection is based on recommendations from J. Taylor, *Decision Management Systems: A Practical Guide to Using Decision Rules and Predictive Analytics*, IBM Press, 2012.

decisions are in this category since the developed models have been generated by a complex AI-based Data Science analytical process.

- *Decisions with measurable business impact.* It must be possible to see the cost of bad decisions and the value of good ones. For a good candidate decision, the business can see the impact of the decision in terms that relate to the measurement framework the business already has. If the business values customer loyalty, then the business impact of a customer retention decision could be measured in terms of an increase in customer loyalty.

It has to be considered that the business impact of a decision may not be immediately apparent. It can take months or even years to see the total impact of an improvement.

In the case of the emissions estimation project, the candidate business rule was based on defining a critical threshold for the production rate of the unit, based on the measured emissions during violation incidents in the last 3 years. The rule defined a conservative high production limit based on the critical threshold minus 5%, which was included in the process control system of the unit. The decision satisfied the criteria for relevance: (1) it was repeatable, being executed every minute by the control system; (2) it was nontrivial, being based on analysis of past trends; and (3) it was measurable, sinse the negative impact of environmental violations was measured by penalties paid.

In the example of the raw materials price negotiation rule, the decision included qualitative suggestions for the amount and price to be negotiated, based on historical data, previous experience, and available reports. The decision also satisfied the criteria for relevance: (1) it was repeatable, being executed periodically during raw materials purchases; (2) it was nontrivial, being based on analysis of past trends and available reports; and (3) it was measurable, since the gain or loss in the negotiated price could be assessed.

Integration of Model Results within Decisions

The next step in defining model-based decisions is integrating model predictions into the selected business rules. Often, in addition to the predictions, their confidence limits with high probability (usually 95%) are used appropriately in the rules. This is one of the biggest benefits of using mathematical models, because it reduces significantly the options in the decision space.

In the emissions estimation project, a new rule, based on the estimated emissions, were added. This includes optimal control of the estimated emissions based on three process variables, Var4, Var6, and Var7 (one of which is the production rate). The initial settings of the controller were based on the model exploration results shown in Fig. 11.7. Another business rule, to allow the execution of the control loop defined in this way, depending on the performance of model, was added as well. This tracked whether the deployed model passed the quarterly tests of accuracy, with an r^2 of 0.85. The control loop was not executed if the model performance was below this limit. A set of corrective actions, discussed in Sect. 11.6, were taken until the model satisfied the quality standard.

In the case of integrating forecasting models, the expected modeling results include the forecasted values with their confidence limits over the defined forecasting horizon.

Communication of Model-Based Decisions

The most important step in defining model-based decisions is communicating them clearly to all potential users and project stakeholders. Some of the defined decisions, after an initial approval, are executed automatically by a decision management or process control system. Many decisions, however, are executed by the users as part of their regular work process. This type of decision requires good communication, with detailed explanations and recommendations. This should include the model predictions, some brief information about the type of model and assessment of its performance, information about discovered influential factors (if available), and a suggested decision. An example of communicating a raw materials price forecasting model is given below.

The document sent to the experts responsible for raw materials purchasing includes the following sections (details are given for some of them):

- *Actual historical trend of prices.* This is shown in Fig. 11.9.
- *Recent trend and price forecast with confidence limits for the next 12 months.* This is shown in Fig. 11.10. Numerical values are given in Table 11.1.
- *Recent trends and forecast of key drivers of the price for next 12 months.* This is shown in Fig. 11.11.

- *Model type and past performance information.* An ARIMAX type of linear forecasting model based on two drivers (crude oil and propylene) was used. The MAPE of the model for the last 12 months was 8.2%, which is in the category of good model accuracy.

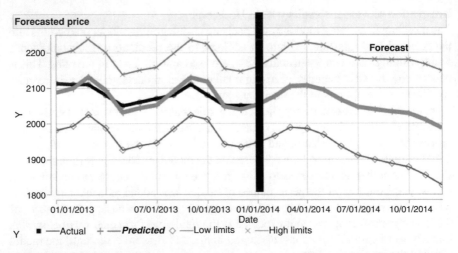

Fig. 11.10 Plot of actual and forecasted price with its confidence limits. The 12 month forecasting horizon is to the right of the vertical line

Table 11.1 Price forecast with low and high confidence limits at 95% probability for 12 month forecasting horizon

Date	Actual	Forecast	Low limits	High limits
1/1/14	2050	2053	1947	2159
2/1/14		2076	1965	2187
3/1/14		2105	1989	2221
4/1/14		2106	1985	2227
5/1/14		2094	1968	2220
6/1/14		2066	1935	2197
7/1/14		2046	1910	2182
8/1/14		2039	1898	2180
9/1/14		2033	1887	2179
10/1/14		2028	1877	2179
11/1/14		2010	1854	2166
12/1/14		1987	1826	2148

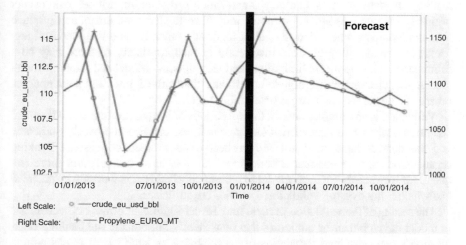

Fig. 11.11 Plot of actual and forecasted key drivers for the predicted price shown in Fig. 11.10. The 12 month forecasting horizon is to the right of the vertical line

- *Summary of past experience.* The price has risen drastically after the recession in 2010 and declined after 2012. Recently the price has been relatively stable and varied around $2050.
- *Summary of forecasting results.* A slight price increase is predicted until April followed by a gradual price decrease until the end of the year. The price decrease is driven by the combined effect of an expected decrease in crude oil prices and an increase in propylene prices in March. These trends are shown in Fig. 11.11.
- *Recommended decision.* Based on the forecasting results and the above analysis, the following decision for purchasing was suggested:

Suggested purchase of 1000 MT (metric tons) in February at a broad price range between \$1965 and \$2187 with most probable price of \$2076. Price increase to around \$2110 is expected in April, followed by a decreasing trend below \$2040 after August. Look at crude oil and propylene price trends as additional indicators.

A comparison between this model-based recommended decision and the subjective one defined in Fig. 11.9 illustrates clearly the advantages of the former: the expert has specific quantitative information about future trends to narrow down the risk and make an appropriate final decision.

Risk Assessment of Model-Based Decisions
Model-based decisions are critical for some sectors of the economy, especially banking. The decision-makers should be attentive to the possible adverse consequences (including financial loss) of decisions based on models that are incorrect or misused, and should address those consequences through active model risk management. Model risk can lead to financial loss, poor business and strategic decision-making, or damage to a banking organization's reputation. Model risk occurs primarily for two reasons: (1) a model may have fundamental errors and produce inaccurate outputs when viewed against its design objective and intended business uses; (2) a model may be used incorrectly or inappropriately or there may be a misunderstanding about its limitations and assumptions. Model risk increases with greater model complexity, higher uncertainty about the inputs and assumptions, broader extent of use, and larger potential impact.

Validation is an important check during periods of benign economic and financial conditions, when estimates of risk and potential loss can become overly optimistic and the data at hand may not fully reflect more stressed conditions. Banking organizations should conduct a periodic review—at least annually but more frequently if warranted—of each model to determine whether it is working as intended and whether the existing validation activities are sufficient.

The results of the validation process may reveal significant errors or inaccuracies in model development or outcomes that consistently fall outside the banking organization's predetermined thresholds of acceptability. In such cases, model adjustment, recalibration, or redevelopment is warranted.[3]

11.3.3   Building a Model-Based Decisions Culture

The ultimate goal is to gradually change the decision-making culture in the organization from "gut"-based to model-based. This is a long and painful process but it is the best way to guarantee consistent results from applied AI-based Data Science solutions. This complex topic is beyond the scope of the book. However, some key steps in this process of cultural change are discussed briefly below.

---

[3]This subsection is based on *Guidance on Model Risk Management*, Federal Reserve System, Washington, DC, 2011

Top Management Support for Using Model-Based Decisions

The first step in opening the door to model-based decision-making in an organization is convincing the key decision makers that forecasting capabilities give them a competitive advantage in the contemporary business world. This can create an environment of strategic management support for applying predictive model-based systems, and such activities are unaffected by future political swings. Some of the arguments that define the benefits of model-based decisions using predictive models are given below:

- *Proactive management*. Predictive models enable model-based decision-making that anticipates future changes and reduces the risk of taking inadequate actions.
- *Improved planning*. A typical case is one of the most popular forecasting applications, demand-driven forecasting, which has significant economic impact. (See the examples given by Chase.[4])
- *Understanding business drivers*. The unique combination of data analysis and forecasting gives insight, based on data and statistical relationships, about the key drivers related to a specific business problem. Model-based decisions consider these relationships.
- *Defining a winning strategy*. Long-term forecasting scenarios based on mathematical models are critical for developing future business strategies.

After convincing the senior leadership about the competitive advantages of using model-based decisions, it is not sufficient to have their support in words only. Top management must define proper model-based KPIs, agree on a new reporting format, establish expectations for information flow and decision-making, require participation from all, and expect actions to be completed on time.

If effectively communicated, formats, expectations, and discipline will cascade from the top. It is more likely that everyone in the organization will fall in line with top management guidance if this communication is clear and consistent.

In the case of the raw materials forecasting project (described in detail in Chap. 15), the top management clearly communicated its long-term support for forecasting-based raw materials purchasing, allocated resources during project development, and expected all related stakeholders to implement the developed model in a well-defined work process.

Introducing a Model-Based Decision-Making Process

Another key step in building a model-based decision culture in a business is defining a work process for consistent implementation of this approach across the organization. The best-case scenario is where this type of decision-making is integrated into existing work processes. A good candidate is the popular Sales and Operations Planning (S&OP) process.[5] The S&OP process includes an updated forecast that leads to a sales plan, a production plan, an inventory plan, a customer lead time

---

[4]C. Chase, *Demand-Driven Forecasting: A Structured Approach for Forecasting*, second edition, Wiley, 2013.

[5]T. Wallace, *Sales and Operations Planning: Beyond the Basics*, T.F. Wallace & Company, 2011.

(backlog) plan, a new product development plan, a strategic initiative plan and a resulting financial plan. The frequency of the planning process and the planning horizon depend on the specifics of the industry. Although invented over 20 years ago, S&OP is currently in a stage of wide popularity and rapid adoption.

Another appropriate popular work process with wide application in industry is Six Sigma. Details on how to integrate AI-based technologies in a Six Sigma are given by Kordon.[6]

In order for the integration of model-based decision-making into these work processes to be successful, the following principles are strongly recommended:

- *Balancing stakeholders' interests.* Understanding the stakeholders' interests and concerns is a winning strategy for avoiding potential political issues. For example, typical business clients for predictive modeling projects expect trustworthy results that are similar to their own expert judgment and can help them make correct decisions and perform profitable planning. They prefer simple explanations and consistent performance. Their main concern is trying to avoid a decision fiasco and its negative impact on their careers.

- *Handling biases.* One of the realities of applying model-based decisions is that users have preliminary opinions about how the future will look based on their knowledge. Handling this biased vision is one of the biggest challenges in managing model-based predictive projects. The most widespread bias is based on overconfidence in the power of modern AI-based Data Science algorithms and ignoring their limitations. Most people are overly optimistic in their forecasts while they significantly underestimate future uncertainty. Often, people replace forecasting with their own wishful thinking.

  Another bias is called the recent event bias (that is, the habit of humans of remembering and being greatly influenced by recent events, and forgetting the lessons from the past). For example, research findings have shown that the largest amount of flood insurance is taken out just after the occurrence of a flood and the smallest amount just before the next flood takes place.[7]

  Some subject matter experts are biased toward their hypotheses about potential process or economic drivers. If their list is not supported by the data and their favorite drivers have not been selected by the variable selection algorithms used, they want detailed explanations. It is strongly recommended that the statistical basis of the variable selection should be described carefully and the supporting data be shown. In some situations, it is possible to reach a compromise and include the expert-defined drivers in the inputs used for modeling and let the variable selection algorithms do the final selection.

- *Avoiding political overrides.* In some cases, statistical forecasts are corrected due to clear departmental political purposes. One extreme is sandbagging, when some sales departments lower the statistical forecast to reduce their sales quota in order

---

[6]A. Kordon, *Applying Computational Intelligence: How to Create Value*, Springer, 2009.
[7]S. Makridakis et al., *Forecasting: Methods and Applications*, Wiley, 1998.

to guarantee their bonuses. The other extreme is sales departments that have losses due to back orders. They prefer to raise the statistical forecast in the hope of managing inventory levels via the sales department is forecast.

Another typical case of political override is "evangelical" forecasting, defined by Gilliland[8]. In this case, top-level management provides numbers, and the forecaster's task is to justify statistically the order given by top-level management.

In the case of the raw materials forecasting project, a work process for using the forecasting models in price negotiation was proposed, discussed, and defined. It included automatic data collection, model execution, report generation, and instructions on how to use the statistical forecasts (similar to the one, given in Sect. 11.3.2).

Performance of Tracking Model-Based Decisions
A decisive step in building trust in model-based decisions is tracking their performance. Of special importance is a comparison with existing heuristic-based decisions. An example of such a comparative study for the raw materials forecasting project is shown in Figs. 11.12 and 11.13.

The objective of the study was, for the same forecasting horizon of 6 months, to compare the forecasting errors between the developed statistical models and the judgmental forecast given by the experts based on the best of their knowledge. The summary results from the model-based forecasts are shown in Fig. 11.12. This figure includes the distribution of MAPEs and summary statistics for 22 deployed models up to six months after moving them into production. As is shown, the range of the MAPEs is between 1.7 and 28% with a mean error of 10.3% (which is on the borderline of a "good" forecast). The corresponding summary results for the experts' judgmental forecasts, shown in Fig. 11.13, illustrate an inferior performance, with a broader error range between 1.7 and 39.3% and a higher mean MAPE of 16.3%. This

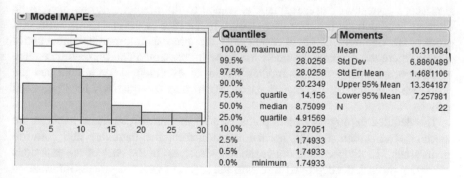

Fig. 11.12 Six-month average MAPEs of forecasting models after model deployment

---

[8]M. Gilliland, *The Business Forecasting Deal*, Wiley, 2010.

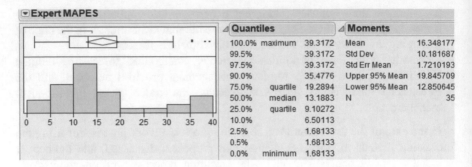

Fig. 11.13 Six-month average MAPEs of expert judgmental forecasting after model deployment

clear advantage of statistical versus heuristic-based forecasting was a critical factor in convincing even the biggest skeptics about using the benefits of model-based decision making for more profitable purchasing.[9]

It is strongly recommended to use a similar approach as an important step in building model-based decision culture.

11.4   Model Deployment

11.4.1   Model Deployment Process

Effective model deployment requires a systematic approach based on well-defined steps. An example of such a work process is shown in the mind map of Fig. 11.14.

It is recommended that the first step, planning the deployment of the models, should be started at the beginning of the project. The deployment environment, user interface, frequency of use, regulations related to the predictions, etc. have to be discussed and defined in the project charter. In the case of the emissions estimation project, the deployment environment was the process information and control system where the developed models would be implemented and become accessible to the process operators. The models would be executed every minute and the predictive accuracy for the emissions needed to pass the quarterly environmental tests.

The detailed deployment plan, prepared after model development, includes the specific requirements for the needed infrastructure: the hardware and software environment, IT support, and organizational structure. In the case of the emissions estimation project, the deployed hardware and software were the process information system. The deployed models need to be supported by the corporate IT division

---

[9]A. Kordon, Applying data mining in raw materials in prices forecasting, invited talk at SAS Advanced Analytics Conference 2012, Las Vegas, October 2012.

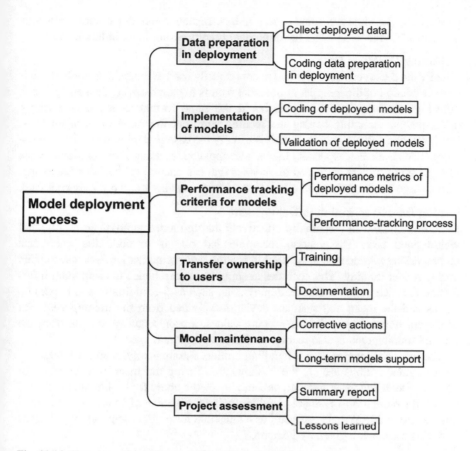

Fig. 11.14 Key steps of a well-organized model deployment process

responsible for process information and control systems. The organizations related to model deployment are the unit operation, unit finance, Data Science, and IT support departments.

Another important section in the deployment plan is devoted to its organization. This includes the specific definition of the tasks to be performed and the necessary resource allocation. The sequence of the tasks for deployment follow the work process in Fig. 11.14. Finally, the deployment plan must include the deployment schedule, based on the defined tasks.

The next steps of the model deployment process (with the exception of the last step, discussed briefly in Sect. 11.1.4) are considered in this chapter.

11.4.2 Data Preparation in Deployment

Dealing with the big differences between development and deployment in data collection and preparation is one of the biggest challenges in moving models into production. The most important actions to be taken are organizing reliable collection of the model-related data and handling data preprocessing in an online mode.

Collecting Deployed Data
Usually the deployed data is limited to data directly used in the models applied and is a small fraction of the long list of potential factors explored during data analysis and model generation. The data must be in the same format as the data used for development, including having the records represent the same units of analysis. For automatic data collection, which is usually the case in deployment, data consistency check is a must, especially in business applications using economic data. Some of these types of data are based on indexes that can change their basis of indexing. This change has to be detected by the data consistency check and properly corrected.

Coding Data Preparation in Deployment
Handling online data preparation effectively during model deployment is still a not well-defined task. The generic recommended rule is to code the same data preprocessing parameters, to be used in scaling and normalization, as those applied during model training. This includes keeping in lookup tables the important information for each variable used in deployment, such as the minimum and maximum values and the mean and standard deviation obtained from the training data. For consistent online imputation of missing values, a representative sample from the original training data is also recommended to be kept and used.[10]

The most challenging task is handling outliers automatically in online mode. The most popular options are (1) the 3 Sigma rule (using the mean of a variable for normal criteria and three standard deviations as a threshold for outlier identification) and (2) the more robust Hampel identifier (using the median of a variable for normal criteria and its MAD scale estimator as a threshold for outlier identification). Details of these methods are given by Pearson.[11]

11.4.3   Implementation of Models in Production

Implementing the developed models in the production environment requires two key steps: (1) transferring the models from the development software setting to the deployment software products and (2) validating the transferred models.

Coding Deployed Models
Transferring developed models to a different deployment software environment can be done automatically in some software tools or manually by data scientists. The software languages used vary from the most popular recent option of Python to the old-fashioned Visual Basic for Applications (VBA). The key options for the

---

[10]D. Abbott, *Applied Predictive Analytics: Principles and Techniques for the Professional Data Analyst*, Wiley, 2014.

[11]R. Pearson, *Mining Imperfect Data*, SIAM, 2005.

deployment software environment of AI-based Data Science models are briefly discussed below:

- *Model development platform.* A possible method of deployment is in the predictive modeling software. This is the simplest of the deployment methods because it does not even require coding of the deployed models. However, this option is limited to small-scale, pilot-type applications. In addition, it requires training of the final user in the basics of the model development platform.
- *Programming language coding.* When models are deployed outside of the model development platform, they need to be coded in another programming language. The most used languages for encoding models are PMML (Predictive Model Markup Language), SAS, Python, and R. Of special importance is the PMML option, defined in 1997. This is based on an XML document designed to support data preparation for predictive modeling, model' representation, and scoring. Most software platforms support the encoding of models into PMML.
- *Process information systems coding.* For real-time applications in manufacturing the common deployment environment is a process information system. Some of these use specialized languages for encoding models but recently most of them have become able to integrate Python-coded models. The complexity and execution speed of the models have to be considered as limiting factors.
- *Enterprise-wide deployment systems.* Complex enterprise-wide systems include a large number of models deployed across distributed hardware. Usually, deployed models are generated and distributed within the software deployed. The cost of these systems is very high.
- *The cloud.* A growing recent trend is to deploy models in the cloud. The key advantage is that the hardware the model run on is in the cloud rather than on a server within the organization. Another advantage is scalability: as the number of usages the model increases, more cloud resources can be purchased to keep up with the demand.
- *IoT.* This new application area requires different model deployment solutions. The coded models should include additional scripts for automatic execution under the proper conditions in the proper segment of IoT.

Validation of Deployed Models
- *Encoding validation models.* This includes validating data flows, model structures and parameters, and the user interface.
- *Validation of performance of deployed models.* This aligns the performance of the developed and corresponding deployed models. The validation is based on both the training and the test data sets used during model development. It is assumed that the deployed model has been validated offline if the performance metrics obtained with the developed model on the same data do not differ significantly. An additional validation of the predictive performance of the deployed model on new data from the production environment is strongly recommended.

11.5   Criteria for Tracking Models Performance

11.5.1   Performance Metrics for Deployed Models[12]

Typically, online models operate at two frequencies: a high frequency with sampling time T_o, driven by the data collection speed for the model inputs x (usually low-cost measurements such as temperatures, flows, and pressures), and low frequency with sampling time T, driven by the data collection speed for model output y (usually high-cost measurements from gas chromatographs and other types of laboratory analysis). For some industrial applications the difference between the sampling times could be significant. The usual sampling time T_o for input data is in the range of minutes to several hours, but the sampling time for the output measurements T could be days or even months. This creates one of the biggest issues in tracking the performance of online models—the lack of model performance information for the majority of the time. As a result, it is necessary to assess the model quality based on measures related only to the available inputs at a sample kT_o and the corresponding model prediction $\hat{y}\,(k + 1)T_o$.

A generic structure of an online model supervisor, which includes the key self-assessment capabilities needed for automatic performance tracking, is shown in Fig. 11.15. The high-frequency real-time data flow is shown above the arrows: x (kT_0) is the vector of measured inputs at sample kT_o $\hat{y}$ $(k + 1)T_o$ is the model prediction with its confidence limits cl and $s(k + 1\ T_0)$ is the state of the model.

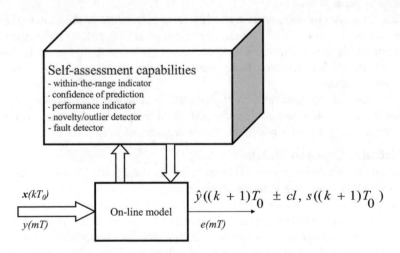

Fig. 11.15   Generic performance-tracking capabilities for self-assessment of models

[12]This subsection is based on the paper A. Kordon et al. Empirical models with self-assessment capabilities for on-line industrial applications, *Proceedings of CEC 2006*, Vancouver, pp. 10463–10,470, 2006.

The low-frequency flow is shown below the arrows and includes the output measurement at sample mT and the corresponding prediction error for that sample, $e(mT)$. The key performance metrics for deployed models are discussed below.

Within-the-Range Indicator

The purpose of this indicator is to evaluate whether the inputs are within the training range of the developed model. The necessity for the within-the-range indicator is due to the limited number of operating conditions usually available for training.

The estimates based on known operating conditions inside the within-the-range indicator are assumed to be reliable. The out-of-range data is explicitly identified by the indicator, and only the trustworthy within-the-range data is presented to the deployed model to calculate the prediction. Another function of the within-the-range indicator is that it can serve to prompt retraining or redesign of the model. If the out-of-range data is more than some acceptable threshold, a new redesign procedure is recommended.

There are different ways to calculate this indicator. One possible way, implemented for performance tracking of a predictive model of a chemical process is discussed below.[13]

In this implementation of the within-the-range indicator, it is assumed that the maximum confidence of 1 is in the middle of the range, and that the confidence decreases linearly to 0 outside the range. The following simple solution was implemented and has proved adequate:

$$\text{Within} - \text{the} - \text{range input x}$$
$$= 1 - \mid (x - x_{\text{middle range}}) \mid) \mid / (0.5 * (x_{\max} - x_{\min})) \tag{11.1}$$

$$\text{If (Within} - \text{the} - \text{range input x)} < 0 \text{ then}$$
$$\text{Within} - \text{the} - \text{range input x} = 0 \tag{11.2}$$

The overall within-the-range indicator of the online model is the product of the within-the-range indicators of all inputs, i.e., if any of the inputs is outside the training range, the predictions are deemed to be unreliable.

Confidence Limits of Prediction

In order to operate reliably in the absence of frequent output measurements, it is highly desirable to have some measure of the confidence of the model's predictions. Different modeling methods offer various ways to estimate confidence limits. The most used option is the confidence limits of a linear model calculated for a selected probability level. An example of predictions with confidence limits with 95% probability, generated by the linear emissions estimation model, is shown in Fig. 11.16.

---

[13] A. Kalos, A. Kordon, G. Smits, and S. Werkmeister, Hybrid model development methodology for industrial soft sensors, in *Proceedings of the ACC'2003,* Denver, CO, pp. 5417–5422, 2003.

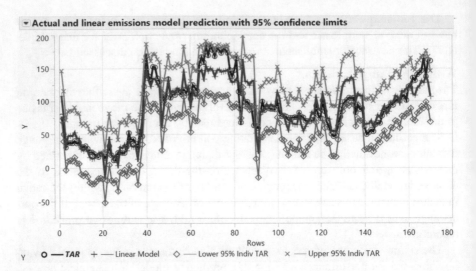

Fig. 11.16 Predictions with confidence limits from the linear emissions model

Prediction Error Performance Indicator

The most used performance tracking metric is based on the prediction error of the deployed model. The prediction error $\varepsilon(mT)$ is defined as:

$$\varepsilon(mT) = y(mT) - \widehat{y}(mT) \tag{11.3}$$

where $y(mT)$ is the measured output and $\widehat{y}(mT)$ is the model prediction at time mT. In most cases there is a time delay between the measured time mT and the current real time kT_o, so actually the model prediction at that moment $\widehat{y}(mT)$ is aligned with previous values. Different statistical measures of this error, such as RMSE, r^2, or MAPE, can be used for assessment of model performance over selected period of time.

Novelty/Outlier Detector

The objective of this indicator is to interpret properly in real time any incoming data that is outside the range used for model development. The key difference of this performance-tracking metric from the *within-the-range* indicator is that the *novelty* detector processes the out-of-range data in order to improve the robustness of the model by either including that data for further adaptation or excluding it as outliers. Of special importance are those data points that differ significantly from the known ranges, since they can influence considerably a model's predictions. In principle, they should not be used for prediction. The interpretation of this data, however, could be entirely different. Some of the data points could be treated as outliers and removed. The assumption is that these data points are based on human errors, measurement faults, or process upsets. Other data points could be interpreted as novelties and kept for future retraining or complete redesign of the models.

Unfortunately, automatic classification of abnormal data points as outliers or novelty data is very difficult and problem-specific. One obvious difference is in their patterns. Usually, outliers appear as individual data points with very large deviations. In contrast, novelty data shows a trend, since there is a transition period for any operating-regime change it may eventually represent.

Fault Detector of Deployed Models

This is the performance indicator that sends critical messages about potential faults in deployed models. There are many ways to organize this performance metric for deployed models. One possible way is by using inputs sensor validation. Both principal component analysis and auto-associative neural networks[14] have been used for sensor validation, fault identification, and reconstruction. If an input sensor fails, it is identified by the PCA or auto-associative neural network model developed, and is reconstructed using the best estimates obtained from its correlation with other sensors.

Another possible way to organize online model fault detection is by constructing an ensemble of symbolic regression predictors with different input sensors. The basis of this approach is the hypothesis that disagreement of model predictions in an ensemble constructed from diverse models indicates some predictive-performance issues. To quantify this idea, a model disagreement indicator is added to the performance assessment metrics.

The role of the model disagreement indicator is to detect areas where the predictions are unacceptable due to degradation of the performance of the ensemble of deployed models outside the training range. The model disagreement indicator is the standard deviation of all models in the ensemble, and a critical limit is defined to quantify the effect. The critical limit is equal to the average plus three times the standard deviation of the model disagreement indicator in the training data.

An example of such an ensemble-based deployed-model fault detector is shown in Fig. 11.17. The ensemble designed used three symbolic regression models with different inputs. To test the robustness of the ensemble in the presence of bad measurements, one of the inputs (variable 18) was fixed between samples 1100 and 1300 in Fig. 11.17. Because variable 18 was used in model 3, but not in models 1 and 2, the median of the three models (which is the predicted Y in Fig. 11.17) was able to give a robust estimate of the true measurement. At the same time, the model disagreement indicator specified abnormal process conditions during this period and raised a flag when the indicator was above the critical limit.

---

[14]Details of auto-associative neural networks are given in S. Heykin, *Neural Networks and Learning Machines*, third edition, Pearson, 2016.

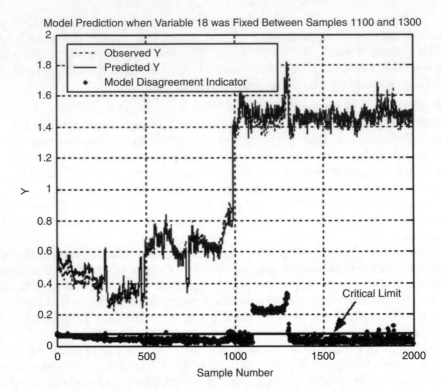

Fig. 11.17 Performance of model disagreement indicator in the case of sensor malfunction

11.5.2   Performance-Tracking Process for Models

The question of whether deployed models are working properly breaks down into three questions (1) whether the hardware is working correctly, (2) whether the models are running, and (3) whether the data being fed into the models is as expected. Because models are often trained on static snapshots of data, their predictions typically become less accurate over time as the environment shifts away from the conditions that were captured in the training data.

After a certain period of time, the error rate on new data exceeds a predefined threshold, and the models must be retrained or replaced. Champion–challenger testing is a common model deployment practice, in which a new, challenger model is compared against a currently deployed model at regular time intervals. When a challenger model outperforms a currently deployed model, the deployed model is replaced by the challenger, and the champion–challenger process is repeated. Yet another approach to refreshing a trained model is through online updates. Online updates continuously change the value of model parameters or rules based on the values of new, streaming data. It is prudent to assess the trustworthiness of real-time data streams before implementing an online modeling system. The most popular statistical method for performing this task is A/B testing.

A/B Testing

A/B testing is often used to answer the question "Is my new model better than the old one?" The performance of both models is split into two groups, A and B. Group A is routed to the old model (the champion model), and group B is routed to the new model (the challenger model). Their performance is compared and a decision is made about whether the new model performs substantially better than the old one. The decision is based on classical statistical hypothesis testing.[15] This decides between a null hypothesis (the new model doesn't change the average value of the metric) and an alternate hypothesis (the new model changes the average value of the metric). Most of the time, A/B tests are formulated to answer the question, "Does this new model lead to a statistically significant change in the key metric?"

A known issue in A/B testing is the level of false positives that can be tolerated. In model deployment A/B test, a false positive might mean switching to a model that should increase revenue when it doesn't. A false negative means missing out on a more beneficial model and losing out on a potential revenue increase. Fortunately, the levels of false positives and false negatives are controlled by the significance level, which by default is 0.05.

The majority of users perform A/B testing to evaluate multiple models in parallel. This means maintaining previous versions of production data with different versions of developed modes and comparing the end performance depending on the model. The decision about model selection is based on the results of the A/B test.[16]

11.6   Model Maintenance

11.6.1   Transfer of Ownership of Models

Many inexperienced data scientists think that their responsibilities end with the completion of model development. In fact, the strategy of "Throw the model and run" is a clear recipe for failure, poisoned connections with the clients and lost opportunities. The right approach is to build long-term relationships with the users based on effective knowledge transfer, collaboration and trust. At the beginning of this process is the proper transfer of ownership of the models from the developers (data scientists) to the final users and the supporting IT organization. It is assumed that this act does not entirely release the model developers from responsibility for the deployed models' performance. Their appropriate intervention in the case of model

---

[15]D. Montgomery, E. Peck, and G. Vining, *Introduction to Linear Regression Analysis*, fourth edition, Wiley, 2006.

[16]T. Dunning and E. Friedman, *Machine Learning Logistics: Model Management in the Real World*, O'Reilly, 2017.

degradation and potential rebuild is needed and should be included in the model maintenance agreement. The responsibilities of the users ownership of the for models, such as training, proper use of the models, maintaining the necessary infrastructure and data quality, have to be clarified as well.

The key steps in the transfer of ownership of models are (1) training of the users and (2) preparing the necessary documentation on how the deployed models can be used.

User Training
It is expected that the users will be trained in the basic manipulations that can be done with the deployed models, such as run the models in manual and automatic mode, understand the results of the models (predictions, confidence limits, forecasting horizons, etc.), checking the performance of the models, generating performance reports on the models and communicating and discussing the models' performance. Managing expectations about the performance of realistic models and its potential deterioration with time is strongly recommended. It is also beneficial to describe briefly the methods by which the models have been built and their limitations.

In the case where ownership is transferred to a specialized IT organization for models maintenance and support, more detailed training is expected from the developers. In addition to the topics discussed above, it may include details of model development, procedures for performance tracking, and a description of recommended corrective actions.

Documentation of Models
All the topics recommended for user training should be documented on paper and in electronic form. The documents must be accessible across the organization to all project stakeholders.

11.6.2   Corrective Actions for Improvement of Model Performance

The key purpose of model maintenance is to counteract the expected performance degradation with time with proper corrective actions. The three main options are discussed briefly below.

Tuning of Model Parameters
The first possible option to improve degrading model performance is to keep the model structure intact but to tune the model parameters. The adjustment should be based on a snapshot of recent data. The size of the data set is problem- and model-specific. It is recommended that the procedure for model readjustment is clarified in advance. It is assumed that it will be performed by the developers offline and the decision about deployment of the candidate model will be made based on the performance improvement. After deployment, both the current (or champion) model and the candidate (or challenger) model are run in parallel until a decision

about replacement or retirement is made based on performance (see Sect. 11.6.3 for details).

Model Redevelopment
The second option to improve degrading performance of a deployed model is to rebuild the model with the selected process variables based on the original data analysis. Obviously, this requires more intensive model development efforts and cooperation with data scientists. It is also based on a more recent data set. The model redevelopment sequence is similar to the model parameter retuning sequence discussed above.

Model Redevelopment with Variable Selection
The most difficult case of improvement of the performance of a deployed model is to begin model development with a new variable selection. It is assumed in this case that process changes have caused significant changes in the correlation among related process variables. As a result, different factors contribute to the target variable and an altered variable selection is required. A new data collection is needed with a revised set of potential drivers, followed by model redevelopment. The new model is deployed according to the process discussed above.

11.6.3   Long-Term Support of Models

In order to deliver effective long-term support for deployed models, creating a container of several models running in parallel in the deployment environment is suggested.[17] A typical container of deployed models includes the currently used model (the champion model) and another model with different parameters or structure (the challenger model). The container also includes a comparison block based on a calculated performance metric for the models. Based on the results of the comparison, a suggestion is made about three possible scenarios: (1) to apply the results from the current model, (2) to replace the champion model with the challenger, and (3) to retire the champion.

This suggested structure is shown in Fig. 11.18, and the nontrivial scenarios (2) and (3) are discussed below.

In Fig. 11.18, the predictions of both the champion and the challenger models are compared with the actual target variable and the prediction errors are calculated. The model selection decision is based on the performance metric using the prediction errors for some period of time.

If the challenger model has shown for a sufficient period of time a competitive accuracy relative to the current champion, it can be gradually put into production. In this case of scenario 2, the new model is progressively used more and more in the

---

[17]T. Dunning and E. Friedman, *Machine Learning Logistics: Model Management in the Real World*, O'Reilly, 2017.

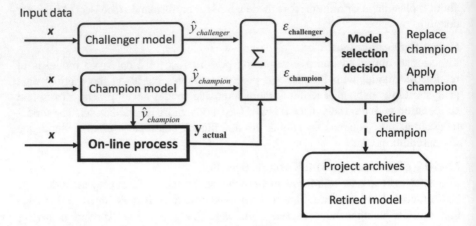

Fig. 11.18 Management sequence of deployed models

production traffic while the previous champion is used less and less. However, after full replacement, it is recommended to keep the old champion still running in case a decision to roll back is made. The dethroned champion model will still handle all the production traffic but its outputs will be ignored. Its performance should still be monitored to verify that it is safe to keep the model as a fallback. Another action item in the model replacement scenario is to select and deploy a new challenger model.

In scenario 3, the former champion is completely removed from the online environment and retires into offline archives. It is recommended to keep all champion models to analyze the progress of their parametric and structural evolution.

11.7   Common Mistakes

11.7.1   Insufficient Model Deployment Efforts

In many cases model deployment is not organized very well and as a result, the value creation opportunities are either reduced or totally missed. One of the reasons for this is the lack of a well-defined process for and literature on this final step of the generic Data Science workflow. Data scientists usually follow some specific process built into the software packages they use for deployment and ignore all other complex issues related to moving models into production.

Another reason for low-quality model deployment effort is a lack of agreement and good coordination between the developers, the IT organization that takes care of the deployed models, and the final users. In order to improve the deployment process, data scientists must lead by offering appropriate information and action items.

11.7.2 No Long-Term Support

One of the symptoms of inexperienced Data Science project leadership is ignoring the critical step of organizing long-term support for the deployed solutions. As a result, the expected value creation flow is limited to the short-term maintenance period initially agreed between the users and the developers. The obvious solution is to organize professional long-term support by an IT organization either inside the business or outside it as a service. This process should be done immediately after finishing the model development step.

11.7.3 Poor Model Performance Tracking

Another common mistake is poorly organized tracking of the performance of deployed models. A typical case is focusing on the prediction error performance indicator only and ignoring the performance metrics of the deployed model discussed in Sect. 11.5.1. As a result, identifying the root causes of model performance degradation is practically impossible.

11.7.4 Low-Quality Training of Final Users

Reducing significantly the effort put into knowledge transfer between the model developers (data scientists) and the users of the deployed models and their supporting IT team has a substantial negative effect on the quality of model use and maintenance. Minimal documentation and short training without the needed details often lead to inappropriate interpretation and manipulation of the modeling results by the users and reduced value creation. This leads to inefficient maintenance and more frequent intervention by data scientists to save the degrading models.

11.7.5 Transferring Ownership Without Model Exploration

Ignoring model exploration before moving the models into production sends a clear message that the data scientists are not interested in the users' feedback about the developed models. As a result, a big opportunity to demonstrate the full potential of the models' performance and give the users a chance to explore it is missed. The negative consequences of this decision are difficult to compensate for during model deployment. In general, unconvinced users question the performance all the time and exaggerate poor results. It takes a lot of effort and almost perfect performance of the models to obtain their firm support. Unfortunately, without the needed backing, the deployed models are not used consistently and the solution is gradually rejected.

11.8   Suggested Reading

- D. Abbott, *Applied Predictive Analytics: Principles and Techniques for the Professional Data Analyst*, Wiley, 2014.
- C. Chase, *Demand-Driven Forecasting: A Structured Approach for Forecasting*, second edition, Wiley, 2013.
- T. Dunning and E. Friedman, *Machine Learning Logistics: Model Management in the Real World*, O'Reilly, 2017.
- M. Gilliland, *The Business Forecasting Deal*, Wiley, 2010.
- R. Pearson, *Mining Imperfect Data*, SIAM, 2005.
- J. Taylor, *Decision Management Systems: A Practical Guide to Using Decision Rules and Predictive Analytics*, IBM Press, 2012.
- T. Wallace, *Sales and Operations Planning: Beyond the Basics*, T.F. Wallace & Company, 2011.
- A. Zheng, *Evaluating Machine Learning Models*, O'Reilly, 2015.

11.9   Questions

Question 1
Is it possible to create value by an AI-based Data Science project without model deployment?

Question 2
When does long-term support for the deployed models have to be organized?

Question 3
What are the advantages of model-based decisions?

Question 4
What is the difference between data preparation in model development and in deployment?

Question 5
What are the key corrective actions to reduce performance degradation of models?

Part III
AI-Based Data Science in Action

Chapter 12
Infrastructure

> *Technology is like a fish. The longer it stays on the shelf, the less desirable it becomes.*
>
> *Andy Rooney*

A typical pattern of the recent AI-driven invasion is the fast pace of technology development. New, spectacular hardware options, big data-processing capabilities, and software "magic" have appeared almost every day. As a result, any attempt to recommend specific solutions or packages at the time of writing this book will be corrected soon by the next wave of technology development. Fast technology dynamics is the only reliable prediction for the AI-based Data Science infrastructure in the foreseeable future. The proposed solution in discussing this topic is to focus on the generic features and options related to Data Science-based hardware, databases, and software and avoid considering concrete products. It is assumed that the reader will search for the best recent options at the time when infrastructure decisions are made.

As we discussed in the Preface, the dominant approach in this book is the integrated view of analyzing Data Science projects, based on three components—scientific methods, people, and infrastructure. This is called the Troika of Applied Data Science, and its key factors are shown in Fig. 12.1.

The core and new AI-driven Data Science methods have been discussed in Chap. 3, and the integration of methods in Chap. 4. This chapter is focused on reviewing the key issues in developing a proper AI-driven Data Science infrastructure. The last component of the Troika, people is discussed in Chap. 13.

The first objective of this short chapter is to clarify the hardware opportunities for developing and deploying AI-driven Data Science systems, while the second is to discuss the possible options for databases. The third objective is to analyze the possibilities for building an adequate AI-driven Data Science software infrastructure.

Due to its size and complexity, the important issue of cyber security infrastructure is not covered in the book.[1]

---

[1] A recommended reference on this topic is W. Stallings, *Effective Cybersecurity: A Guide to Using Best Practices and Standards*, Addison-Wesley, 2018.

© Springer Nature Switzerland AG 2020
A. K. Kordon, *Applying Data Science*,
https://doi.org/10.1007/978-3-030-36375-8_12

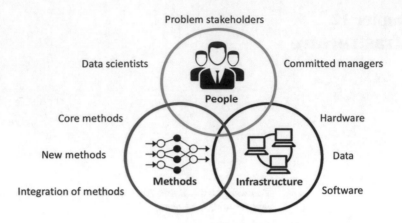

Fig. 12.1 The key factors of the Troika of Applied Data Science

12.1   Hardware

In general, AI-driven Data Science systems require high computational power, especially when evolutionary computation and deep learning methods are used. There are three key options for building a hardware infrastructure to fulfill the expected number-crunching needs of AI-driven Data Science: (1) specialized chips, (2) distributed computing, and (3) cloud computing. These are discussed briefly below.

12.1.1   Specialized Chips

A well-known recent trend in hardware is that the growth of computational power of the computer "brain"—its CPU—has slowed down. Obviously, the design of computer chips is close to the saturation curve of the famous Moore's law[2] due to reaching the physical limits of transistor density. The key option for increasing computational power is by accelerating specific calculations using specialized chips. The most popular solution is to use graphical process units.

A GPU is a specialized type of microprocessor, primarily designed for quick image processing. GPUs appeared as a response to graphically intense applications that put a burden on the CPU and degraded computer performance. They became a way to offload those tasks from CPUs, but modern graphics processors are powerful enough to perform rapid mathematical calculations for many other purposes, including for AI-based technologies.

---

[2]Moore's law is the observation that the number of transistors in a dense integrated circuit doubles about every 2 years.

GPUs are not replacements for the CPU architecture. Rather, they are powerful accelerators for the existing infrastructure. GPU-accelerated computing offloads computanionally intensive portions of the application to the GPU, while the remainder of the code still runs on the CPU. From a user's perspective, applications just run much faster.

Modern GPUs provide superior processing power, memory bandwidth, and efficiency over their CPU counterparts. They are 50–100 times faster in tasks that require multiple parallel processes, such as machine learning and big data analysis. The key vendors of this hardware are continuously developing new, improved versions of GPUs with different architectures and configurations. They ccan be used in graphic cards with single or multiple GPUs, linked to workstations or as a cloud service.

Another potential hardware solution for increasing computational power is by using specialized chips for fast computation in selected AI methods. An example is the current development by some key chip vendors of so-called neuromorphic chips, which attempt to more closely resemble how a real brain functions. Similarly to how neuroscientists think the brain functions, the chip transmits data through patterns of pulses (or spikes) between neurons. These chips are based on a fully asynchronous neuromorphic many-core mesh that supports a wide range of sparse, hierarchical, and recurrent neural network topologies with each neuron capable of communicating with thousands of other neurons. Each neuromorphic core includes a learning engine that can be programmed to adapt the network parameters during operation, supporting supervised, unsupervised, reinforcement, and other learning paradigms.

12.1.2   Distributed Computing

The next option for increasing the computational power of hardware is by parallelizing appropriate calculations and data and distributing them across multiple computers. The key databases and software techniques for accomplishing this are discussed in the next section. However, this approach will not work for some AI-based techniques that are not inherently parallel.

12.1.3   Cloud Computing

Cloud computing, or what is simply referred to as the cloud, can be defined as an Internet-based computing model that largely offers on-demand access to computing resources. These resources comprise many things, such as application software, computer memory resources, and servers and data centers. Cloud service providers usually adopt a 'pay-as-you-go' model, something that allows companies to scale their costs according to need. It permits businesses to bypass infrastructural setup costs, which were a part of the only solution prior to the advent of the cloud.

From modeling perspective, cloud computing gives opportunities to use almost unlimited computational power and to interact with big data, with easy deployment on a large scale. It is expected that the share of this hardware option in building AI-based Data Science businesses will significantly grow in the future.

12.2   Databases

Organizing access to data is a key objective in building an AI-driven Data Science infrastructure. Most businesses have inherited classical relational databases with predominantly structural numerical and textual data. The current trend toward analyzing unstructured heterogeneous data, such as images, video, and audio, requires a different organization of big data. The two key options for building a database infrastructure are discussed briefly below.[3]

12.2.1   Relational Databases

A relational database is based on the relational model, in which the raw data is organized into sets of tuples, and the tuples are organized into relations. This relational model imposes structure on its contents, in contrast to the unstructured or semistructured data of the various NoSQL architectures.

Database management systems and their relational counterparts (RDBMs) were the most widely used database systems for a long time, beginning in the 1980s. They are generally very good for transaction-based operations. These databases have the following limitations: they are relatively static and biased heavily toward structured data, they represent data in nonintuitive and nonnatural ways, and they are computationally inefficient. Another downside is that the table-based stored data does not usually represent the actual data (i.e., domain/business objects) very well. This is known as the object-relational impedance mismatch, and it requires a mapping between the table-based data and the actual objects of the problem domain. Database management systems usually include Microsoft SQL Server, Oracle, MySQL, etc.

The alternative to the classical relational databases is the NoSQL database technology. NoSQL is a term used to describe database systems that are non relational, are highly scalable, allow dynamic schemas, and can handle large volumes of data accesses of high frequency. They also represent data in a more natural way and are largely used for high-scale transactions.

---

[3]A recommended reference for understanding databases and building a proper infrastructure is W. Lemahieu, B. Baesens, and S. vanden Broucke, *Principles of Database Management*, Cambridge University Press, 2018.

NewSQL is a relatively new type of database management system. Such systems try to blend the best characteristics and the querying language of relational database management systems with the highly scalable performance of NoSQL databases. It is still unknown as to whether NewSQL will gain enough popularity for adoption and acceptance like relational and NoSQL databases have done.

Practitioners in big data, however, need entirely new approaches for high-scale data storage, processing capabilities, and analytics of enormous amounts of data. The principles of one of the most popular such systems (Apache Hadoop) are discussed briefy below.

12.2.2   Big Data Frameworks

Traditional database management systems are used to store and process relational and structured data only. However, recently business applications have needed to deal with lots of unstructured data such as images, audio files, and videos. Unfortunately, traditional system are unable to store and process these kinds of data. There are several frameworks available to solve this problem, and the most popular is Hadoop.

Hadoop is a very powerful open source platform built on Java technologies and is capable of processing huge volume of heterogeneous data in a distributed, clustered environment. The Hadoop framework consists of the Hadoop core components and other associated tools. In the core components, Hadoop Distributed File System (HDFS) and the MapReduce programming model are the two most important concepts. The generic structure of Hadoop is shown in Fig. 12.2.

HDFS takes care of the storage part of the Hadoop architecture. HDFS stores files across many nodes in a cluster. Each file is stored in HDFS as blocks. The default size of each block is 128 MB in Apache Hadoop 2.9.

An example of how the data is distributed is shown in Fig. 12.3. After the original file of 320 MB has been divided into data blocks, as shown in Fig. 12.3, these data blocks are then distributed across all the data nodes present in the Hadoop cluster. NameNodes and DataNodes are the core components of HDFS. NameNodes manage DataNodes, which store the data in specific data blocks.

Fig. 12.2 Generic Hadoop framework

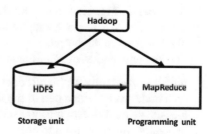

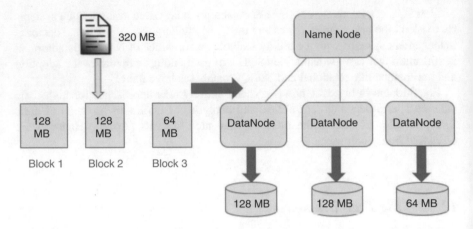

Fig. 12.3 Example of data distribution in HDFS

What are the advantages of HDFS, and what makes it ideal for distributed systems?

- *Fault tolerance*. Each data block is replicated three times in the cluster (everything is stored on three machines/DataNodes by default.) This helps to protect the data against DataNode (machine) failure.
- *Disk space*. The size of the needed disk space is flexible and can be changed by adding or removing data nodes.
- *Scalability*. Unlike traditional database systems, which cannot scale to process large data sets, HDFS is highly scalable because it can store and distribute very large data sets across many nodes that can operate in parallel.
- *Generic*. HDFS can store any kind of data, whether structured, semistructured, or unstructured.
- Cost-effective. HDFS has direct attached storage and shares the cost of the network and computers it runs on with MapReduce. It is also open source software.

The MapReduce paradigm has two main components. One is the Map() method, which performs filtering and sorting. The other is the Reduce() part, designed to make a summary of the output from the Map part. The generic MapReduce structure is shown in Fig. 12.4.

As shown in Fig. 12.4, the input data is divided into partitions that are Mapped (transformed) by a mapper and Reduced (aggregated) by defined functions, and finally delivered to the output.

The key advantages of MapReduce are:

- *Parallel processing*. In MapReduce, jobs are divided among multiple nodes, and each node works with a part of the job simultaneously and hence helps to process the data using different machines. As the data is processed by different machines in parallel, the time taken to process the data is reduced by a tremendous amount.

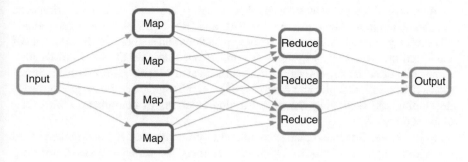

Fig. 12.4 MapReduce structure

- *Cost-effective solution*. Hadoop's highly scalable structure also implies that it comes across as a very cost-effective solution for businesses that need to store the ever-growing data dictated by today's requirements.
- *Fast*. MapReduce is often on the same servers where the data is located, resulting in faster data processing.

While Hadoop brings the above-mentioned benefits, it comes with the disadvantages of a steep learning curve and a lot of cost to architect the solution and hire additional staff to create it. Regular businesses with traditional RDBMS data warehouses or data appliances have to hire or train to build entirely new skills. Hadoop is an entire ecosystem on its own and staff is typically just too costly and complex for most businesses, especially those not using big data.

12.3   Software

The biggest challenge in building an AI-driven Data Science infrastructure is choosing the proper software tools. There are many different options and the number is growing. Two key selections should be made by data scientists: (1) appropriate software platforms that include a variety of algorithms and data analysis capabilities, and (2) appropriate programming languages for implementing the AI-based Data Science work process.

12.3.1   Integrated Data Science Software Platforms

As we have discussed in the book, solving complex business issues requires a broad selection of AI-based Data Science methods and the integration of many of these. Several integrated software platforms offer the capabilities needed to develop and (for some of them) to deploy AI-based Data Science models.

At one end of the spectrum in this category are high-cost, enterprise-wide commercial systems (examples are the SAS and IBM SPSS platforms) that include all necessary tools for large-scale global applications and support. At the other end of the spectrum are low-cost systems (examples are RapidMiner and KNIME plat-forms) that offer the key algorithmic capabilities at affordable prices. Three industry giants (Microsoft, Google, and Amazon) also offer integrated software platforms for developing and deploying machine learning solutions in combination with their cloud services.

Most of these integrated platforms allow fast prototype model development based on built-in algorithmic blocks. Often programming languages and skills, nor deep knowledge of the selected algorithmic blocks, are not needed for developing the solution. Another advantage is the minimal training of data scientists required.

It is strongly recommended that a proper integrated Data Science software package should be a part of the infrastructure built. A comparative analysis of the other recent alternative—open source AI-based Data Science algorithmic pack-ages—is given in Sect. 12.3.3.

12.3.2   Data Science-Related Programming Languages

The other key selection to be made for the software infrastructure is about the programming languages that will be mostly used. The industrial reality is that developing and deploying AI-based Data Science solutions cannot be done with integrated software platforms only. Data scientists need to use coding in corrections to systems, integration, support, and many situations outside the automatic solution generation offered by integrated packages. Without programming skills, data scien-tists become prisoners of the built-in capabilities of the integrated platforms, with limited degrees of freedom to improve model performance and introduce new methods.

Two programming languages, Python and R are the most popular choices from the long list of potential options for AI-based Data Science infrastructures. Python is a general-purpose language that balances the flexibility of conventional scripting with the numerical power of a good mathematical package. Its programming style is close to other popular languages such as PHP, C#, and Java, and data scientists with programming skills in them can easily transfer to Python. There is a recent trend toward growing use of Python and openness of the integrated packages to Python scripts and to models coded in Python. Most of the open source software related to AI-based Data Science is also based on Python.

The other popular programming language option for AI-based Data Science development is R. This language was designed by and for statisticians, and is natively integrated with a graphics capability and extensive statistical functions. However, the syntax is clunky, the support for strings is limited, and the type system is obsolete. While its popularity among data scientists is declining, R is still the main tool used by the statistical community.

Businesses also use other programming languages related to AI-based Data Science. Some of them, such as MATLAB (Octave) and Mathematica, are of use to generic number crunching and have excellent packages for the most of AI-based Data Science methods. Others, such as SAS, are specific to the vendor and include rich libraries related to statistics and machine learning.

12.3.3   Recommendations for AI-Based Data Science Software Selection

A key dilemma, especially for businesses introducing AI-based Data Science, is wether to invest in commercial software or to start with open platforms. Both options are available, and each has its strengths and weaknesses. Open-source software is free and has all the latest algorithms, but may not scale well, or be as maintainable as the commercial tools over time. Commercial software is often expensive and may be slightly behind in implementing new, unexplored methods, but has known quality, is more maintainable, and is more readily accepted by IT support than open ource tools are.

It has to be considered that there is a cost associated with the "free" open source tools that is not obvious up front—for example, the cost of integration, training and maintenance.

Another factor that has to be considered in software selection is the capacity of the vendor to introduce, develop, and validate new AI-based Data Science methods. Only the Olympians and the big vendors, such as SAS, have the R&D capacity to allocate the necessary R&D resources to deliver reliable new, complex algorithms. On the opposite side, most of the open source methods offered by individuals are not well documented or tested. Using these scripts without comprehensive validation is a risky business.

12.4   Infrastructure as a Service

For many businesses, AI-driven Data Science seems close to rocket science, appearing expensive and talent demanding. They understand the potential value that this set of technologies may create for the business but do not want to allocate resources for building the needed infrastructure. Fortunately, the trend toward making everything-as-a-service gives them an opportunity. One of the available services covers machine learning methods and is called machine learning as a service (MLaaS).

Machine learning as a service is an umbrella definition of automated and semi-automated cloud platforms that cover most of the infrastructure issues such as data preprocessing, model training, and model evaluation, with generation of prediction or classifications. Prediction results can be bridged with the business's internal IT infrastructure through appropriate APIs.

The Amazon Machine Learning service, Azure Machine Learning, and Google Cloud AI are three leading cloud MLaaS platforms that allow fast model training and deployment with little to no data science expertise. It is expected that these types of services will grow and offer a broad range of capabilities in the future.

12.5   Common Mistakes

12.5.1   Investing in an Inappropriate Infrastructure

The most common mistake is either investing in a larger infrastructure than the business needs or the opposite, investing in a smaller infrastructure. Usually this mistake is made when the decision-makers do not know or estimate the expected demand for AI-based Data Science capabilities. In order to avoid this misstep, it is recommended to first analyze the business needs and to learn from the experience of similar companies in applying these technologies. Another piece of advice is to be careful about overoptimistic offers from vendors, which usually exaggerate the size of the infrastructure.

12.5.2   Ignoring Maintenance Infrastructure

Another frequent mistake is not to include investment in a maintenance infrastructure. Mostly this issue relates to integrating the model development infrastructure with the existing business infrastructure in which the models will be deployed. Usually this topic needs good coordination with the corresponding IT department in the business.

12.5.3   Using Only One Software Tool/Programming Language

Some organizations, especially those invested in integrated software platforms, use only the capabilities and the programming language of that specific tool. This solution reduces the cost of introducing new methods and of training in another programming language. However, it limits the capabilities for solving complex AI-based Data Science problems that may require a broader set of algorithms.

12.5.4 *Selecting an Inappropriate Software Tool/ Programming Language*

The most frequent mistake in this category is to minimize the investment in software infrastructure and rely only on open source packages. The recent wave of data scientists just released from the academic world are big proponents of this approach, based on their experience during education. For large-scale industrial applications, however, such an infrastructure is not adequate for the size and complexity of the problems.

The opposite mistake in this category is to invest in a large enterprise-wide integrated software platform for a small or medium-size business. A combination of small-scale solutions and open source software packages is more appropriate in this case.

12.6 Suggested Reading

- F. Cady, *The Data Science Handbook*, Wiley, 2017.
- W. Lemahieu, B. Baesens, and S. vanden Broucke, *Principles of Database Management*, Cambridge University Press, 2018.
- M. Minelli, M. Chambers, and A. Dhiraj, *Big Data Big Analytics*, Wiley, 2013.

12.7 Questions

Question 1
Which AI-based Data Science methods require high computational power?

Question 2
What are the advantages of cloud computing?

Question 3
What are the key limitations of relational databases?

Question 4
What are the advantages and limitations of open source software packages?

Question 5
What is the key factor in selecting the size of an AI-based Data Science infrastructure?

Chapter 13
People

> *AI is not about replacing the human with a robot. It is about taking the robot out of the human.*
>
> *Usama Fayyad*

Effectively involving human intelligence in AI-generated solutions is one of the biggest challenges in the field of AI-based Data Science. The third component of the Troika of Applied Data Science, the people, is the most neglected topic of consideration in Data Science community, ignoring the fact that this is the critical factor for final success in real-world applications. A business can invest in the most powerful hardware and sophisticated tools and use the most advanced algorithms, but all these efforts are doomed to failure without the support of key categories of people related to AI-based Data Science, such as data scientists, problem stakeholders, and managers.

Behavior change driven by AI-based Data Science is often the hardest part of improving a company's performance on any dimension. This is the reason why only 12% of applied AI-based Data Science projects achieve or exceed a company's expectations and 38% fail by a wide margin. Many of the companies making huge investments in this area will be disappointed to discover that software tools and modern methods alone aren't enough to grow a company's fortunes.[1]

The focus of this chapter is to discuss briefly some important topics in this very broad and unexplored area. This is expected to help the reader to understand the significance of analyzing the human factor in developing AI-based Data Science solutions. The chapter includes some recommendations for evaluating behavior patterns and for improved communication between categories of people related to AI-based Data Science, by effective team building and marketing.

The first objective of this chapter is to analyze the main types of response of human intelligence to applied AI-based Data Science systems, while the second is to summarize the behavior of the key categories of people related to AI-based Data Science projects. The third objective is to discuss the issue of creating and operating

---

[1] J. Deal and G. Pilcher, *Mining Your Own Business*, Data Science Publishing, 2016.

© Springer Nature Switzerland AG 2020
A. K. Kordon, *Applying Data Science*,
https://doi.org/10.1007/978-3-030-36375-8_13

teams in this area, and the fourth is to give some guidance about marketing AI-based Data Science to technical and nontechnical audiences.

13.1   Human vs. Artificial Intelligence

There is a famous story about the two founding fathers of AI, Marvin Minsky and Douglas Engelbart. Minsky declared, "We're going to make machines intelligent. We are going to make them conscious!" To which Engelbart reportedly replied, "You're going to do all that for the machines? What are you going to do for the people?" [2]

Answering this question is one of the key driving forces of applied AI. It requires a very good understanding of people's expectations from AI, and their attitudes to and interactions with applied AI-based systems. One lesson learned from industry is that, due to their built-in intelligent component, applied AI-based solutions can have different impacts on a business. The key distinction is in the way they challenge to some extent the human intelligence of the users. As a result, the success or failure of these applications depends not only on their technical accomplishments but also on the complex interactions between human and artificial intelligence.

We'll focus on three practical issues related to these interactions: (1) which key weaknesses of human intelligence can be compensated by AI, (2) how human intelligence can benefit from AI, and (3) why human intelligence resists AI.

13.1.1   Weaknesses of Human Intelligence

The Wikipedia definition of human intelligence is:

> Human intelligence is the intellectual prowess of humans, which is marked by complex cognitive feats and high levels of motivation and self-awareness. Through their intelligence, humans possess the cognitive abilities to learn, form concepts, understand, apply logic, and reason, including the capacities to recognize patterns, comprehend ideas, plan, solve problems, make decisions, retain information, and use language to communicate.

These extraordinary abilities are not unlimited, however. Above all, human intelligence has weaknesses due to its biological limitations. Some of them, which are potential candidates for improvement with AI technologies, are shown in Fig. 13.1 and discussed briefly below.

---

[2] A. Burgess, *The Executive Guide to Artificial Intelligence: How to Identify and Implement Applications for AI in Your organization*, Palgrave Macmillan, 2018.

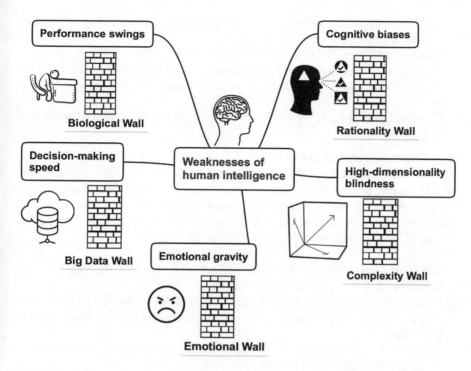

Fig. 13.1 The key weaknesses of human intelligence

Cognitive Biases

According to Wikipedia, "A cognitive bias is a systematic pattern of deviation from norm or rationality in judgment." Individuals create their own "subjective social reality" from their perception of the input. An individual's construction of social reality, not the objective input, may dictate their behavior in the social world. Thus, cognitive biases may sometimes lead to perceptual distortion, inaccurate judgment, illogical interpretation, or what is broadly called irrationality.

This process is visualized in Fig. 13.1 as hitting the Rationality Wall.

The top 10 cognitive biases, selected by the author, which are defined in Wikipedia, are listed below:

- *Dunning–Kruger effect.* The tendency for unskilled individuals to overestimate their own ability and the tendency for experts to underestimate their own ability.
- *Bandwagon effect.* The tendency to do (or believe) things because many other people do (or believe) the same.
- *Recency effect.* The tendency to weigh the latest information more heavily than older data.
- *Confirmation bias.* The tendency to search for, interpret, focus on, and remember information in a way that confirms one's preconceptions.
- *Ostrich effect.* The tendency to ignore an obvious (negative) situation.

- *Ambiguity effect*. The tendency to avoid options for which missing information makes the probability seem "unknown."
- *Clustering illusion*. The tendency to see patterns in random events (data).
- *Status quo bias*. The tendency to like things to stay relatively the same.
- *Weber–Fechner law*. The difficulty in comparing small differences in large quantities.
- *Choice-supportive bias*. The tendency to remember one's choices as better than they actually were.

High-Dimensionality Blindness

The limitations of human intelligence related to high-dimensional visualization, perception, attention, etc. are well known. The total impact is that human intelligence is "blind" when we wish to analyze phenomena that have many interactive factors and beyond some level of complication, i.e., it hits the Complexity Wall. Some of the key contributors to this type of weakness are given below:

- *3D visualization limit*. The constraints of human intelligence on imagining and identifying patterns beyond three dimensions.
- *Multivariate relationships limit*. The difficulties of human intelligence in discovering, defining, and understanding the dependencies among many factors (variables).
- *Attention limit*. The limitations of human intelligence in paying attention to or tracking a large number of factors.
- *Combinatorial limit*. Human intelligence is not very impressive in tasks that require evaluation of many alternative solutions. As IBM's Big Blue has shown long ago, even the best chess players cannot compete with computers...

Emotional Gravity

Probably the best-known limitation of human intelligence is that the rationality of human thought is entrapped in the emotional "gravity" of human behavior. If a person cannot handle the negative influence of her/his emotions and hits the Emotional Wall, the consequences can be damaging. Fortunately, the broad area of emotional intelligence delivers a popular framework and methods for controlling human emotions and reducing the effect of emotional gravity.[3]

Decision-Making Speed

Human intelligence depends on the way our brain operates. Unfortunately, the brain's processing speed is very slow in comparison with contemporary computers. In addition, its memory capabilities are not impressive and degrade with age or illness. Both of these limitations become an important obstacle when optimal decisions, based on millions of records, hundreds of diverse factors, and thousands of conditions have to be made in seconds in real time. It is practically impossible for human intelligence to accomplish this task alone, since it hits the Big Data Wall.

---

[3]The classical book on this topic is D. Goleman, *Emotional Intelligence*, Bantam Books, 1995.

Performance Swings

Another well-known weakness of human intelligence is its changing and unpredictable performance. Even the best and the brightest have their bad times due to gazillions of reasons. In general, performance swings are a result of the biological limitations of human bodies. They cannot operate continuously at top performance, since they hit the Biological Wall.

13.1.2   Benefits to Human Intelligence from AI

The majority of people in businesses look at the capabilities of AI-based Data Science as a unique opportunity to compensate for these limitations and to enhance human intelligence and improve productivity. The term "augmented intelligence" has even been used to define this cooperation between human intelligence and AI technologies. A search engine can be viewed as an example of augmented intelligence (it augments human memory and factual knowledge), as can natural language translation (it augments the ability of a human to communicate).

Most of the weaknesses of human intelligence can be compensated for by augmented intelligence. Cognitive biases can be reduced or eliminated by using model-based decisions.[4] AI-based Data Science gives human intelligence the missing high-dimensionality capabilities, as well as the ability to make fast, emotionless decisions using big data. Most of these systems raise significantly the performance level of all users and make it consistent all the time.

The positive impact of AI-based Data Science technologies on human intelligence can be a big driver in people's career development. Combining the benefits of augmented intelligence with appropriate incentives for the most productive users is the best recipe for success in applying these systems in business.

13.1.3   Resistance of Human Intelligence toward AI

Not everybody is happy with and supportive of the new opportunities offered by AI, however. For some people, the new, powerful capabilities of AI-based Data Science technologies are a direct threat to their jobs and current competencies. They look at AI as a negative amplifier, i.e., it enhances their weaknesses and ignorance, and raises the chance of decreased performance due to new skillset gaps. This attitude may evolve into an AI-driven inferiority complex with different forms of resistance, such as questioning the recommended decisions and results, exaggerating cases with wrong predictions, incorrectly using deployed solutions, or not using them at all.

---

[4]However, math model-related biases have to be considered and properly handled.

Analyzing the key factors that contribute to the resistance of human intelligence toward AI-based Data Science is of big importance for the successful application of this technology in industry.[5] Some of them, observed in many real-world applications, are discussed briefly below.

- *The business fails to reinforce AI-based Data Science applications properly.* The leadership has not sent a clear message to all employees that AI-based Data Science will not endanger their jobs, and encouraged users of new technology with corresponding benefits, proportional to the productivity gain.
- *Business users are not prepared to change their behavior.* Adopting new processes and tools can disturb even the most enthusiastic employees. It is strongly recommended that companies deploying AI-based Data Science solutions should provide constant training and coaching of employees not only in how to use the new technology but also in understanding the implications for decision-making at every level of the company. Without this, the common response is to reject and resist change with all possible means available.
- *Experts are not committed to using AI-based Data Science.* Business experts weren't engaged from the beginning of the company's raids into AI-based Data Science or they don't see the value of the technology. In either case, it is likely that the business leadership has not communicated the company's vision of how the experts and their careers would benefit from AI-based Data Science.
- *Data scientists and business teams are not communicating effectively.* Too often, data scientists "throw the models and run" and let the users in the business struggle with all the pains of the application of the models. That approach rarely works, and triggers complaints from the users. Most of them will resist any future attempts to implement such technology.
- *The delivered solutions are not user-friendly.* A specific issue in this category is the case of "black-box" or purely academic AI-based Data Science solutions. They provide clunky, overly complicated insights that are impossible for business users to fully understand. As a result, users do not trust this "academic abracadabra" and question its application.

Most of the above factors contributing to the resistance toward AI-based Data Science can be resolved with an improved business strategy, proper incentives, communication, and project management.

---

[5]The key driving forces of resistance toward AI-driven Data Science were discussed in Chap. 1.

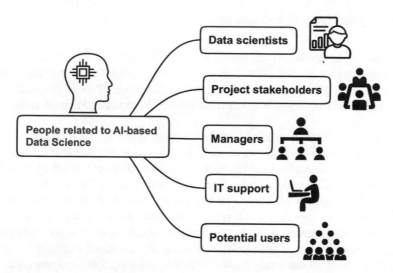

Fig. 13.2   Key categories of people related to AI-based Data Science

13.2   Key Categories of People Related to AI-Based Data Science

The growing popularity of AI-based Data Science is affecting a large number of people in industry, government, academia, the military, etc. Some of them are directly involved in the development, deployment, support, and management of the solutions generated by this spectacular technology. In this section we'll focus on the key categories of people, related to AI-based Data Science, which are shown in Fig. 13.2 and discuss next.

13.2.1   Data Scientists

Data scientists are responsible for everything that AI-based Data Science is used to do and they must also implement the resulting models in production systems. That is, data scientists own the entire workflow, from the formulation of a business objective and goals to the accomplishment of those goals and delivering the expected value.

The needed skillset and responsibilities of data scientists in a business are discussed in Chap. 17.

13.2.2   Problem Stakeholders

This is the category who decide, within a specific business, the fate of each project in particular and of AI-based Data Science in general. Above all, the problem stakeholders own the deployed solutions and extract the value. Their collaboration in all steps of the workflow, especially in problem definition and knowledge acquisition, is critical.

Having the problem stakeholders' full support is the ultimate task for the data scientist' members of the project team. In addition to good generic communication skills, some techniques for establishing an effective dialog and avoiding the lost-in-translation trap discussed in Chap. 5 are recommended for data scientists. Understanding the business problem and presenting related AI-based Data Science methods for its solution in plain English is the first necessary step for an effective dialog. Involving the stakeholders and regularly reporting the project's progress to them is another important good practice. Giving the developed solutions to the final users for exploration has a very positive effect by showing that one appreciates their valuable feedback, and increases the trust.

A critical step in the collaboration between data scientists and problem stakeholders is the transfer of ownership the developed solutions during the model deployment phase. It has to be properly done, according to the recommendations in Chap. 11.

13.2.3   Managers

Managers are the critical people component for the starting, operating, and survival of AI-based Data Science in a business. The vision of the top-level leadership is a necessary precondition for opening the door to the technology in the organization, but the continuous and enthusiastic support of low-level management is the key factor to making AI-based Data Science a profitable reality.

Allocating resources to introduce AI-based Data Science is not an easy decision for top leadership. A 2017 Deloitte survey of executives with high AI awareness found that 37% of the respondents said the primary AI challenge was that "managers don't understand AI-based technologies and how they work." [6]

Only the most visionary leaders have the guts to take the risk. Based on the data from the survey, the following seven attributes of AI-driven executives were defined:

- *They learn the technologies.* It helps a lot in leading with AI to know what AI is and does. AI-driven organizations typically want to explore a wide variety of

---

[6]A significant part of this subsection is based on reference T. Davenport and J. Foutty, AI-driven Leadership, *MITSloan Management Review*, 08/10/2018.

technologies, and leaders need to know enough about them to be able to weigh in on which ones will be most critical to their organization's success.

- *They establish clear business objectives.* As with any other technology, it is important to have clear objectives for using AI. The same Deloitte survey of US executives with a high level of AI awareness found that the most popular objectives involved using AI to improve existing products and services, make better decisions, create new products, and optimize business processes.
- *They set an appropriate level of ambition.* Some organizations have difficulties in pulling off highly ambitious goals and may set back AI initiatives overall by years if they fail. The alternative is to undertake a series of less ambitious projects—often called the low-hanging fruit. While less transformational individually, a series of such projects can add up to major change in a product or process.
- *They look beyond pilots and proofs of concept.* Usually AI projects start as pilots. But to improve productivity and achieve the needed ROI, leaders need to push their companies to scale up these projects to full production status. This means identifying process improvements before applying technology and figuring out how to integrate AI technologies with existing applications and IT architectures.
- *They prepare people for the journey.* Most AI projects will involve "augmentation"—smart people working in collaboration with smart machines—rather than large-scale automation. That means that employees will have to learn new skills and adopt new roles, which won't happen overnight. Good leaders are already preparing their people for AI by developing training programs, recruiting for new skills when necessary, and integrating continuous learning into their models.
- *They get the necessary data.* AI-driven leaders know that data is a very important asset if they want to do substantial work in AI. Many organizations will need to turn to external data to augment their internal sources, while others will need to improve the quality and integration of their data before they can use it with their AI projects.
- *They orchestrate collaborative organizations.* Often C-level executives—CEOs and heads of operations, IT, etc.—do not collaborate closely on initiatives involving technology. But these groups need to work together in AI-driven organizations to establish priorities, determine the implications for technology architectures and human skills, and assess the implications for key functions such as marketing and supply chains.

It is not a bad idea, before starting any activities for introducing AI-based Data Science in an organization, to check if their executives have these attributes.

13.2.4 IT Support

In order to function effectively, an AI-based Data Science organization needs IT support. In cases of enterprise-wide development packages, this support is needed during model development. Often the IT organization is also responsible for data collection. As we discussed in Chap. 11, IT support is critical for model deployment and long-term maintenance.

Tight collaboration between data scientists and the corresponding IT groups in the business is a must. It is of special importance to coordinate the integration of new software, including that from open sources, into the company's software architecture. It is good practice to train the IT colleagues related to deployed projects in the technical details of the developed solutions.

13.2.5   Potential Users

The majority of people in businesses are not involved in AI-based Data Science projects or initiatives. However, with a good business strategy, visionary leadership, and experienced data scientists, they could become potential users. A critical step in this process is establishing high credibility of the technology by a sequence of successful applications. Communicating the success stories and recognizing the contribution of technology pioneers across the organization may encourage others to follow their path.

It is also recommended to explain AI-based Data Science in appropriate parts of the organization. Some suggestions are given in Sect. 13.4.

13.3   Teamwork

AI-based Data Science requires a peculiar combination of skills—mathematics, statistics, programming, storytelling, project management and domain knowledge. This combination of skills is not easily available in the existing talent pool. Most people will have one or more of these skills but not all of them. Only a well-selected team can successfully accomplish the defined projects. A good team consists of more than a few data scientists and a champion. It includes people who understand the business and the data, and people who are going to be using the model. Involving a wide range of people from the beginning increases buy-in, which increases the probability of the results being used, and therefore the probability of success. Some suggestions for how to organize such a team for AI-based Data Science projects were discussed in Chap. 6.

Assuming that an effective project team has been built, it is important to transform effectively this group of talented individuals into a high-functioning team. Among the many issues in making this happen, we'll focus on two: (1) identifying and leveraging different work styles, and (2) understanding the team development process.[7]

13.3.1   Identifying Teamwork Styles

Just as important as understanding the technical skillset of the selected team is understanding how each personality type drives performance and behavior. Understanding why team members may react differently to a given situation is important to managing the team's success. The most popular identification is of the following four work styles:

- *Pioneers* are innovators. They don't mind taking risks and are excited by the idea of trying out a new, untested technique or technology.
- *Guardians* are detail-oriented and methodical. They want to know the facts and are risk-averse.
- *Integrators* help bring a team together. They care about people and the relationships within a group and prefer to make decisions by consensus.
- *Drivers* are workhorses. They value getting the job done and seeing results.

Most team members will identify with one or two of the work styles above. It is essential to note that there is no one right work style for an AI-based Data Science project. The characteristics of each could be a positive or a negative, depending on the situation. Individuals with any work style can come together to create a winning solution, but people with different styles often take different approaches to or have different insights about solving the same problem. Having access to different points of view can lead to more productive team collaboration.

However, a team with different work styles can also create conflict. To work together most effectively, team members need to understand (and respect) the work styles of their colleagues. While it is often easiest to work with people of like style, the advantages of diversity will be lost.

Conflict arises when team members do not adapt their styles to benefit their team, or fail to recognize what everyone brings to the table. For example, a guardian might stress that a new technology proposed by a pioneer is too risky whereas the pioneer may believe their ideas are being dismissed when the guardian highlights potential problems that might arise. Without understanding the benefits of differing points of view, both parties could come away feeling dissatisfied. One potential solution in this example would be for the guardian to propose boundaries around the time spent investigating the new technology.

Another typical example is when a driver is acting as a singular force on the team, essentially attempting to do all the work. Such projects rarely turn out as great as

---

[7]A significant part of this subsection is based on C. Everington, *Building a High-Functional Analytics Team*, Elder Research, 03/16/2018.

they could. The lack of respect for and trust in the team shown by trying to shoulder all the work deteriorates relationships, prevent others in the team from contributing as much as they could and learning new skills, and blocks everyone from bringing their full knowledge to the table. The act of one person trying to be a "hero" hurts the immediate project but also has a lasting negative impact on the team's overall growth and professional development—which is a big loss for the company.

The first step in applying knowledge of work styles is to have each team member reflect on which style(s) applies to them. Then schedule a team meeting to share which work style each member most closely identifies with. This is especially important when working together for the first time. This dialog about work styles enables members to understand how to make each member feel valued and how complementary styles, when brought to light and planned for, can enhance team performance.

13.3.2   Bringing the Team Together

Once a team understands these work style dynamics, they must embrace style diversity on each project to deliver the best business outcomes for their stakeholders. The next key topic for the team's success is handling team dynamics and performance in the best possible way. The most popular model of group development, first proposed by Bruce Tuckman in 1965, can be used. He proposed that groups undergo four phases (forming storming norming and performing) as they eventually grow into a high-functioning team. These phases are shown in Fig. 13.3 and discussed briefly below.

- *Forming*. This occurs when the group first comes together to work on a project. The group becomes familiar with the project itself, as well as with one another. It is a key objective for the group to form a common understanding of the project goals. Individual tasks are assigned, and group members act independently. The team is really just a set of individuals during this stage. To progress to the next stage, each member must embrace the possibility of conflict.
- *Storming*. The team likely experiences conflict due to differing opinions on how to approach a particular problem. Storming can occur during a single period of time or on a reccurring basis when new challenges arise. During this phase disagreements can potentially make individuals stronger, more versatile, and able to work more effectively as a team. To move to the next phase, it is important that the team learns what each member brings to the table in terms of work styles and skills and begins to establish trust.
- *Norming*. This takes advantage of the trust built during storming. Norming encompasses both conflict resolution and the development of a team mentality—putting the goals of the team and the project before any one individual's pursuits. Group norms are established as individuals learn to understand other team members' tendencies and abilities. The biggest risk during this stage is that

Fig. 13.3 Tuckman's
phases of team development

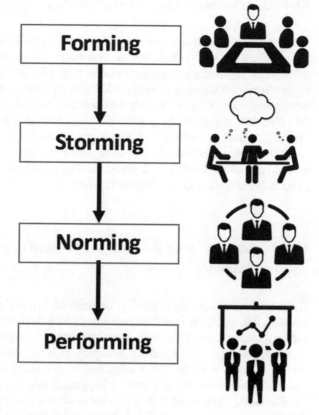

members become so averse to conflict that they suppress controversial opinions
or ideas that could benefit the project.

- *Performing.* The team members are highly functional and collaborate to achieve
  their goals. Any differences of opinion are handled in a nonstorming fashion,
  allowing the group to quickly and effectively develop solutions. Members of
  performing groups are beyond the phase of understanding their teammates; these
  groups use their inherent knowledge of work styles and team dynamics to assign
  tasks, make decisions, and complete projects successfully together.

It has to be considered that becoming a team, especially in such a complex area as
AI-based Data Science, is a process. Every team will move through challenging
times on the way to becoming a cohesive unit that effectively delivers successful
solutions. It takes both time and awareness for a group to become a high-
functioning team.

13.4   Marketing AI-Based Data Science

The majority of people in a business are not familiar with the capabilities of AI-based Data Science. There is a high chance that their view of this topic is formed by the media or nontechnical references. In order to minimize the hype and raise realistic expectations it is strongly recommended that AI-based Data Science is introduced into an organization by technically competent experts. Usually this role is taken by data scientists who don't have specialized marketing skills. It is not expected that they will explain the technology like advertising agents or car dealers, but it is good practice if data scientists are aware of some relevant marketing techniques. Of special importance for them are some guidelines on how to market AI-based Data Science to technical and nontechnical audiences.

13.4.1   Marketing AI-Based Data Science to a Technical Audience

The success of technology marketing depends on two key audiences—technical and business. The technical audience includes those potential users with an engineering and mathematical background who are interested in the principles and functionality of the technology. The business audience includes those potential users of the applied systems who are more interested in the value creation capability of the technology and how easily it can be integrated into the existing work processes. The marketing approach for the technical audience is discussed in this section, and that for the nontechnical audience is presented in Sect. 13.4.2. We'll concentrate on two topics: (1) how to prepare an effective presentation for an audience with a technical background, and (2) how to approach technical gurus.

Guidelines for Preparing Technical Presentations on AI-Based Data Science
The main objective of a good technical presentation is to demonstrate the technical competitive advantage of the proposed approaches. The key expectations of a technical introduction to a new technology are the main principles and features well-explained, a definition of the competitive advantage it offers; the potential application areas within the users' area of interest; and an assessment of the implementation effort required. One of the challenges in the case of AI-based Data Science is that the technology is virtually unknown to the technical community at large. The assumption has to be that this technology must be introduced from scratch with minimal technical details. In order to satisfy the different levels of knowledge in the audience, a second, more detailed presentation can be offered offline to those who are interested.

One of the key topics in the presentation is the differences between the proposed methods and the most popular approaches, especially those used frequently by the target audience, such as statistical methods. The examples given in Chaps. 3 and 4

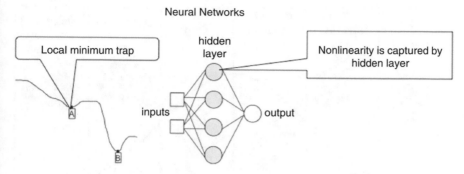

Fig. 13.4   An example of a slide introducing neural networks to a technical audience

are a good reference for preparing of this type of presentation. The audience must be convinced about the technical competitive advantage of a given approach relative to another technology with arguments based on scientific principles, simulation examples, and application capabilities. A realistic assessment, with a balance between the strengths and weaknesses is strongly recommended. An example of an introductory slide for neural networks is shown in Fig. 13.4.

Two central topics of neural networks are addressed here: (1) the capability of a neural network to generate nonlinear models; and (2) the danger that these solutions are inefficient due to local optimization. The first topic is demonstrated by showing how nonlinearity is captured by the hidden layer of a three-layer neural network. The second topic, generation of nonoptimal models due to the potential of the learning algorithm to be entrapped in a local minimum, is represented visually in the left part of the slide.

Another, more complicated type of slide for introducing a new method to a technical audience, called the "kitchen slide" includes information about the technology "kitchen" used in generating the final solutions. Examples of such slides for related AI-based technologies were shown in Chap. 5.

Key Target Audience for Technical Presentations
A typical technical audience includes data scientists, technical subject matter experts, practitioners, system integrators, and software developers. Their incentives to break the cozy status quo and introduce a new technology depend on the benefits they'll get as a result of increased productivity, ease of use, and their risk-taking being rewarding with potential career development and bonuses. In general, it is very difficult to expect that introducing AI-based Data Science will satisfy all of these incentives. Thus, management support is required to start the effort. The critical factor in gaining the needed support is the opinion of the technical leaders (gurus) in the organization. It is practically impossible to open the door to the new technology

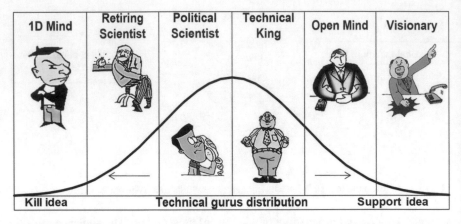

Fig. 13.5 Distribution of technical guru based on their support for a new idea

without their blessing. That is why, in the rest of this section, we'll focus our attention on approaching effectively this key category of the technical audience.[8]

In order to specify the marketing efforts as much as possible, we divide technical gurus into six categories based on their support for a new idea. The bell curve of the different types is shown in Fig. 13.5, where the level of support increases to the right and the level of rejection of the new idea increases to the left.

Two types of guru—the Visionary and the Open Mind—can fully support the new technology. On the opposite side are the 1D Mind and the Retiring Scientist gurus, who will try by all means available to kill or postpone the introduction of the new idea. Most technical gurus are neutral to a new technology and their support will depend on increasing their personal technical influence (in the case of the Technical King) or gaining political benefit (in the case of the Political Scientist). The typical behavior of the different gurus and the recommended approaches for presenting a new idea to them are given below.

Visionary Guru

Typical behavior. Having the blessing of a Visionary Guru is the best-case scenario of full, enthusiastic support for the new idea. The Visionary Guru is the driving force for innovation in an organization. She/he is technically sharp and has an abundance of ideas, i.e. shares her/his own ideas and enjoys exploring new ones. Other important characteristics of the Visionary Guru are an outside focus with respect to external expertise, risk taking, and the political skills to convince management. The Visionary Guru is well informed, and her/his office is full of bookshelves containing technical journals and books from diverse scientific areas.

---

[8] A very good source for analyzing target audiences for high-tech products is the classic book by G. Moore, *Crossing the Chasm: Marketing and Selling Disruptive Products to Mainstream Customers*, HarperCollins, 2002.

Recommended approach. Prepare an inspirational talk with a very solid technical presentation. Be ready to answer detailed technical questions and to demonstrate competence about the scientific basis of the idea. Convincing the Visionary Guru is not easy, but if successful, the chance of a positive decision to apply the proposed approach is almost 100%.

Open Mind Guru

Typical behavior. The Open Mind Guru accepts new ideas but will not initiate change without serious technical arguments. She/he is more willing to support methods that require gradual changes or have an established application record. The risk-taking level and enthusiasm are moderate. Before supporting a new idea, the Open Mind Guru will carefully seek the opinions of key technical experts and, especially, managers. Her/his bookshelf is half the size of the Visionary Guru is.

Recommended approach. Focus on detailed technical analysis that demonstrates competitive advantages over the methods closer to the Open Mind Guru's personal experience. Show specific use cases or impressive simulations in areas similar to the targeted business. It is also recommended to discuss the synergetic options between the new method and the approaches most used in the targeted business. An important factor for gaining the Open Mind Guru's support is if the new method is implemented by an established software vendor.

Technical King Guru

Typical behavior. The Technical King Guru dominates an organization with her/his own ideas, projects, and cronies. The key factor in her/his behavior is gaining power, i.e. this person is most likely to give a new idea her/his blessing if it will increase her/his technical influence. In this case, the Technical King Guru will fully support the idea and will use all of her/his influence to apply it. Otherwise, the chances of success depend only on top management push. The Technical King Guru shines in her/his area of expertise and has a good internal network and business support. On her/his bookshelf one can see only books related to her/his favorite technologies.

Recommended approach. Understand the key areas of expertise of the Technical King Guru and prepare a presentation that links the proposed new approach with the areas identified. Recognize the importance of the Technical King and her/his contribution and try to demonstrate how the new idea will fit in with her/his technical kingdom and will increase her/his glory and power. Independently, top management support could be pursued to counteract possible rejection of the idea.

Political Scientist Guru

Typical behavior. The Political Scientist Guru is on the negative side of the "new idea support" distribution shown in Fig. 13.5. By default, she/he rejects new approaches, since they increase the risk of potential technical failure with corresponding negative administrative consequences. Technically, the Political Scientist Guru is not on the list of "the best and the brightest," and this is another reason for looking suspiciously at any new technology idea. The real power of the Political Scientist is in using political means effectively to achieve technical objectives. From that perspective, support for new ideas depends on purely political factors, such as top management opinion, current corporate initiatives, and the balance of interests between the different parts of the organization related to the new technology. On her/his bookshelves one can see a blend of books on technical and social sciences with the person's favorite author, Machiavelli.

Recommended approach. The marketing effort must include a broader audience than technical experts. It is critical to have a separate presentation to top management first, which emphasizes the competitive advantages of the proposed approach. The ideal argument will be that some of the competitors are using the new methods. The presentations must have minimal technical details, be very visual, and be application-oriented.

Retiring Scientist Guru

Typical behavior. The Retiring Scientist Guru is counting the remaining time to retirement and trying to operate in safe mode with maximum political loyalty and minimum technical effort. In order to mimic activity, she/he uses a sophisticated rejection technique known as "killing a new idea by embracing it." The Retired Scientist is a master of freezing time by combining bureaucratic gridlock with departmental feuds. As a result, the new idea is buried in an infinite interdepartmental decision-making loop. Often the possible solution is in her/his potential career plans as a consultant after retirement. Don't expect bookshelves in her/his office. Only outdated technical manuals occupy the desk.

Recommended approach. Broaden marketing efforts to several departments. Sell the new idea directly to top management first. Be careful to avoid the "embrace and kill" strategy. Try to understand if the new technology may fit in with the Retiring Scientist's consultancy plans after leaving office.

1D Mind Guru

Typical behavior. The 1D Mind Guru is the worst-case scenario, with almost guaranteed rejection of new ideas. She/he has built her/his career on one approach only, which has created value. This mode of idea deficiency creates a fear of novelty and aggressive job protection. Any new idea is treated as a threat that must be eliminated. The 1D Mind Guru is well informed on the weaknesses of any new approach and actively spreads negative information, especially to top management. Bookshelves are absent in the 1D Mind Guru's office. However, the walls are often filled with certificates and company awards.

Recommended approach. Don't waste your time. Try other options.

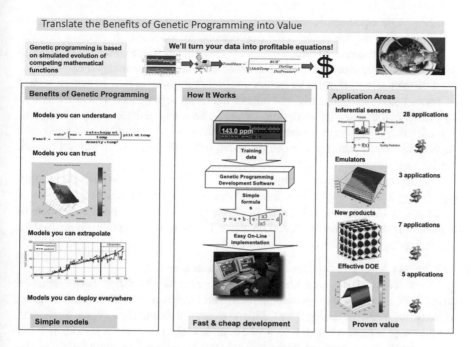

Fig. 13.6   An example of a slide introducing genetic programming to a business audience

13.4.2   Marketing AI-Based Data Science to a Nontechnical Audience

Guidelines for Preparing Nontechnical Presentations on AI-Based Data Science
The main objective of a good nontechnical presentation is to demonstrate the value creation capabilities of AI-based Data Science. The key expectations of a nontechnical introduction to a new technology, such as AI-based Data Science, are the benefits of the technology well-explained; a definition of the business competitive advantage, that it offers to the businesses; suggested potential application areas within the users' area of interest, and an assessment of the implementation effort required.

An example of a nontechnical introduction to one of the AI-based Data Science technologies, genetic programming, is given in Fig. 13.6. This is a typical slide for introducing a new method to nontechnical audiences. We call it the "dish slide."

In contrast to the "kitchen slide," shown in Fig. 5.11, it does not include information about the principles of the technology. The focus is on the benefits of the final product from the technology—the "dish." The slide template is organized in the following way. The title section includes a condensed message that represents the source of value creation, supported by an appropriate visualization. In the case of genetic programming, the source of value creation is defined as "We'll turn your data into *profitable* equations!", visualized by icons showing the transformation of the data into equations that create value (represented by a dollar sign).

The slide is divided into three sections: (1) benefits, (2) deployment, and (3) application areas. The benefits section focuses on the key advantages of symbolic regression models, generated by genetic programming—interpretable models trusted by users, increased robustness, and almost universal deployment. The interpretability benefit is illustrated by an equation that can be easily understood by the experts. The advantage of having a trustable model is shown by a "what-if" scenario response surface. Increased robustness is demonstrated by performance of the model outside the training range.

The deployment section is illustrated by a simple sequence on how the technology is applied. The application area section focuses on the key implementation successes of the method. It is strongly recommended to give specific numbers of applications and the value created, if possible. The proposed template can be used for any technology.

Key Target Audience for Nontechnical Presentations

A typical nontechnical audience includes professionals in different nontechnical areas, such as accounting, planning, sales, and human resources. Their incentives to introduce a new technology are the same as for the technical audience. In general, this audience is more resistant to new and complex technologies. An important factor for support is the ease of use of the technology and minimal required training.

The final fate of the new technology, however, will be decided by the management. That is why, in the rest of this section, we'll focus our attention on approaching effectively this key category of nontechnical audience. We have to make an important clarification, though. The critical factor in the management decision-making process is the opinion of technical gurus. Very often, it automatically becomes the official management position. Smart managers even go outside the organization for the advice of external technical gurus before making such decisions.

The hypothetical bell curve distribution for managers, supporting or rejecting a new technology, is similar to the one for technical gurus shown in Fig. 13.5. At the edges of the distribution we have the two extremes: (1) on the positive side, visionary managers who initiate and embrace change, and (2) on the negative side, low-risk "party line soldiers" who shine in following orders and hate any novelty. In the same way as the size and the content of these bookshelves is a good indicator of the type of a technical guru, the frequency of use of the word *value* can be used for categorization of managers into types. Visionary managers who embrace novelty use it appropriately. The other extreme, innovation-phobic managers who are resistant to supporting new technologies, use the word *value* two to three times in a sentence.

The majority of managers, however, are willing to accept innovation if it fits into the business strategy and their career development. We'll focus on some recommendations on how to be prepared for approaching this type of manager:

- *Understand business priorities.* Do your homework and collect sufficient information about relevant business priorities. Proposing technologies that are outside the scope of the business strategy and the manager's goals is marketing suicide.
- *Understand management style.* Information about the style of the manager is very important. Is it a hands-on, detail-oriented, need-to-be-involved-in-every-decision approach, or a hands-off, bring-me-a-solution-with-results style? Is she/he a quick learner? How rigid or flexible is she/he?
- *Understand the paramount measure of success in the targeted business.* The proposed new technology must contribute to the key business metric of success. If on-time delivery is the paramount measure of success in the business, any solutions in this direction will be fully supported. If low-cost development is the primary goal, any suggestion that is seen as adding cost will not be accepted. Promises of future savings that offset a current addition may fall on deaf ears.
- *Avoid the problem saturation index (PSI).* The meaning of PSI is that management is saturated with today's problems. Any attempt to promise results that will occur 3 years from now is a clear recipe for rejection.
- *Offer solutions, not technologies.* Managers are more open to solutions that address current problems. For example, if one of the known problems in the targeted business is a need for improved quality, it is better to offer online quality estimators using robust inferential sensors (the solution) rather than symbolic regression models generated by genetic programming (the technology).

13.5 Common Mistakes

13.5.1 Not Analyzing the Effect of AI on Human Intelligence

Assuming automatically that an AI-based Data Science solution will be enthusiastically accepted by all users can bring unpleasant surprises. Ignoring the potential negative response of some users who feel threatened can lead to their resistance, and application failure. This can be prevented with proper identification of the users, addressing their fears, and inclusion of them in the model development process from the beginning.

13.5.2 Not Understanding some Categories of People Related to AI-Based Data Science

A well known weakness of just-come-from-university data scientists is their lack of knowledge of office politics in general and the different categories of people, discussed in Sect. 13.2 in particular. They are unprepared for effective communication and interaction with their business partners. As a result, projects are not organized and do not operate in the most effective way.

13.5.3   Ineffective Teamwork

Ignoring effective team building and operation significantly reduces project efficiency. Some techniques for selecting the composition of the team, suggested in Chap. 2, and for team operation, suggested in this chapter, can help to correct this mistake.

13.5.4   No Marketing Skills on AI-Based Data Science

Due to their complex nature, introducing the capabilities of AI-based Data Science into a business requires more popular style. Usually the data scientists who take care of this process do not have the necessary marketing skills. They can use the guidelines on how to communicate the technology to technical and nontechnical audiences, given in this chapter.

13.6   Suggested Reading

- A. Burgess, *The Executive Guide to Artificial Intelligence: How to Identify and Implement Applications for AI in Your Organization*, Palgrave Macmillan, 2018.
- J. Deal and G. Pilcher, *Mining Your Own Business*, Data Science Publishing, 2016.

13.7   Questions

Question 1
What are the weaknesses of human intelligence?

Question 2
How can AI compensate for the weaknesses of human intelligence?

Question 3
What are the key categories of people related to AI-based Data Science?

Question 4
What are the key phases of team development?

Question 5
What are the key types of technical gurus?

Chapter 14
Applications of AI-Based Data Science in Manufacturing

> *Theory is knowledge that doesn't work. Practice is when everything works and you don't know why.*
>
> Hermann Hesse

If we follow the proverb that "the proof is in the pudding," the industrial applications of AI-based Data Science presented in this and next chapter are the "pudding" of this book. Most of the examples described reflect the personal experience of the author in applying AI-based Data Science in large corporations, such as The Dow Chemical Company. However, they are typical of the chemical industry in particular, and of manufacturing and business problems in general.

Since manufacturing and business problems are very different, we will divide the "pudding" into two parts. The first part, described in this chapter, is devoted to AI-based applications in manufacturing. The first objective of the chapter is to clarify the specifics of manufacturing applications and the second is to discuss the key directions of the manufacturing of the future, driven by digital technologies. The third and most important objective is to give examples of successful applications of AI-based Data Science in manufacturing.

For each example, first the business case will be described briefly, followed by the technical solution. Some of the technical details require more specialized knowledge than the average level and may be skipped by nontechnical readers. However, technically savvy readers may find the specific implementation details very helpful.

14.1 Manufacturing Analytics

Manufacturing is one of the industrial areas where AI-based Data Science has been explored and used with growing impact. For the success of application efforts, however, more detailed knowledge of the features of manufacturing is needed. Some specifics of the analytical process in manufacturing and the data that it uses are discussed briefly below.

© Springer Nature Switzerland AG 2020
A. K. Kordon, *Applying Data Science*,
https://doi.org/10.1007/978-3-030-36375-8_14

14.1.1 Specifics of Manufacturing Analytics

Math modeling and statistics have been used for improving and optimizing manufacturing processes for more than 50 years. However, the capabilities of AI-driven Data Science offer new opportunities to target more complex systems that are dependent on big data and distributed as part of the IoT. Although the analytics for manufacturing and business use similar methods for data analysis and model development, manufacturing applications have several specific features that have to be considered separately. They are shown in Fig. 14.1 and discussed below:

- *High level of automation.* Most of modern manufacturing processes include process control and information systems with a large number of various types of sensors. Many of these processes are partially or fully automated. Some operations are performed by robots.
- *Key users are engineers.* The majority of potential clients for AI-based Data Science solutions in manufacturing have a technical and engineering background. On average, they have a good level of understanding of physical and chemical relationships. Some of them have tremendous knowledge about related manufacturing processes that can be used in the developed solutions.
- *Most problems are technically driven.* Usually the defined problems in manufacturing are based on issues with physical assets (broken equipment, faulty sensors, poorly tuned control loops, frequent shutdowns, etc.) The economic impact of the technical solution, however, is the final criterion for project success.
- *Model deployment is in real time.* Most of the derived AI-based Data Science solutions create value operating in real time. This is a very important constraint on model complexity and execution speed for deployed solutions.

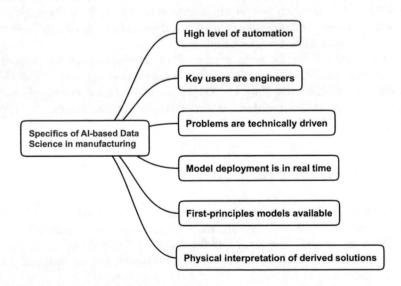

Fig. 14.1 Key features of an applied AI-based Data Science system in manufacturing

- *First-principles models available.* Some manufacturing processes operate based on complex first-principles models (recently called digital twins of the corresponding physical assets). They have a high model development cost but have more robust performance over a broad-range of operating conditions than the cheaper data-driven models have.
- *Derived solutions require physical interpretation.* As a result of their engineering background and good process understanding, the users of manufacturing models expect transparency and potential interpretation. Black boxes are not very welcome in process control rooms ...

14.1.2   Features of Manufacturing Data

The value of data was understood much earlier in manufacturing than in other types of business. An obvious reason is that process control and optimization are not possible without high-quality data. Many manufacturing units have process information systems with a long history of plant data. This is a very good basis for effective data analysis and modeling. Some specific features of manufacturing data are also a factor for an efficient model development process. These features are shown in the mind map in Fig. 14.2 and discussed briefly below:

- *Most of the data is numeric.* The majority of manufacturing data includes process measurements and calculations based on them. Most of these are numeric. There is a small fraction of textual data, however, linked to process and alarm logs.

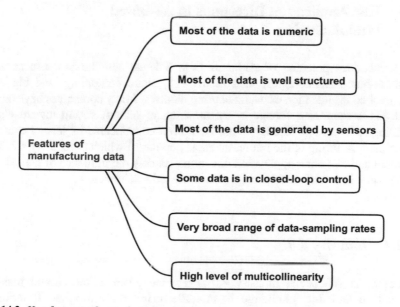

Fig. 14.2 Key features of manufacturing data

- *Well-structured data.* Most manufacturing data is in the category of structured data. An additional advantage is that, once defined, the data structure is constant, in contrast to the dynamic changes in the structure of businesses and their related data.
- *Most of the data is generated by sensors.* An advantage of sensor-based data is that it is objective and depends on the state and accuracy of measurement instruments. In the worst-case scenario of a broken sensor, this state is recorded as a fault in the process historian, and can be identified as an outlier and handled as such for the related period.
- *Some data is in closed-loop control.* A part of the process data is included in control loops. It has to be considered that the responses of these loops depend on the tuning of controller parameters, which are changed often.
- *Data is collected at different sampling rates.* The range of data collection frequency in manufacturing is very broad—from milliseconds to days. As a result, defining the proper sampling time for data analysis and model building is critical for project success.
- *High level of multicollinearity.* Due to the process physics, some manufacturing data is highly correlated. A typical case is of several temperatures in different sections of a chemical reactor. A representative sensor from such a highly correlated group needs to be selected by the SMEs.

14.2   Key Application Directions in AI-Based Manufacturing

The tremendous potential of AI-based Data Science and the progress in other computer-based technologies, such as the IoT, cloud computing, and big data, have grabbed the attention of manufacturing industry. Many companies have understood that a significant change in the direction of manufacturing toward digital technologies is needed. The key ideas have been consolidated in several concepts for the manufacturing of the future, the most popular of which is Industry 4.0. This paradigm and two other application directions, namely the influence of the IoT and digital twins in smart manufacturing, are discussed briefly below.

14.2.1   Industry 4.0

According to Wikipedia, Industry 4.0 is a name given to the current trend of automation and data exchange in manufacturing technologies. It includes

Fig. 14.3 Key components
of the manufacturing of the
future

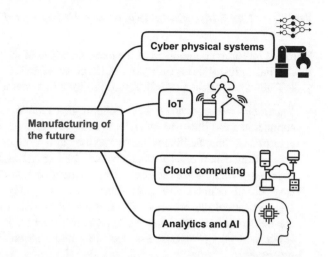

cyber-physical systems, the Internet of things, cloud computing and cognitive computing. Industry 4.0 is commonly referred to as the fourth industrial revolution.

Industry 4.0 fosters what has been called a "smart factory." Within modular structured smart factories, cyber-physical systems monitor physical processes, create a virtual copy of the physical world and make decentralized decisions. Over the Internet of Things, cyber-physical systems communicate and cooperate with each other and with humans in real-time both internally and across organizational services offered and used by participants of the value chain.

The key components of this version of manufacturing of the future are shown in Fig. 14.3.

In this subsection, we'll focus on the most important component—cyber physical systems.

Cyber physical systems are the basis of and enable new capabilities in areas such as product design, prototyping and development, remote control, services and diagnosis, predictive maintenance, systems health monitoring, planning, innovation capability, and real-time applications.

Cyber physical systems can communicate. They have intelligent control systems, embedded software, and communication capabilities as they can be connected into a network of cyber physical systems.

Cyber physical systems can be uniquely identified. They have an IP (Internet Protocol) address and are part of an IoT in which they can be uniquely addressed (each system has an identifier). Cyber physical systems have controllers, sensors, and actuators. This was already the case in previous stages before cyber physical systems; however, as part of the IoT, their importance is growing significantly.[1]

---

[1]This and the next two subsections are based on Industry 4.0: The fourth industrial revolution—guide to Industrie 4.0, https://www.i-scoop.eu/industry-4-0

14.2.2   The Manufacturing of the Future and the IoT

The next important component of the manufacturing of the future is the IoT. It is safe to say that Industry 4.0 is only possible because of IoT.

IoT has three distinct uses in future production systems:

- *Smart enterprise control.* IoT technologies enable tight integration of smart connected machines and smart connected manufacturing assets with the wider enterprise. This facilitates more flexible and efficient, and hence profitable, production. Smart enterprise control can be viewed as a mid-to-long-term trend.
- *Asset performance management.* Deployment of cost-effective wireless sensors, easy cloud connectivity (including wide area network), and data analytics improves asset performance. These tools allow data to be gathered easily from the field and converted into actionable information in real time. The expected result will be better business decisions and forward-looking decision-making processes.
- *Augmented operators.* Future employees will use mobile devices, data analytics, augmented reality, and transparent connectivity to increase productivity. As fewer skilled workers will be left to run core operations due to a rapid increase in baby boomer retirement, younger replacement plant workers will need information at their fingertips. This will be delivered in a real-time format that is familiar to them. Thus, plants will evolve to be more user-centric and less machine-centric.

It is expected that two AI-based Data Science approaches will be used in most IoT manufacturing applications: (1) complex-event processing (CEP) and (2) streaming analytics.

Complex-Event Processing

In general, event processing is a method of tracking and analyzing streams of data and deriving a conclusion from them. Complex-event processing is event processing that combines data from multiple sources to infer events or patterns that suggest more complicated circumstances. The goal of CEP is to identify meaningful events (such as opportunities or threats) and respond to them as quickly as possible.

CEP relies on a number of techniques, such as pattern recognition, abstraction, filtering, aggregation and transformation. CEP algorithms build hierarchies of model events and detect relationships (such as causality, membership, or timing) between events. They create an abstraction of event-driven processes. Thus, typically, CEP engines act as event correlation detectors, where they analyze a mass of events, pinpoint the most significant ones, and trigger actions.

Streaming Analytics

Real-time systems differ from CEP in the way they perform analytics. Specifically, real-time systems perform analytics on short data streams. Hence, the scope of real-time analytics is a "window" which typically comprises the last few time slots. Making predictions on real-time data streams involves building an offline model and applying it to a stream. Models incorporate one or more AI-based Data Science algorithms, which are trained using the training data. Models are first built offline based on historical data. Once built, a model can be validated against a real-time system to find deviations in the real time stream data. Deviations beyond a certain threshold are tagged as anomalies.

14.2.3   The Manufacturing of the Future and Digital Twins

Digital Twins

A digital twin is intended to be a digital replica of a physical asset, process, or system, in other words, a first-principles model. The simplest definition of a digital twin is an image connected to a physical object by a steady stream of data flowing from sensors, thereby reflecting the real-time status of the object. The data stream connecting the physical and digital is called a digital thread. In some implementations, a digital twin not only reflects the current state, but also stores the historical digital profile of the object.

What is a digital twin?

As a virtual representation of a physical asset, a digital twin enables an asset operator to derive actionable insights about both the performance and the health of the asset. These insights can result in reduced costs, new revenue opportunities, and improved overall business operations. Sensors provide data about the asset's operating conditions and the key performance parameters that describe the asset's real-world behavior. When both real-time and historical operational data are combined with physics-based scientific insights from the design of the asset, a unique digital representation of the asset emerges: the digital twin. The digital twin is a confluence of physics, sensors, and related data.

First-principles-based models, along with the insights derived from them, constitute the first building block of a digital twin. Sensor-enabled data and associated insights constitute the second building block.

Manufacturing processes and supply chains can have corresponding digital twins of critical assets that participate in the process or chain.

The digital twin of an asset or system has several benefits:

- Preempting asset maintenance needs, thereby reducing costs.
- Reducing asset downtime.
- Improving plant efficiency.
- Optimizing process times.
- Reducing time to market.

Digital Twins Versus Empirical Emulators

There is one big obstacle to mass-scale development of digital twins, however—their high cost. First, their offline model development requires a team of SMEs with high expertise in the related technical domains, process knowledge, and expensive modeling packages. In addition, offline validation of a first-principles model needs a lot of accurate process data that is not always available. Often the deployment of

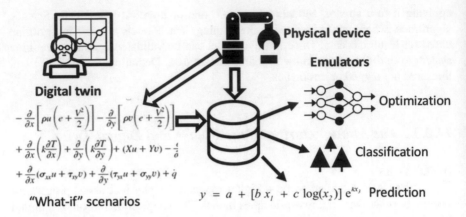

Fig. 14.4 The options of a digital twin and an empirical emulator for digital representation of a physical device

fundamental model requires special software licenses and fast hardware to operate in real time. Model maintenance requires deep knowledge at Ph.D. level of the first principles used in building the model. All of this combined significantly increases the total cost of ownership of digital twins.

Another issue is that some manufacturing processes require several complex first-principles modeling approaches to represent adequately the physical devices. A typical example is provided by the chemical industry, where many processes are mathematically described by a combination of kinetic, fluid dynamics, thermal, and mass balance models. Developing such a monstrous digital twin is extremely difficult and costly. Deploying it for real-time operation would require significant computer power and high speed. Maintaining the digital twin would be practically impossible due to the frequent changes in process conditions, typical in this industry.

An alternative solution for the digital representation of physical assets is the concept of empirical emulators. The idea is to partially replace the first-principles model with a set of data-driven models, called empirical emulators, that capture different features of the physical device. The concept of both a digital tween and an empirical emulators are illustrated in Fig. 14.4.

A popular use of empirical emulators is as a proxy for a complex first-principles model, i.e., a digital twin. This option for empirical emulators was described in Chap. 4 and illustrated in Fig. 4.7. An example of an industrial application of this type of emulator is given in Sect. 14.3.1. The key reason for developing this type of model is to resolve the issue of slow calculations in complex digital twins by developing simple data-driven models and using them for online process control and optimization.

A more effective option, as shown in Fig. 14.4, is to replace the digital twin by a set of different empirical emulators based on the available data. The coordinated responses of these models for prediction of the behavior of the physical device or finding some optimal model of operation should be similar to the responses of an equivalent digital twin. AI-based Data Science methods are very appropriate for generating such empirical emulators.

Developing an effective alternative to digital twins, however, requires careful integration of the potential empirical emulators to achieve the same level of prediction accuracy. Some of the relevant model integration approaches that could help the reader to accomplish this task were discussed in Chap. 4.

14.3 Examples of Applications in Manufacturing

The four examples, discussed in this chapter, are linked to the following key AI-based Data Science problems in manufacturing, shown in Fig. 2.7: enhanced process monitoring, process optimization, automated operating disciplines, and predictive maintenance. The first application covers the broad area of inferential sensors and focuses on robust solutions, based on symbolic regression. The second application is in the area of online process optimization and uses emulators of complex first-principles models (digital twins). Both neural networks and symbolic regression were used in this application. The third manufacturing application is in one critical area for successful plant operation—an automated operating discipline. The following AI-based Data Science technologies were integrated into the application: expert systems, genetic programming, and fuzzy logic. The fourth application is in predictive maintenance, where a paper machine health monitor was developed to predict potential machine breakdowns and reduce process shutdowns.

14.3.1 Robust Inferential Sensors

Inferential Sensor Technology
Some critical parameters in chemical processes are not measured online (composition, molecular distribution, density, viscosity, etc.), and their values are instead captured either from lab samples or by offline analysis. However, for process monitoring and quality supervision, the response time of these relatively low-frequency (several hours or even days) measurements is very slow and may cause loss of production due to poor quality control. When critical parameters are not available online in situations with potential for alarm "showers," the negative impact could be significant and eventually could lead to shutdowns. One of the ways to address this issue is through the development and installation of expensive hardware online analyzers. Another solution is to use soft, or inferential, sensors that infer the

critical parameters from other, easy-to-measure variables such as temperatures, pressures, and flows.[2]

Several AI-based Data Science methods are used to extract relevant information from historical manufacturing data to develop inferential sensors. In the case of linear relationships between the process and quality variables, multivariate statistical regression models such as partial least squares can serve to find the empirical correlations. Since the early 1990s, a more generic approach that captures nonlinear relationships based on artificial neural networks has been used. Artificial neural networks have several features that are very appropriate for the design of inferential sensors, such as universal approximation, and that models are developed by learning from data and can be implemented online. Due to these features, many applied inferential sensors are based on neural networks. However, neural networks have some limitations, such as low performance outside the ranges of the process inputs used for model development. Model development and maintenance require specialized training—and frequent retraining, which significantly increases the maintenance cost. In addition, model deployment demands specialized run-time licenses.

An alternative technology—the robust inferential sensor—has been under development at The Dow Chemical Company since 1997.[3] It is based on genetic programming and resolves most of the issues of neural-network-based inferential sensors. These robust inferential sensors are in the form of explicit algebraic equations that are automatically generated, with an optimal trade-off between accurate predictions and simple expressions. As a result, they offer more robust performance in the face of minor process changes.

Robust inferential sensors provide the following economic benefits:

- Inferential sensors allow tighter control of the most critical parameters for final product quality and, as a result, enable significant improvement in product consistency.
- Online estimates of critical parameters reduce process upsets through early detection of problems.
- The sensors improve working conditions by decreasing or eliminating the need for laboratory measurements in a dangerous environment.
- Very often such sensors provide optimum economics. Their development and maintenance costs are much lower than those of first-principles models and less than the cost of buying and maintaining hardware sensors.
- Inferential sensors can be used not only for estimating parameters but also for running "what-if" scenarios in production planning.

---

[2]The state of the art in soft sensors was presented in the book by L. Fortuna, S. Graziani, A. Rizzo, and M. Xibilia, *Soft Sensors for Monitoring and Control of Industrial Processes,* Springer, 2007.

[3]A current survey of robust inferential sensors was given in A. Kordon, et al., Consider robust inferential sensors, *Chemical Processing*, September 2014.

Application Areas of Inferential Sensors

Manufacturing industry rapidly realized the economic benefits of inferential sensors. From the early 1990s on, vendors and articles in the literature have reported a spectacular record of successful applications.

Environmental emissions monitoring epitomizes the role that inferential sensors can play. Traditionally, analytical instruments with high maintenance costs have performed such monitoring. The inferential-sensor alternative, implemented as a classical neural network, is much cheaper and provides an accuracy acceptable for federal, state, and local regulations in the United States and the European Union. Process variables enable the level of NOx emissions in burners, heaters, incinerators, etc. to be inferred.

One of the first popular applications of inferential sensors was for estimating product composition in distillation columns. However, the most widespread implementation in the chemical industry is for predicting polymer quality. Several polymer quality parameters, such as melt index, polymerization rate and conversion, are deduced from the reactor temperature, the jacket inlet and outlet temperatures, and the coolant flow rate through the jacket. Of special interest are nonlinear control systems that optimize the transitions between different polymer products.

A number of well-established vendors, such as Rockwell Automation, Aspen Technology, Siemens and Honeywell, have already implemented thousands of inferential sensors in a wide variety of industries. The benefit from improved product quality and reduced process upsets is estimated to be in the hundreds of millions of dollars but the potential market is much bigger.

We'll illustrate the potential of robust inferential sensors with two applications in different types of manufacturing process—continuous and batch processes.

Robust Inferential Sensor for Alarm Detection (Continuous Process)

The objective of this robust inferential sensor was the early detection of complex alarms in a chemical reactor.[4] In most cases, late detection of such alarms causes unit shutdown, with significant losses. The alternative solution of developing hardware sensors was very costly and required several months of experimental work. An attempt to build a neural-network-based inferential sensor was unsuccessful, due to the frequent changes in operating condition.

Twenty-five potential inputs (hourly averaged reactor temperatures, flows, and pressures) were selected for model development. The output was a critical parameter of the chemical reactor, measured by lab analysis of a grab sample every 8 h. The selected model was an analytical function of the type

---

[4]A. Kordon and G. Smits, Soft sensor development using genetic programming, in *Proceedings of GECCO 2001*, San Francisco, pp. 1446–1451, 2001.

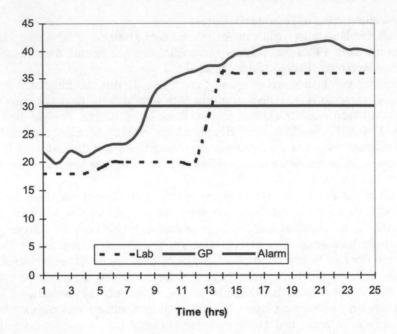

Fig. 14.5 Successful alarm detection in real time by the inferential sensor. The sensor (continuous line) triggered an alarm (the alarm level was at 30) several hours in advance of the laboratory measurements (dashed line)

$$y = a + b\left[e\left(\frac{x_3}{x_5}\right) - d\right]^c \qquad (14.1)$$

where x_3 and x_5 are two temperatures in the reactor, y is the predicted output, and a, b, c, d, and e are adjustment parameters.

The simplicity of the selected function for prediction of the critical reactor parameter allowed its implementation directly in the process control system as a critical alarm indicator in the automated operating discipline system, described in Sect. 14.3.3. The system initially included a robust inferential sensor for one reactor. An example of a successful alarm detection several hours before the lab sample is shown in Fig. 14.5.

The robust performance over the first 6 months gave the process operation the confidence to ask for the solution to be leveraged to all three similar chemical reactors in the unit. No new modeling efforts were necessary for model scale-up, and the only procedure needed to fulfill this task was to fit the parameters a, b, c, d, and e of the GP-generated function (14.1) to the data sets from the other two reactors.

The robust long-term performance of the GP-generated soft sensor convinced the process operation to reduce the lab sampling frequency from once a shift to once a day. The maintenance cost after the implementation was minimal and covered only the efforts required to monitor the performance of the three inferential sensors.

Robust Inferential Sensor for Biomass Estimation (Batch Process)

Biomass monitoring is fundamental to tracking cell growth and performance in bacterial fermentation processes. During the growth phase, determination of the biomass over time allows calculation of the growth rate. A slow growth rate can indicate nonoptimal fermentation conditions, which can then be a basis for optimization of the growth medium or conditions. In fed-batch fermentations, biomass data can also be used to determine feed rates of growth substrates when the yield coefficients are known.

Usually the biomass concentration is determined offline by lab analysis every 2–4 h. This low measurement frequency, however, can lead to poor control, low quality, and production losses, i.e., online estimates are needed. Several neural network-based soft sensors have been implemented since the early 1990s. Unfortunately, due to the batch-to-batch variations, it is difficult for a single neural-network-based inferential sensor to guarantee robust predictions over the whole spectrum of potential operating conditions. As an alternative, an ensemble of GP-generated predictors was developed and tested for a real fermentation process.[5]

Data from eight batches was used for model development (training data), and the test data included three batches. Seven process parameters including pressure, agitation, oxygen uptake rate, and carbon dioxide evolution rate, were used as inputs to the model. The output was the measured optical density (OD), which was proportional to the biomass. Several thousand candidate models were generated by 20 runs of GP with different numbers of generations. An ensemble of five models with an average r^2 performance above 0.94 was selected. The prediction of the ensemble was defined as the average of the predictions of the five individual models. The accuracy requirements for the ensemble were to predict the OD to within 15% of the observed OD level at the end of the growth phase. The performance of the ensemble on the training and test data can be seen in Fig. 14.6.

In Fig. 14.6(a), the OD level is plotted against the sample number, which corresponds to the time from the beginning of the fermentation. The solid line indicates the observed OD level. The dashed line corresponds to the prediction of the ensemble. In Fig. 14.6(b) the residuals of the ensemble's prediction with respect to the observed OD are shown. The 15% error bound is also shown. For the training data, one sees that for three batches (B_3, B_4, and B_8), the ensemble predicts outside the required accuracy. For batch B_3, it was known that the run was not consistent with the other batches. However, this batch was added in order to increase the range of operating conditions captured in the training set. The performance of the ensemble at the end of the run for all the batches in the test data (batches B_9, B_{10}, and B_{11}) is within the required error bound.

[5]A. Kordon, E. Jordaan, L. Chew, G. Smits, T. Bruck, K. Haney, and A. Jenings, Biomass inferential sensor based on ensemble of models generated by genetic programming, in *Proceedings of GECCO 2004*, Seattle, WA, pp. 1078–1089, 2004.

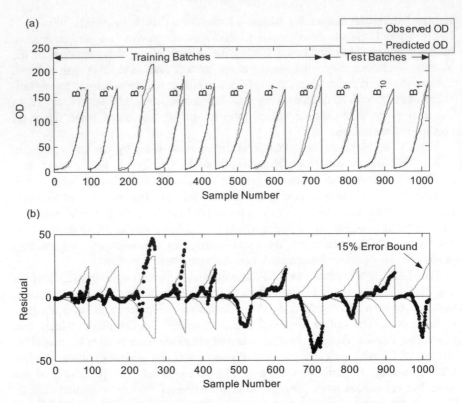

Fig. 14.6 Performance of the ensemble of biomass predictors on the training and test data. (**a**) Performance of ensemble; (**b**) accuracy of ensemble

14.3.2 Empirical Emulators for Online Optimization

Empirical emulators mimic the performance of first-principles models by using various data-driven modeling techniques. The driving force to develop empirical emulators is the push to reduce the time and cost of development of new products or processes. Empirical emulators are especially effective when hard real-time optimization of a variety of complex fundamental models (digital twins) is needed. As was discussed in Sect. 14.2.3, a set of appropriate empirical emulators can represent a digital twin.

Motivation for Developing Empirical Emulators

The primary motivation for developing an empirical emulator of a first-principles model is to facilitate the online implementation of a model for process monitoring and control. Often it may prove difficult or impractical to incorporate a first-principles model directly within an optimization framework. For example, the complexity of the model may preclude wrapping an optimization layer around it. Or, the model may be implemented on a different software/hardware platform

Fig. 14.7 Hybrid scheme of empirical emulators and fundamental models

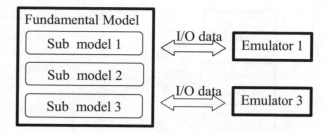

than the process control system, again preventing its online use. On other occasions, the source code of the model may not even be available. In such circumstances, an empirical emulator of the fundamental model can be an attractive alternative. An additional benefit is the significant acceleration of the execution speed of the online model (10^3–10^4 times faster).

Structures of Empirical Emulators

The most obvious scheme for utilization of empirical emulators is for the complete "replacement" of a fundamental model. The key feature of this scheme, shown in Fig. 4.17 in Chap. 4, is that the emulator represents the fundamental model entirely and is used as a stand-alone online application. This scheme is appropriate when the fundamental model does not include too many input variables and a robust and parsimonious empirical model can be built from the available data, generated by design of experiments.

In the case of higher dimensionality and more complex models, a hybrid scheme that integrates fundamental models and emulators is recommended (Fig. 14.7). Emulators based only on submodels with a high computational load are developed offline using different training data sets. These emulators substitute for the related submodels in online operation and enhance the speed of execution of the original fundamental model. This scheme is of particular interest when process dynamics has to be considered in the modeling process.

Finally, an item of special importance to online optimization is a scheme (shown in Fig. 14.8) where the empirical emulator is used as an integrator of different types of fundamental models (steady-state, dynamic, fluid, kinetic, and thermal).

In this structure, data from several fundamental models can be merged and a single empirical model can be developed on the combined data. The empirical emulator, as an integrator of different fundamental models, offers two main advantages for online implementation. The first advantage is that it is simpler to interface only the inputs and outputs from the models than the models themselves. More importantly, when constructing the data sets using DOE, the developer selects only those inputs/outputs that are significant for the optimization. Hence, the emulator is a compact empirical representation of only the information that is pertinent to the optimization.

The second advantage is that one optimizer can address the whole problem, rather than trying to interface several separate optimizers. The objectives, costs,

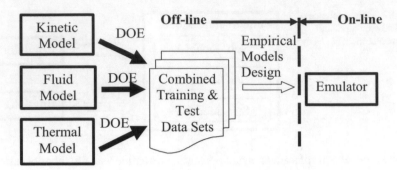

Fig. 14.8 Empirical emulator as integrator and accelerator of fundamental models

constraints, algorithm, and parameters of the optimization are more consistent, allowing the multimodel problem to be solved more efficiently.

Case Study: An Empirical Emulator for Optimization of an Industrial Chemical Process[6]

Problem Definition

A significant problem in the chemical industry is the optimal handling of intermediate products. Of special interest are cases where intermediate products from one process can be used as raw materials for another process in a different geographical location. This case study is based on a real industrial application of optimization of intermediate products between two plants belonging to The Dow Chemical Company, one in Freeport, Texas and the other in Plaquemine, Louisiana. The objective was to maximize the intermediate-product flow from the plant in Texas and to use it effectively as a feed in the plant in Louisiana. The experience of using a huge fundamental model for "what-if" scenarios in planning the production schedule was not favorable because of the specialized knowledge required and the slow execution speed (~20–25 min/prediction). Empirical emulators were a viable alternative for solving this problem. The objective was to develop an empirical model which emulated the existing fundamental model with an acceptable accuracy ($r^2 \sim 0.9$) and calculation time (<1 s).

Data Preparation

Ten input variables (different product flows) were selected by the experts from several hundred parameters in the fundamental model. There were 12 output variables (Y_1 to Y_{12}) that needed to be predicted and used in process optimization. The assumption was that the behavior of the process could be captured with these most significant variables and that a representative empirical model could be built for each output. A specialized 32-run Plackett–Burman experimental design with 10 factors

---

[6]A. Kordon, A. Kalos, and B. Adams, Empirical emulators for process monitoring and optimization, in *Proceedings of the IEEE 11thConference on Control and Automation, MED 2003*, Rhodes, Greece, p. 111, 2003.

Table 14.1 Performance of all emulators on training and test data

Output	r^2 for neural network on training data	r^2 for neural network on test data	Number of hidden nodes
Y_1	0.910	0.890	30
Y_2	0.994	0.989	20
Y_3	0.984	0.979	20
Y_4	0.987	0.981	20
Y_5	0.991	0.967	30
Y_6	0.999	0.999	1
Y_7	0.995	0.999	1
Y_8	0.995	0.993	10
Y_9	0.994	0.992	10
Y_{10}	0.992	0.993	1
Y_{11}	1.000	1.000	1
Y_{12}	0.997	0.989	20

at four levels was used as the DOE strategy. The training data set consisted of 320 data points. For 15 of these cases, the fundamental model did not converge for three of the outputs. The test data set included 275 data points, where the inputs were randomly generated within the training range.

Empirical Emulator Based on Neural Networks

Several runs with different numbers of hidden nodes were done and the results for all 12 neural-network-based emulators are summarized in Table 14.1.

The structure of the neural network for each emulator included 10 inputs and one output. The same set of inputs was used for all emulators. A number of different structures (with between 1 and 50 hidden nodes) were developed. The optimal number of hidden nodes was then selected by applying each neural network to the test data set and selecting the structure with the minimum r^2 value. This procedure was repeated for each emulator.

As shown in Table 14.1, all emulators have acceptable accuracy on the training and test data. The emulators' performance on the test data varies between $r^2 = 0.89$ for Y_1 and a perfect fit for Y_{11}. The complexity of the neural network also varied, from an almost linear structure with one hidden node for Y_6, Y_7, Y_{10}, and Y_{11} to a structure with 30 hidden nodes for Y_1 and Y_5. The prediction quality is good in all ranges. An example of the performance of emulator Y_5 is shown in Fig. 14.9 for the test data set.

Empirical emulators, based on GP-generated equations, were also developed. They had slightly lower performance and were not used.

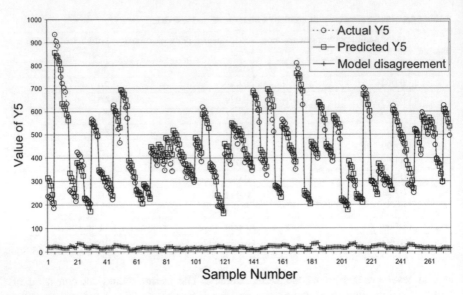

Fig. 14.9 Actual and predicted values for emulator Y_5 on test data

14.3.3 Automated Operating Discipline

The operating discipline is a key factor for competitive manufacturing. Its main goal is to provide a consistent process for handling all possible situations in the plant. It is the biggest knowledge repository for plant operation. However, the documentation is static and is detached from the real-time data for the process. The missing link between the dynamic nature of process operation and the static nature of operating discipline documents is traditionally bridged by the operating personnel. It is the responsibility of the process operators and engineers to assess the state of the plant, to detect problems, and to find and follow the recommended actions. The critical part in this process is problem recognition. The capability to detect a complex problem depends strongly on the experience of the process operators. This makes the existing operating discipline process very sensitive to human error, competence, inattention, and lack of time.

One approach to solving the problems associated with the operating discipline and making it adaptive to the changing operating environment is to use real-time intelligent systems. Such a system runs in parallel with the chemical manufacturing process, independently monitoring events in the process, analyzing the state of the process, and performing fault detection. If a known problem is detected, which is not accommodated by the existing control system and which requires human intervention, the system can automatically suggest the appropriate corrective actions.

Fig. 14.10 Example of an alarm case

```
          Alarm Case "Excess feed pre-cooling"

IF
          High value of critical variable
AND
          Rapid temperature decrease in sections 3-4
OR
          Fast change of reactor pressure
AND
          Decreased flow to unit 12
THEN
          Raise alarm and follow corrective actions for
          "Excess feed pre-cooling"
```

The advantages of the proposed approach are illustrated below with an industrial application for automating the operating discipline in a large-scale chemical plant.[7]

The specific application was based on using alarm troubleshooting as the key factor in establishing a consistent operating discipline for handling difficult operating conditions. The objectives of the system were as follows:

- Automatically identify the root causes of complex alarms.
- Reduce the number of unplanned unit shutdowns.
- Provide proper advice to the operators about the root cause of alarms in real time.
- Streamline the operating discipline through automatically linking the corrective actions to the identified problems.
- Assist the training of new operators to detect problems and carry out corrective actions.

The integration of the various intelligent systems techniques into the hybrid intelligent system was done in the following sequence.

Knowledge Acquisition from Experts

The knowledge acquisition was carried out in two main steps: (1) definition of alarm cases by experts and (2) validation of these alarm cases by a knowledge engineer through analysis of the potential causes of chemical process upsets. After discussions between the experts, the knowledge about detecting and handling process upsets was organized into building blocks called "alarm cases." An alarm case defines a complex root cause and usually contains several low-level alarms, linked with logical operators. The description of the alarm case is given in natural language. An example of an alarm case is given in Fig. 14.10.

---

[7]A. Kordon, A. Kalos, and G. Smits, Real-time hybrid intelligent systems for automating operating discipline in manufacturing, in *Artificial Intelligence in Manufacturing Workshop Proceedings of the 17th International Joint Conference on Artificial Intelligence, IJCAI 2001*, pp. 81–87, 2001.

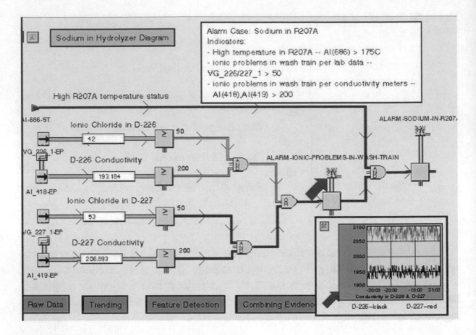

Fig. 14.11 Example of an alarm logic diagram

Part of the knowledge acquisition was linked to fuzzy logic in order to quantify with membership functions expressions like "fast temperature change" and "not enough liquid in the reactor." Of special importance was the validation process, where the defined alarm cases were evaluated with the operating history of the plant.

Organization of the Knowledge Base

The alarm logic that was defined during the knowledge acquisition phase was implemented using logic diagrams. These diagrams were organized in the following manner (Fig. 14.11).

First, the raw data from the chemical process was linked to the diagram, followed by trending. The processed data was then supplied to a feature detection component, by testing against a threshold or a range. Every threshold contained a membership function, as defined by fuzzy logic. The values of the parameters of the membership function were based either on the experts' assessment or on statistical analysis of the data. The "combining evidence" phase of the diagram included the necessary logic to implement the alarm logic that triggered the alarm block.

Implementation of Prototype for One Process Unit

This phase included the full integration of all components. A critical task for validation of the knowledge base was the adjustment of the alarm threshold. This process is essential for establishing credibility of tan expert alarm system because incorrect thresholds have a negative effect on alarm triggering. They either trigger the alarm too often (nuisance alarms) or respond to the alarm condition too late. In order to improve this process, it is necessary to verify the thresholds defined by the experts with statistical information. If significant differences are detected, they need to be resolved before putting the alarm system into operation. Otherwise, it will lose credibility at the beginning of online operation, and this may have a negative psychological effect on the process operators.

After 6 months of operation and corresponding tuning of the alarm logic and thresholds, the system was scaled up to all process units.

Operators' Involvement

Two operators from each shift were trained in how to interact with the system. At the end of the training, the operators knew how to respond to an alarm case, how to access and print the corrective actions, how to access the alarm log file, and how to start up and shut down the intelligent system. Most of the operators did not have problems interacting with the system after training. Unfortunately, some of them did not use the system regularly and their experience in its use gradually decreased. However, other operators used the system regularly and even suggested important improvements in the logic diagrams and in the corrective actions.

Value Evaluation

After an evaluation period of 6 months, the operations personnel identified the following items that they believed provided value to the plant:

- Detecting problems the operators might have missed.
- Eliminating upsets or at least minimizing the number of upsets to just a few each year.
- Catching operator mistakes, specifically in complex alarm cases that included parameters from different process units.
- Training new operators in problem detection and corrective actions.
- Fast problem recognition due to reliable inferential sensors.

The main issue in value assessment of this class of systems is the time span for performance evaluation. The key source of value is prevention of major upsets. These are rare events in general and usually are not regular. That is why it is very difficult to assess the cost/benefits of such an intelligent system on a short time-scale (less than 1 year). A reasonable time period for cost evaluation is 3–5 years.

14.3.4 Process Health Monitoring[8]

A Case Study: A Health Monitor for Paper Machine Breakdowns
Paper machine breakdowns due to sheet breaks are a significant contributor to lost market opportunity, as well as increased downtime and greater operating expenses, in the pulp and paper industry. Sheet breaks can occur at different stages of processing and may have different root causes. Some of the causes of sheet breaks build up slowly, ultimately leading to paper machine breakdowns, while other causes are quite abrupt. The abrupt faults generally come without any prior indication and are often difficult to detect in advance. However, some of the faults in the press section and in the dryer section occur because of slow changes in the process and are believed to be predictable in advance.

In a recent study it was reported that paper machine breakdowns led to 1.6 h average loss of production time per day, which amounts to \$6–8 million per year for each production line.[9] The early detection and diagnosis of the root causes of many of the sheet breaks would bring substantial value to the pulp and paper industry.

Therefore, there is a need to develop monitoring schemes for the health of paper machines which are simple in application, easily maintained, and of manageable size. In this case study, the development and implementation of a PCA-based sheet-break-monitoring scheme that was carried out for a major paper mill is described. The main objective was to detect and isolate the root cause of sheet breaks well ahead of the break, so that corrective actions could be taken to prevent a paper machine breakdown. An important objective of this application was to develop a monitoring scheme that was fairly sensitive to detecting sheet-break faults and yet had as few false alarms as possible. Though the theory of the use of PCA for fault detection and isolation is well known, the successful implementation of PCA-based monitoring partly depends on the preprocessing of the data. In addition to building a multivariate monitoring scheme, an extensive root cause analysis was carried out by applying data mining techniques combined with process knowledge and engineering judgment.

Health Monitor for Paper Machines Breakdowns Based on Multivariate Statistics
Since paper machine breakdowns are caused by multiple factors in different sections, the appropriate modeling method should be multivariate. In this case, PCA was used to model the process and a combined index based on Hoteling's T^2 and the squared prediction error (SPE) was developed as a sheet-break detection indicator.[10] As the

---

[8]This subsection is based on S. Imtiaz et al. Detection, diagnosis and root cause analysis of sheet-break in a pulp and paper mill with economic impact analysis, *Canadian Journal of Chemical Engineering*, **85**, 8, pp. 512–525, 2007.

[9]P. Bonissone and K. Goebel, When will it break? A hybrid soft computing model to predict time-to-break in paper machine, *General Electric Global Research Centre Report*, 2002.

[10]Details about these metrics can be found in S. J. Qin, Statistical process monitoring: basics and beyond, *Journal of Chemometrics* **17**(8–9), pp. 480–502, 2003.

process is subject to external disturbances, changes and frequent interruptions, preprocessing of the data played an important role in getting consistent results. Several methods for normalizing and filtering the data were explored before achieving the needed data quality.

Another technical issue in effectively detecting the faults was finding the best combination of the two statistical metrics T^2 and SPE. Unfortunately, each metric demonstrates unacceptable performance for process fault detection. Because the pulp and paper process is subject to many external disturbances, the SPE metric generated several false alarms even during normal operation. On the other hand, the T^2 metric was below the threshold and indicated normal operation very distinctly. But, at the same time, it was not as sensitive to sheet-break faults as the SPE. Therefore, it was evident that T^2 and the SPE complemented each other in fault detection. Combining T^2 with SPE helped to suppress the false positives during normal operation, but at the same time the result was also sensitive to faults. The fault sensitivity of the combined index (CI) could be increased or decreased by changing the weighing parameter k. A value of 0.5 (equal weight) was used in the application.

A critical factor for robust performance of the paper machine breakdowns monitor was variable selection. Consequently, 39 variables (from the initial list of 1024 process inputs) were chosen using variable selection methods and engineering judgment. The T^2 and SPE values were calculated based on 10 principal components.

An example of early detection of a paper machine breakdown by the health index developed is shown in Fig. 14.12. The plot covers the progress of the machine health monitor between two consecutive machine breakdowns. The threshold for the normalized index CI', shown by the dashed line in Fig. 14.12, was one. The alarm for a potential machine breakdown was raised when the index is above the threshold (shown by an arrow in Fig. 14.12) for some period of time. As is shown in Fig. 14.12, the paper machine health monitor sent a reliable warning message before the breakdown.

The models detected 63% of the faults successfully from the selected test cases with very few or no false alarms. If the machine breakdowns health indicator gave a consistent warning 15 min prior to the sheet-break event, then this was considered to be a good detection. This was considered quite satisfactory by the plant personnel, considering that a small number of variables were included in the model and, among the test cases many of the faults were abrupt and hence undetectable until the very end.

Value Evaluation

An important result of the developed solution was that the root causes of sheet breaks were identified and recommendations were made. Implementation of changes made on the basis of these recommendations reduced the frequency of sheet breaks significantly. The economic impact of the changes was evaluated. Savings of more than $1 million in terms of fewer sheet breaks and reduced downtime per year were realized as a result of implementing the suggested changes.

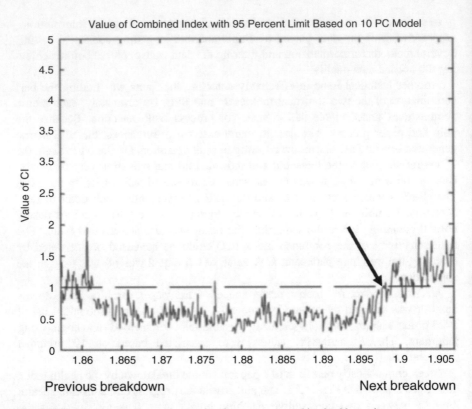

Value of Combined Index with 95 Percent Limit Based on 10 PC Model

Previous breakdown   Next breakdown

Fig. 14.12 An example of an early warning from a paper machine health monitor

In order to quantify the benefits, several key performance indicators were calculated from the historical data before and after the changes were made. The performance indices of the plant prior to the changes were an average of 13 months of production history, and the performance indices after the changes were calculated from 9 months of data. These two phases were termed as "pre project" and "post project", respectively.

An example of the progress of one of the most important KPIs, the average production rate in these two phases, is shown in Fig. 14.13. The y-axis values of the plot have been masked in order to maintain confidentiality. However, the relative changes have been stated. The key result from the changes was that the paper machines were operated at significantly higher speeds. This together with the increase in operation time, gave a boost to the production. The average production rate of the plant increased by 35.1 tons/day, as shown in Fig. 14.13.

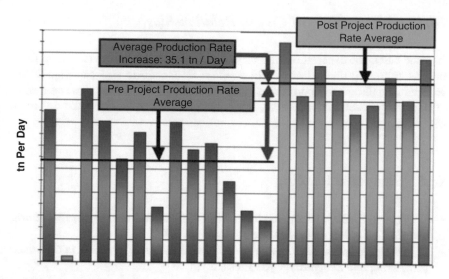

Fig. 14.13 Production rate improvement based on reduced machine breakdowns

14.4   Common Mistakes

14.4.1   No Need for AI-Based Data Science

Many manufacturing units don't open the door to the new AI-based Data Science technologies. The key argument is that the current sophisticated process control systems are already optimizing the plant and there is no need to overload the plant operators with more complex technologies. Most of process operators also see AI as a direct threat to their jobs. They create an image of themselves as the ultimate experts of all process knowledge and convince the management that no AI can fully replace their "bag of tricks" for operating the plant.

14.4.2   No Links to Business-Related AI-Based Data Science

Another issue in the current state of industrial applications of AI-based Data Science is that manufacturing and business systems are separated. Often the division is at the organizational and software levels. It is true that the objective differences in the problems to be solved require different skillsets in the project team composition (engineering in manufacturing and finance/economics in business applications). However, looking to build bridges and find appropriate joint problems will benefit both key groups of data scientists. Let's not forget that the problem spaces are different but the solution spaces (i.e., AI-driven Data Science methods) are almost the same.

14.4.3 Expectations of Causal Relationships

In some cases, the expectation of a clear interpretation and full cause–effect explanations from any derived empirical solution is unrealistic and can significantly
reduce the application opportunities in manufacturing. On occasions when a causal
"white-box" model cannot be built, it is better to use an "inferior black-box" model
than nothing.

14.5 Suggested Reading

- L. Fortuna, S. Graziani, A. Rizzo, and M. Xibilia, *Soft Sensors for Monitoring
  and Control of Industrial Processes,* Springer, 2007.
- D. Pascual, *Artificial Intelligence Tools: Decision Support Systems in Conditions
  Monitoring and Diagnosis*, CRC Press, 2015.

14.6 Questions

Question 1
What are the specific features of manufacturing applications?

Question 2
What are the key features of manufacturing data?

Question 3
What is the difference between a digital twin and an empirical emulator?

Question 4
What are the key components of the manufacturing of the future?

Question 5
Rank the most important application areas of AI-based Data Science in
manufacturing.

Chapter 15
Applications of AI-Based Data Science in Business

Big breakthroughs happen when what is suddenly possible meets what is desperately needed.

Thomas Friedman

The second part of the "pudding," described in this chapter, is devoted to AI-based Data Science applications in business. They include the broad range of business activities, such as: defining business strategy, handling business finance, planning, procurement, marketing, and human resources. The first objective of the chapter is to clarify the specifics of business applications, and the second is to discuss the key application directions in business driven by digital technologies and AI. The third objective is to give examples of successful AI-based Data Science applications in business.

15.1 Business Analytics

For more than a decade the business world has understood the value of using AI-driven Data Science. As a result, the number of applications in different business activities is growing fast. Based on accumulated experience, some specific features of the analytical process for developing business systems and the related data have been defined. The key learnings are discussed briefly below.

15.1.1 Specifics of Business Analytics

Math modeling and statistics have not been frequently used in operating businesses in the past. Relative to manufacturing analytics applications, the culture of model acceptance in business settings is in an early phase. This is one of the observed differences of business applications. The key distinctions are shown in Fig. 15.1 and discussed below:

© Springer Nature Switzerland AG 2020
A. K. Kordon, *Applying Data Science*,
https://doi.org/10.1007/978-3-030-36375-8_15

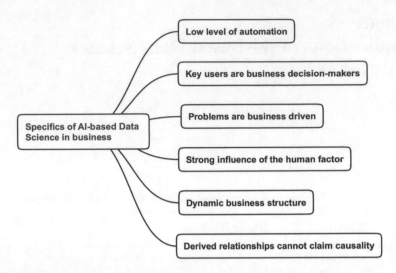

Fig. 15.1   Key features of an applied AI-based Data Science system in business

- *Low level of automation.* In contrast to manufacturing systems, business pro-cesses are paper-hungry and have a very low level of automation. One of the reasons is a resistance to computerizing business decision-making. Some routine business operations, however, such as elementary customer support and adver-tising, are in a process of growing replacement by automatic solutions.
- *Key users are business decision-makers.* The majority of the potential users of AI-based Data Science systems in business are decision-makers with an econom-ics or financial background. On average, they have minimal statistical knowledge and no math modeling skills. Most of them expect analytics deliverables in the form of Excel spreadsheets, reports, and dashboards.
- *Most problems are business driven.* Usually the defined problems are based on issues in business operation, such as demand forecasting, fraud detection, and customer churn. The economic impact is measured by increased efficiency, reduced decision uncertainty, lower cost, etc.
- *Strong influence of the human factor.* In contrast to manufacturing analytics where the derived insight is based on objective technical relationships and laws of nature, business analytics generates solutions based on subjective dependen-cies with people in the loop.
- *Dynamic business structure.* Due to frequent mergers and acquisitions some businesses are in a mode of continuous structural change. As a result, most of the insight and models delivered need to be redeveloped and adapted to the latest business structure. Aligning historical data from different business sources which are part of the new structure, can be very challenging, though.
- *Derived relationships cannot claim causality.* Most of the insight and models in business analytics are based on discovered statistical patterns and relationships between economic or financial factors. The basis of these models is correlations,

not laws of nature. As correlations do not imply causation, it is not expected that the derived solutions will be causal.

15.1.2 Features of Business Data

Most of businesses keep internal data for reporting purposes in their local databases. Unfortunately, the databases for the different parts of the business process are not integrated and are difficult to synchronize. Recently, the value of external macro-economic and microeconomic data as an important factor for business forecasting has been recognized. As a result, such data complements the internal business data for developing AI-based Data Science models. Aligning all related data with different data collection periods (monthly, quarterly, or annual) is one of the known requirements for business data preparation. Some specific features of business data that need to be considered in modeling and project planning are shown in the mind map in Fig. 15.2 and discussed briefly below.

- *Most of the data is categorical.* The majority of business data includes textual information, such as customer category, product type, and payment option. Due to the recent trend toward using social media data in model building, the share of textual data is significantly growing. Unfortunately, this is a severe limitation on using modeling approaches that require numerical data. A procedure for the transformation of the categorical variables into numeric ones is needed. Some modeling packages have this option.
- *Unstructured data.* Most business data is in the category of unstructured data. As a result, the deliverable of many business problems is a combination of classification and prediction or forecasting models.

Fig. 15.2 Key features of business data

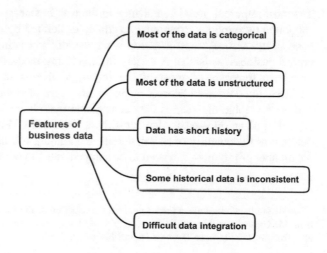

- *Most of the data has a short history.* Due to relatively late introduction of digital technologies and frequent changes in business structure, some businesses do not have an acceptable set of historical data. This may be an issue in developing forecasting models.
- *Some historical data is inconsistent.* Changing business rules and regulations can lead to diverse business metrics for different time periods. Different companies in the process of mergers and acquisitions may have used diverse data structures and KPIs. As a result, historical data could be inconsistent for some periods of time.
- *Difficult data integration.* In contrast to manufacturing data which is well synchronized in real time in process information systems, business data is gathered asynchronously from different sources. Integration and synchronization of all the pieces in the puzzle is a more challenging task than with manufacturing data.

15.2   Key Application Directions in AI-Based Business

In general, the key directions of the Industry 4.0 paradigm are valid for the business applications of AI-based Data Science as well. However, there are specific topics that are not covered by this paradigm. Some of these particular future trends in business applications have been defined by leading analytical groups.[1] A summarized view of the key components of the business of the future, based on these analyses, is shown in Fig. 15.3 and discussed below.

15.2.1   Forecasting-Driven Decisions

The most important trend is a change in the way businesses operate from decisions supported by previous experience to strategies defined by predictive analytics. The basis of this critical transformation is the use of forecasting models in most of the critical business decisions. A variety of forecasting methods can generate short and long-term forecasts of most of the important factors in and parameters of the business. In addition to internal business data, a lot of external economic, financial, and demographic time series data is available that allows building such models. A typical application area is demand-driven forecasting. Predicting future demand determines the quantity of raw materials, the amount of finished goods inventories, the number of products that need to be shipped, the number of people to hire, and the

---

[1]Two recommended references with detailed descriptions of the trends are the following reports from McKinsey Global Institute: *The Age of Analytics: Competing in a Data Driven World*, 2016, and *Artificial Intelligence: The next Digital Frontier?*, 2017.

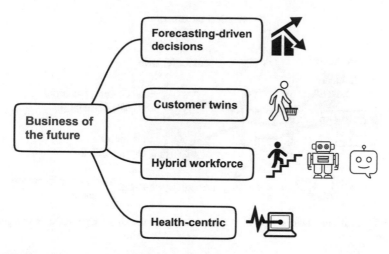

Fig. 15.3 Key components of the business of the future

number of plants to build, right down to the office supplies that should be purchased.[2]

Since running complex businesses involves a wide range of information and possible responses, data scientists need to handle a large, complex set of interdependent variables and deploy a range of forecasting methodologies, including judgmental forecasting. A short summary of the most important techniques for business forecasting is given in Sect. 15.3, as well as an example of a large-scale corporate forecasting project for predicting raw materials prices.

15.2.2   Customer Twins

If we could borrow the concept of a digital twin from manufacturing analytics and apply it to business problems, the proper analogy would be the development of a customer twin. In the same way as a digital twin is a digital replica of a physical asset, a customer twin is a digital duplication of a customer. The key difference, however, is in the nature of the digital representation. Digital twins are based on first-principles models, while customer twins are not. A digital customer is a set of different empirical emulators, developed from diverse customer data from her/his purchasing history, financial activities, social media activities, etc. Each empirical emulator characterizes a specific feature of the customer, for example her/his customer life value or engagement. Some emulators, such as those for estimating customer churn and loan return potential, are based on classification algorithms.

---

[2]C. Chase, *Demand-Driven Forecasting: A Structured Approach to Forecasting*, second edition, Wiley, 2013.

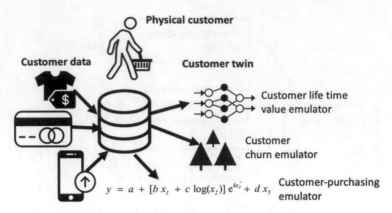

Fig. 15.4 A customer twin as a set of empirical emulators for digital representation of a physical customer

Others, such as those for evaluating investment decisions or forecast purchasing, require predictive models.

A generic structure of a customer digital twin based on a set of empirical emulators is shown in Fig. 15.4.

A high volume and quality of customer data are a key condition for developing these diverse emulators. The greater the number of data sources used, the more complete customer twin that is developed. The recommended list includes:

- *Transactional data* (the products the customer buys each time, the time of purchase, etc.).
- *Demographic data* (customer gender, age, family size, etc.).
- *Financial data* (income, loans, debt).
- *Web behavior data* (the products puts into her/his basket when she/he shops online).
- *Data from customer-created texts* (comments about the retailer that the customer leaves on the internet).

Big retailers have some of this data already and are using it for more and more detailed customer segmentation in their marketing strategies. The ultimate goal is to develop an effective individual-customer marketing strategy. The best possible way to accomplish this ambitious objective is to develop a customer twin. This process will include gradually building different empirical emulators related to the specific activities and behavior of the customer. It is possible that at some point these emulators will be integrated into a coordinated set, leading to a "unified" digital replica of the customer. It is assumed that as a result of the integration, the accuracy of predicting the customer's responses in different marketing situations will be higher.

In contrast to the digital twins in manufacturing, which are broadly used in industry, customer twins are at the beginning of demonstrating their big potential. The computer giant Cisco, for example, went from generating tens of "artisanal" or human-created propensity models to about 60,000 autonomously generated ones. A small group of analysts and data scientists in a group called Global Customer Insights generates these models each quarter using machine learning. Every potential customer–more than a hundred million of them–for every Cisco product in every country is represented in those models, which is why so many different models are required. Cisco's sales and marketing teams use these models to decide which products to offer to which customers.[3]

There are a lot of issues, however, related to data privacy, the ownership of customer twins, and customer' rights over their digital replicas.

15.2.3 Hybrid Workforce

Another key direction of the business of the future is a significant change in the structure of the labor force. The more efficient AI-based Data Science technologies not only will transform the way humans work and develop their careers but also will add new active participants to the work process, such as robots and chatbots. The first group includes all devices, programs, or business processes with the common name "robots" that replace humans and operate autonomously. The second group includes the different type of interactive agents with the common name "chatbots" that participate actively in the business process via communication with humans.

This hybrid workforce—a combination of humans, robots, and chatbots—will require different management and control. The division of labor will be one of the key challenges. Of special interest is the growing influence of chatbots that help customers transact or solve problems. Usually they use NLP to interact with customers in natural language (by text or voice), and use machine learning algorithms to improve it over time. Chatbots are being introduced by a range of financial services firms, often in their mobile apps or in social media. While many are still in the trial phase, there is potential for growth as chatbots become more sophisticated and gain increasing usage, especially among millennials. The current generation of chatbots in use is simple, generally providing routine information or alerts to customers, or answering simple questions. It is worth observing that the rising usage of chatbots is correlated with the increased use of messaging applications.

Chatbots are gradually moving toward giving advice and prompting customers to act. In addition to using chatbots to assist customers in making financial decisions, financial institutions can benefit by gaining information about their customers based on interactions with chatbots. The insurance industry has also explored the use of chatbots to provide real-time insurance advice.[4]

---

[3] This example was given in: T. Davenport, *The AI Advantage: How to Put the Artificial Intelligence Revolution to Work*, MIT Press, 2018.

[4] Financial Stability Board, *Artificial Intelligence and Machine Learning in Financial Services*, 11/01/2017,

Chatbots have begun to be used even in human resources. For example, talent acquisition and new-hire onboarding are ripe areas where intelligent assistants can tap multiple data sources to develop candidate profiles, schedule interviews, and make decisions about prospective job candidates.

The capabilities of chatbots and AI-based Data Science can increase the level of human employee's collaboration and effectiveness as well. The new AI-based human resources tools should be able to evaluate employee's performance more objectively, based on quantitative criteria. The balance between the three components of the hybrid workforce should not be at the expense of humans. It should not reduce the number of people needed, but make the business more profitable by using efficiently the new capabilities of robots and chatbots.

15.2.4 Health-Centric Business

The consensus of the analysists of the business of the future is that one of the key directions will be healthcare. Some important areas where AI-based Data Science technologies can be used to improve health care performance are discussed briefly below.[5]

Performance Evaluation
Keeping tabs on hospital activities by maintaining relevant databases can help administrators find inefficiencies in service provision. Based on the results found from data analysis, specific actions can be taken to reduce the overall costs of a healthcare facility. Reduced costs may be reflected in the form of a reduced cost burden on the consumers of such healthcare facilities.

According to U.S. News and World Report Beaufort Memorial Hospital in South Carolina found that it could save approximately $435,000 annually by implementing the simple act of discharging patients half a day earlier. Hospital administration didn't just make a random decision. They reached their conclusion after carefully analyzing the data.

Patient Satisfaction
By incorporating AI-based Data Science into healthcare systems we can improve the efficiency of the organization in terms of administrative tasks. These technologies have not only proven to be beneficial for the mundane tasks related to administration, but also to have a positive impact on a patient's overall experience. By maintaining a database of patients' records and medical histories, a hospital facility can cut the cost of unnecessary, repetitive processes.

In addition, AI-based Data Science can help in keeping an updated record of a patient's health. With the adoption of more advanced analytics techniques,

---

[5]This subsection is based on R. van Loon, *Data Analytics Is Transforming Healthcare Systems*, 05/09/2017.

healthcare facilities can even remind patients to maintain a healthy lifestyle and provide lifestyle choices based on their medical conditions. After all, the whole purpose of introducing digital technology into the health sector is to make sure that people are getting the best facilities at a subsidized cost.

Healthcare Management
Using AI-based Data Science technologies can help healthcare organizations to obtain useful metrics about the population. It can reveal information such as if a certain segment of the population is more prone to a certain disease. Moreover, if a healthcare facility is being operated in multiple units, analytics can prove beneficial for ensuring the consistency across all facilities and specific departments. For example, Blue Cross of Idaho used Pyramid Analytics BI Office to create a population health program. The results were evident in the form of reduced ER visits and emergency cases.

Labor Utilization
In addition to patient's data, hospitals and healthcare facilities can also store staff data. They can observe staff performance and find any loopholes or inefficiencies in the system. Based on the results, they can arrange for staff to be divided strategically across departments. Some healthcare units call for more staff than others. Failure to understand the organization's requirements will lead to a loss for both patients and staff.

15.3 Examples of Applications in Business

The examples in this section represent two of the key types of business problem shown in Fig. 2.1—price forecasting and customer churn. First, a more general view of the problem of business forecasting is presented, followed by a specific application for large-scale forecasting of raw materials prices in a global corporation. The first two examples presented are based on the author's experience at The Dow Chemical Company. The third example is related to reducing customer churn in an energy distribution company.

15.3.1 Business Forecasting[6]

The ultimate objective of business forecasting is to deliver to the key decision-makers a reliable forecast of specific economic variables, such as product demand,

---

[6]This subsection is based on A. Kordon, Applying genetic programming in business forecasting, *Genetic Programming Theory and Practice XI* (eds: R. Riolo, T. McConaghy, and E. Vladislavleva), *Springer, pp. 101–117, 2014.*

raw materials prices, and labor costs. Two technical factors have contributed to the increased use of business forecasting, especially after the recession in 2008–2009: (1) improved internal discipline and processes for collection of time series data on important economic variables within the businesses, and (2) the explosion of available business-related time series data. For example, the leading source of this kind of information, IHS Markit, offers more than 30 million time series of economic data across the globe (www.ihsmarkit.com). The key issue is how to use these sources of information for more reliable prediction, having in mind the available forecasting methods and their limitations.

Key Forecasting Methods
Classical time series forecasting methods can be subdivided into the following categories:

- exponential smoothing methods;
- regression methods;
- ARIMA methods;
- threshold methods;
- generalized autoregressive conditionally heteroskedastic (GARCH) methods.

The first three can be considered as linear methods, i.e., methods that employ a linear functional form for modeling time series, and the last two as nonlinear methods. For example, in exponential smoothing, a forecast is given as a weighted moving average of recent time series observations. In regression, a forecast is given as a linear function of one or more explanatory variables. ARIMA methods, presented by Box and Jenkins in the 1970s, give a forecast as a linear function of past observations (or differences of past observations) and the error values of the time series itself. They are the most popular and used forecasting methods for univariate forecasting of a single variable based on time-related patterns identified from historical data. When a model includes explanatory variables with their past values, it is called an ARIMAX model. Often ARIMAX models are referred to as dynamic regression. They are a typical example of multivariate forecasting approaches. The method has been widely used in the financial, economic, and social sciences fields.[7]

Issues in Business Forecasting
All the above-mentioned linear forecasting methods assume a functional form which may not be appropriate for many real-world time series. Linear models cannot capture some features that commonly occur in actual data such as asymmetric cycles and occasional outlying observations. Regression methods often deal with nonlinear time series by logarithmic or power transformation of the data, however, this technique does not account for asymmetric cycles and outliers.

In principle, no univariate forecasting method is capable of accurately predicting the future when history does not repeat itself. As is well known, however, this is a

---

[7]T. Rey, A. Kordon, and C. Wells, *Applied Data Mining for Forecasting*, SAS Press, 2012.

very strong condition. Due to social, technological, and structural economic changes, history never exactly repeats itself. Another limiting factor that has to be considered in business forecasting is the level of randomness of complex social systems. For example, efficient markets have a random-walk-type behavior, which reduces their predictability. The generic limitations of time series forecasting and the constraints of the different forecasting horizons are discussed briefly below and in detail in (Rey et al., 2012).

Generic Limitations of Time Series Forecasting

Univariate forecasting requires a representative historical data set that accurately captures the trend, seasonality, and cyclicality in the data. Collecting such data sets is one of the biggest challenges for the development of forecasting models. The most difficult requirement is representing the cyclicality, which is business-specific, and may require a long history of several years, containing at least several business cycles. Very often internal business data sets have lengths that are too short to satisfy this condition and to represent cyclicality.

When there are fundamental changes in the subject of forecasting, such as the introduction of a revolutionary new technology or significant business transformation, time series forecasting cannot identify automatically the new pattern.

Univariate time series forecasting assumes no change in the environment—in a similar fashion, changes in the economic environment, such as the introduction of new taxes or environmental regulations, cannot be captured automatically, and model redevelopment is needed.

Another issue with univariate forecasting is its limited capability for anticipating rare events. Some forecasting software allows defining and adding events, which improves the forecast accuracy after the event. Unfortunately, it is of limited value in the forecasting of a future event before it occurs.

Limitations of Forecasting Time Horizon

Another important factor that has to be considered in defining realistic expectations of business forecasting is the length of forecasting time horizon. Obviously, the longer the time horizon, the higher the uncertainty in the forecast and the requirement for a long historical data set. Forecasting time horizons can be defined into three categories—short-term (up to three time samples in the future), medium-term (up to 18 time samples in the future), and long-term (more than 18 time samples in the future). The limitations of the short-medium- and long-term time horizons are discussed briefly below:

* _Limitations of short-term forecasting horizon._ This is the best-case scenario for forecasting since the key economic phenomena that contribute to forecasting models (trend and seasonality) do not change much or frequently over a short time span.[8] Most of the available forecasting methods can be used for short-term forecasting with different degrees of success based on their specific limitations.

---

[8]S. Makridakis, S. Wheelwright, and R. Hyndman, _Forecasting: Methods and Applications_, Wiley, 1998.

The key issue in short-term forecasting is the possibility of appearance of an unexpected event, such as a weather disaster or an unplanned shutdown.

- *Limitations of medium-term forecasting horizon.* The key limitation of this forecasting category is the influence of business cycles and major economic movement, such as recessions and booms. Only a number of limited univariate forecasting methods, such as ARIMA, are trustable in this case. Complementary estimation of business cycles is strongly recommended. The recommended method is to use multivariate ARIMAX-based forecasting methods.
- *Limitations of long-term forecasting horizon.* In most cases the available time series forecasting methods are unreliable over a long time span. In addition to time series forecasting models, it is recommended to identify and extrapolate megatrends, and use analogies and scenario-based forecasting.

Modern Forecasting Approaches

Some AI-based Data Science methods can overcome the shortcomings of the classical forecasting approaches discribed above. The focus in modern business forecasting is on two topics: (1) multivariate forecasting based on selection of appropriate economic drivers and (2) nonlinear forecasting dealing with the nonlinear dependencies of the selected drivers. In both cases, the critical factor for success is dynamic variable selection, based on lagged economic drivers. Usually the selection is based on appropriate macroeconomic and microeconomic indicators and specific factors related to the target variable. Many AI-based Data Science methods can be used for this task.[9]

Nonlinear methods, such as SVMs, genetic programming, and neural networks have been used for forecasting in academic studies. Of special interest is the new option of long-short-term-memory recurrent deep neural networks with their capability to represent long-term time patterns (see the description Sect. 3.3.2). Unfortunately, these academic studies have not yet achieved a sufficient level of maturity to be accepted in industry. An important obstacle is the lack of statistical confidence limits in the forecasting horizon, which are a critical piece of information for most of the decisions, based on the forecast.

Recently, an approach of integrating genetic programing with the popular ARIMAX forecasting method has been attracting attention with its potential in business forecasting. The idea is to combine the advantages of both methods— generation of nonlinear transforms by genetic programming, and forecasting by an established technology, namely ARIMAX. The nonlinearity is represented with transforms generated by GP and used as explanatory variables in ARIMAX models. The final forecast is generated by these ARIMAX models. The nonlinear equations generated by GP can be based on both contemporaneous and dynamic relationships.

The development process of this integrated GP-ARIMAX forecasting system includes the following steps:

- *Step 1: Preparation of time series data.* In order to apply GP for variable selection and model generation in a time series modeling framework, some preparation of

---

[9]*See details in* T. Rey, A. Kordon, and C. Wells, *Applied Data Mining for Forecasting*, SAS Press, 2012.

Fig. 15.5 Key areas of Pareto front for selection of GP-generated transforms

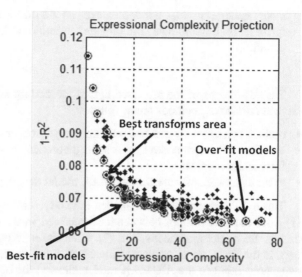

the available time series data is needed. It is crucial to include the lags for each input when selecting the variables because the relationships among the variables with the lags could be important. The selection of the number of lags is a trade-off between fully representing the dynamics and the increased dimensionality of the search space for variable selection. The key methods for preparing the data for applying GP include: stationarity and seasonality analysis, exploring cross-correlation lags, differencing, and detrending (see details in Rey et al., 2012).

- *Step 2: Model generation by Pareto-front GP*. Among the many versions of GP, it is recommended that model generation is based on Pareto-front GP, discussed in Sect. 3.3.5.
- *Step 3: Selection of nonlinear transforms*. The selection of nonlinear transforms is still an art rather than a science and is usually done around a special area on the Pareto front. An example is given in Fig. 15.5, where the area which is the focus of attention for the selection of appropriate nonlinear transforms, contains the simple models that occupy the upper left part of the Pareto front. In contrast to the case of generation of nonlinear models, the best-fit models at the knee of the Pareto front are too complex for integration with the ARIMAX framework. For the proposed integrated approach, model selection is reduced to a small number of candidate models from the first area of the best transforms with low-complexity models.
- *Step 4: Generation of ARIMAX models*. The selected transforms are used as explanatory variables in ARIMAX models. It is recommended to develop alternative ARIMAX models based on the different transforms, as well as a linear benchmark model with the original explanatory variables. The selection of the final ARIMAX model is based on the forecasting accuracy on holdout data. It has to be considered that a better fit of a transform on historical data used for GP

model generation does not "guarantee" an ARIMAX model with more accurate forecasts on holdout data. Building an ensemble of ARIMAX models is also an option.

The advantages of the proposed hybrid forecasting system based on integrating GP and ARIMAX methods are as follows:

- optimal synergy between two well-known approaches (GP and ARIMAX);
- avoiding the need to develop a solid theoretical basis for GP-based nonlinear forecasting models;
- using available software for ARIMAX model development.

This hybrid system has been applied in two typical cases in business forecasting: (1) when the relationships between the forecasted variable and the related economic drivers are contemporaneous, and (2) when the relationships are dynamic due to lags. In the first case, a simple nonlinear transform, generated by GP and used as an exogenous input in the ARIMAX model, showed the best *ex-ante* performance. In the second case, GP generated a dynamic model with accurate lags. This model was compared with another contemporaneous nonlinear model, generated by GP, and the best linear ARIMAX model. The *ex-ante* performance of the dynamic model was the best for the period of time tested. These encouraging results based on real-world applications demonstrate the big potential of the proposed approach.

15.3.2   Raw Materials Price Forecasting[10]

As we discussed in Chap. 2, price forecasting is one of the key application areas of AI-based Data Science. For some manufacturing companies, especially in the processing industry, forecasting raw materials prices is of critical importance. Some of these prices show significant variations and different patterns. An example of such a price, with 28% variation in a couple of months, is shown in Fig. 15.6. Predicting this price with univariate forecasting methods is difficult because of the changing patterns in time after the recession in 2008.

As the largest US chemical company with a broad range of products, Dow Chemical is using many different raw materials. Their total cost is in the range of billions of dollars. Reducing this cost with reliable forecasting models can deliver significant value and directly influence the company's profit.

In order to accomplish this goal, a large-scale forecasting project for the development and deployment of predictive models for the top 51 costliest raw materials prices was initiated with the support of the purchasing department. The project,

---

[10]This subsection is based on A. Kordon, Applying data mining in raw materials forecasting, in *SAS Analytics 2012 Conference*, Las Vegas, 2012.

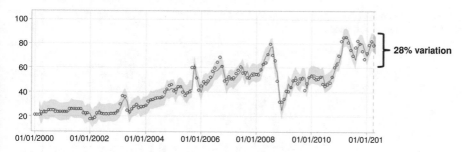

Fig. 15.6 Example of variation of a raw material price

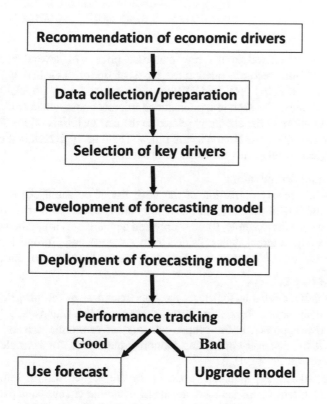

Fig. 15.7 Steps of the project for forecasting raw materials prices

based on the generic methodology given in Chap. 6, included the key steps shown in Fig. 15.7 and discussed briefly below.

Recommendation of Economic Drivers

The project included several different teams, related to selected raw materials prices in the corresponding geographic regions. Each team involved the most experienced SMEs familiar with the specific requirements of price negotiation and with the purchasing history. The key objective of this step is to define the list of potential

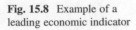

Fig. 15.8 Example of a leading economic indicator

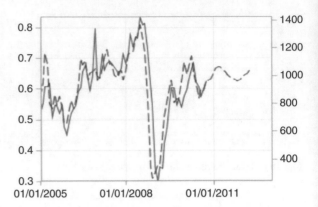

economic drivers related to the raw materials prices. In several brainstorming sessions, the SMEs recommended three types of driver. The first type included macroeconomic factors, such as GDP, consumer confidence, population growth, and exchange rates. The second type included microeconomic data from the supply and demand sectors of the economy related to the raw materials prices, for example the construction index and personal care index. The third type included competitive or related prices of other raw materials or products.

Data Collection/Preparation
This step is the reality correction of the wish list of potential economic drivers, defined by the SMEs. Usually not all of the data is available or the frequency of update is too low (for example, some supply and demand data is updated only once a year). Collecting external data outside North America and Europe is not always possible, or the available time series may be too short for historical data.

Selection of Key Drivers
In order to deliver reliable forecasts, the data scientists in the project decided to implement multivariate forecasting based on the ARIMAX approach. The key to success in this approach is the proper selection of economic drivers. Of special importance is the discovery of leading economic indicators. An example of such an indicator is shown in Fig. 15.8.

The selected leading indicator is shown by the dashed line in Fig. 15.8. The changes and patterns in this indicator are ahead of similar changes and patterns in the raw materials prices that we want to predict (shown by the continuous line in Fig. 15.8). The estimated leading lag is a month, i.e., the changes in the leading indicator influence the forecasted raw materials price a month later. Another big advantage of this leading indicator is that a forecast for it is available and can be purchased from the vendor supplying the indicator.

Development of Forecasting Models
A total of 51 ARIMAX forecasting models were developed based on the most important economic drivers, some of them leading indicators. The last 18 months of actual data were used as holdout data for validating model performance. The rest of the data was used for model development. The performance metric was the mean

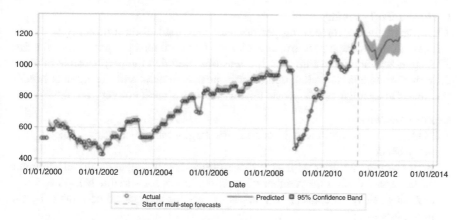

Fig. 15.9 Raw materials price forecast from the best forecasting model

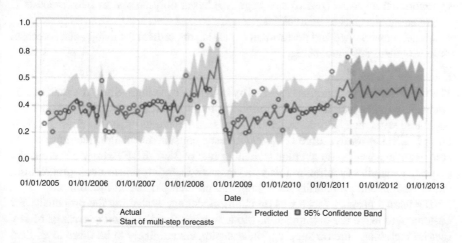

Fig. 15.10 Raw materials price forecast from the worst forecasting model

absolute percentage error. The forecasting horizon included 12 monthly forecasts. Different raw materials prices have different time series histories and forecastability, however. As a result, the performance of the models varied significantly, with MAPEs between 1.7 and 18.5 on holdout data. The forecasting plots of the best and worst raw materials models are shown in Figs. 15.9 and 15.10, respectively.

Deployment of Forecasting Models

The model deployment included an automatic data update and model execution on a monthly basis by the development software (SAS Forecast Studio). The forecasted raw materials prices and plots were delivered to price managers in Excel spreadsheets. They also included plots and values of all economic drivers related to the corresponding models.

Performance Tracking

The progress of the forecasting performance was tracked on a monthly basis. An example of the average performance after the first 6 months in operation of the first 22 models was shown in Fig. 11.12. It was observed that the performance of about 20% of the models degraded significantly after 3 months and the models needed retuning. On average, the forecasts beyond 6 months were not very reliable.

Lessons Learned

- Management commitment is critical for large-scale forecasting projects with many users.
- Brainstorming sessions with experts are critical for finding appropriate economic drivers and reducing the number of potential inputs for variable selection.
- Selecting relevant economic drivers is critical for acceptance of models by the final users.
- Managing realistic expectations of forecasting is critical for final success.
- Emotional response (positive or negative) based only on one or two forecasts is premature and has to be avoided.
- Model maintenance and performance tracking are critical for using such forecasts in raw materials purchasing.

Customer Churn[11]

The UK energy market is characterized by high levels of churn. Since the market was completely opened up to competition in 1999, 19 million customers have changed suppliers, with many domestic consumers seeing savings of up to £150 on their energy bills. Currently, some 160,000 domestic consumers change electricity or gas supplier every week—an average switching rate of 38%. EDF Energy has created a dedicated analytics function to focus on key areas including: customer segmentation, churn assessment, probability modeling, and product placement modeling.

The team's primary focus was on churn modeling: evaluating the propensity for customers to leave the organization. Modelling involved examining a range of key metrics including "life on supply" (older customers are likely to be more loyal than younger ones) and overall interaction levels. The first stage of the churn modeling process was to ensure the team had as much relevant data as possible. To support this, EDF Energy bought in third-party data sets: attitudinal data to better understand customer attitudes, and lifestyle data with demographic details, including people's locations and household arrangements. Bringing these diverse data sources together was critical to optimizing its modeling toolset.

With appropriate modeling software used to classify customer data and identify which customers were most likely to leave, EDF Energy then deployed a validation data set to monitor those customers. If they subsequently defected the Customer Insight team took the "evidence" to the EDF marketing department and so helped ensure that it could prioritize customer communications based on identified risk. An

---

[11]This subsection is based on *Understand the Propensity of Your Customers to Defect*, https://www.sas.com/en_us/customers/edf-energy.html, 2017.

additional focus was the company's "dual fuel upgrade to reduce churn" program. The aim was to understand which customers were currently "electricity only" but also mains gas capable: they could take dual fuel.

EDF Energy built customer models that predicted that the top 25% of customers covered were four times more likely to take dual fuel—and are therefore far less likely to churn. Equally, by identifying those customers least likely to take action, EDF Energy could save money that might otherwise have been wasted trying to sell upgrades to them. The company estimated the potential value from this project to be between £15 and £30 million a year.

15.4   Common Mistakes

15.4.1   Modeling with Insufficient or Low-Quality Data

Short time series, incomplete or low-quality data are more the rule than the exception in business settings. The mindset that this crappy data can be compensated with complex AI-based Data Science algorithms (referred to as the GIGO 2.0 effect) is one of the biggest mistakes in business-related applications. Especially vulnerable are forecasting projects based on short internal time-series data. Data scientists have to be very firm in explaining to clients that developing reliable forecasting models without the collection of relevant historical data is a waste of time.

15.4.2   Businesses Not Prepared for Cultural Change

AI-based Data Science business applications require changes in existing work processes and modes of operation. It is recommended that these changes are well defined and communicated to all related stakeholders. Unfortunately, this is a task of low priority to management and neglected by data scientists. As a result, effective technical solutions may sink in an inappropriate organizational structure.

15.4.3   No Links to Manufacturing-Related AI-Based Data Science

This is the same mistake that manufacturing-related businesses make. All recommendations, given in Sect. 14.4.2 are valid for this case as well.

15.5   Suggested Reading

- C. Chase, *Demand-Driven Forecasting: A Structured Approach to Forecasting*, second Edition, Wiley, 2013.
- T. Davenport, *The AI Advantage: How to Put the Artificial Intelligence Revolution to Work*, MIT Press, 2018.
- M. Gilliland, L. Tashman, and U. Sglavo, *Business Forecasting: Practical Problems and Solutions*, Wiley, 2015.
- S. Makridakis, S. Wheelwright, and R. Hyndman, *Forecasting: Methods and Applications*, Wiley, 1998.
- T. Rey, A. Kordon, and C. Wells, *Applied Data Mining for Forecasting*, SAS Press, 2012.

15.6   Questions

Question 1
What are the specific features of business applications?

Question 2
What are the key features of business data?

Question 3
What is a customer twin?

Question 4
What are the advantages of business forecasting?

Question 5
Rank the most important application areas of AI-based Data Science in business?

Chapter 16
How to Operate AI-Based Data Science in a Business

> *Use AI or die!*
>
> *Anonymous*

AI-based Data Science is rapidly becoming a core enterprise competency. Organizations slow to implement it risk falling behind competitors. Companies need quick and reliable insight into the current and future performance of their processes, as well as the evolving needs of customers. AI-based Data Science is no longer the purview of the Olympian companies like Google or Amazon. It is a critical competitive differentiator for any modern business.

Introducing and operating these technologies in a business is not a trivial process, however. It requires some level of digital maturity and risk-taking culture in the business. Above all, it is a long and challenging process that is business-specific. There are some generic steps that can be defined, based on the experience of leading companies in applying AI-based Data Science, though.[1]

The focus of this chapter is on giving some guiding principles to help the reader to navigate the complex process of introducing and operating AI-based Data Science in a business. The first objective of the chapter is to suggest proper criteria to evaluate when a business is ready to start this long journey. The second objective is to propose a roadmap for operating AI-based Data Science, with three key steps of introducing, applying, and leveraging the technology in a business, while the third is to discuss the possible organizational structures for the related capabilities.

---

[1] A recommended reference that summarizes experience in applying AI is T. Davenport, *The AI Advantage: How to Put the Artificial Intelligence Revolution to Work*, MIT Press, 2018.

© Springer Nature Switzerland AG 2020
A. K. Kordon, *Applying Data Science*,
https://doi.org/10.1007/978-3-030-36375-8_16

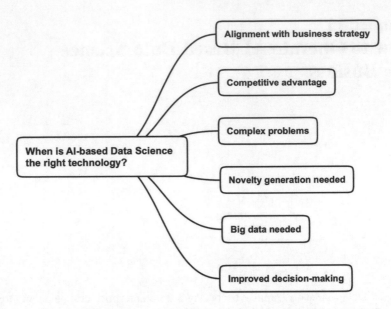

Fig. 16.1 Criteria for evaluating potential applications of AI-based Data Science technology

16.1   Is a Business Ready for AI-Based Data Science?

The first step in the process of applying these technologies in a business is answering this question as clearly as possible. In many cases, however, this is not an easy task. It requires convincing arguments that AI-based Data Science fits into the business strategy and will increase the company's competitive advantage. This is only one side of the question, though. The other side depends on the assessment of both the technical and the organizational readiness of the business to satisfy the requirements of the technology. Both of these issues are discussed briefly below.

16.1.1   When Is AI-Based Data Science the Right Technology?

Some criteria to be used in answering this question are captured in the mind-map in Fig. 16.1 and discussed shortly below:

- *Alignment with business strategy*. The litmus test for the future need for applying AI-based Data Science is whether this technology fits into the business strategy. It is not recommended to start these efforts when the answer is no. It is suggested, however, to discuss the new opportunities that this technology may bring to the company and revise the business strategy accordingly.

- *Competitive advantage.* The most important criterion for applying AI-based Data Science is the competitive advantage it gives to the business. It will be extremely difficult to sustain long-term success if there is no clear vision of the sources of competitive advantage in implementing this emerging technology in specific business areas. Two of the most frequent sources of competitive advantage from AI-based Data Science, dealing with complex problems and novelty generation, discussed in detail in Chap. 1, are highlighted below.
- *Complex problems.* Often businesses look at AI-based Data Science after exhausting the capabilities of existing technologies. The key reason is the high level of complexity of problems in the business, especially those related to a large number of factors with nonlinear dependencies. This category also includes problems with very high dimensionality and any type of complex social interaction. The methods that contribute the most to the modeling of complexity are multivariate statistics, machine learning, deep learning, and evolutionary computation.
- *Novelty generation.* AI-based Data Science is very effective for businesses where innovation is critical. Intelligent agents can capture emergent behavior from local interactions, which could be identified as novelty. Evolutionary computation automatically generates novel structures, such as the areas of electric circuits, optical lenses, and control systems. Neural networks, support vector machines, and evolutionary computation capture unknown dependencies between variables that could be used for process monitoring and new product design.
- *Big data needed.* A clear indicator of potential AI-based Data Science applications is a growing use of various types of data in the business. It is especially critical if image, video, audio, or text data from social networks is needed. In general, the more data is expected to be used, the more the benefits there will be from implementing this technology by effectively translating the available data into insight and value.
- *Improved decision-making.* Another important criterion for selecting AI-based Data Science is the importance and complexity of business decisions. This technology can significantly increase the speed of decision-making and even automate it (as is the case with fraud detection in bank applications.) Business forecasting is a must in making correct business decisions based on data and appropriate predictive algorithms. In general, the overall business decision-making process in an organization can benefit by reducing the number of subjective gut-based decisions with objective analytically driven decisions.

16.1.2   Estimating the Readiness of a Business for AI-Based Data Science

If the business decides that it will benefit from implementing this technology, estimating its readiness to start this process is strongly recommended. Some important criteria to be checked are shown in Fig. 16.2 and discussed briefly below:

- *Leadership support.* The most important thing to be done before estimating the readiness of the organization for applying AI-based Data Science is to grab the top management's attention and make them aware of and excited about the potential of this technology. Executives should approach AI-based Data Science as an instrument to expand their business—creating new products or services, increasing productivity, or winning more market share—as much as a tool to cut costs. The expected leadership support is defined as assigning an exploratory team to study the technology, approval of initial goals, justifying the projects needed and securing adequate funding.
- *Identified business needs.* The opportunities are identified by considering a number of different business-specific criteria. The potential needs are brought

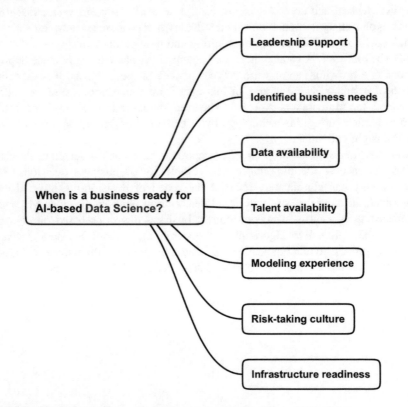

Fig. 16.2   Criteria for readiness of a business to implement AI-based Data Science

to the surface through interviews with relevant managers in the areas being assessed, and these interviews should be done by people who have a sound understanding of the framework for the AI-based Data Science capability as well as of business-related used cases. Usually the starting list of potential projects is concerned with improving the processes that the business already has, without necessarily changing the way the business works. The most explored areas are saving employee hours, decreasing response times, and saving expense on repair and replacement of material.

- *Data availability.* The most important technical criterion for assessing the readiness of the business to apply AI-based Data Science is the status of its data-handling capabilities. This includes an evaluation of the completeness of the current internally collected data and an estimate of the external data for potential future use in terms of documents, images, video, audio, and system logs. Another action item is switching from legacy data systems to a more nimble and flexible cloud architecture to store and harness big data. For some targeted applications in business forecasting, an assessment of the length of the internal historical data is required.
- *Talent availability.* It is assumed that, at the beginning, the business is lacking the talent needed to move effectively into this technology. Assessing the level of knowledge or awareness of statistics, analytics, or AI among the available in-house specialists is the first recommended step. Another option is to use external experience from consultants or vendors to begin filling the talent gap. An initial target number of data scientists, based on projected demand, should be defined and hired as soon as possible. It is not recommended to start any activities in applying AI-based Data Science without a small team of data scientists and external experts in the field.
- *Modeling experience.* Any previous experience in using statistical methods or any different type of models should be considered. This is a critical factor for accepting the new technology. The high complexity of AI-based Data Science is a high cultural obstacle for users. The more experience they have from previous use of math and statistics, the less the cultural shock.
- *Risk-taking culture.* Applying this complex technology requires accepting a higher level of risk than using established, well-defined solutions. Some organizations are not ready or willing to take such risks. It is very important to identify the groups or sections in the business with a low-risk culture and to avoid starting implementation efforts with them.
- *Infrastructure readiness.* A summary assessment of how the capabilities of the business infrastructure can satisfy the expected requirements of AI-based Data Science applications is also suggested. This includes evaluating the gaps in computing power, needed data, software, and work processes. Some action items, such as establishing a cloud service, upgrading databases, and purchasing appropriate software, are recommended.

16.2   Roadmap for Operating AI-Based Data Science in a Business

At the beginning of the long journey to implementing AI-based Data Science in a business, all stakeholders must align their expectations. Of critical importance is understanding that it takes time to get AI-based Data Science right. It took even the leaders, such as Google and Amazon, years to integrate AI into their business DNA. The recommended time range for implementing the technology in big enterprises is from 3 to 5 years, dependent on their size, the complexity of their dominant AI-based Data Science applications, their infrastructure capabilities, and the maturity of the talent available.

AI-based Data Science is not a plug-and-play exercise. It requires well-coordinated joint efforts to integrate big data with complex software, and well-educated data scientists collaborating with business experts. The final success and real speed of introducing AI-based Data Science depends strongly on the efficiency of this integration.

There are many ways to introduce and integrate a new technology such as AI-based Data Science in a business.[2] These ways also depend on the size of the business and its capacity to allocate resources for internal exploration of the technology and for project development. At one end are the small businesses that cannot afford any development efforts and are interested in cheap and simple solutions with minimal capital and training cost. At the other extreme are the large international corporations with tremendous capabilities to use internal and external resources for development and deployment of the technology across the globe.

A roadmap showing several generic phases in applying AI-based Data Science, summarizing experience from big businesses, is shown in Fig. 16.3. While it best fits big enterprises, the proposed approach can be used, on a different scale, in most businesses. The roadmap separates the application sequence into three distinct phases with growing impact: (1) an introduction phase, (2) an application phase, and (3) a leveraging phase.

The purpose of the introduction phase is to validate the potential of the technology based on well-selected pilot projects. Since there is a deficit of knowledge about AI-based Data Science, external resources from universities, vendors, and consultants are used in marketing, training, and project development. The objective of the application phase is to validate the value creation potential of the technology in several high-value business projects. Very often, external resources from vendors are used in project development and support. The goal of the leveraging phase is to maximize the benefits of AI-based Data Science through its integration into existing work processes. A more detailed description of each phase follows next.

---

[2]A recommended reference discussing this issue is T. Davenport and J. Harris, *Competing on Analytics: The New Science of Winning*, Harvard Business School Press, 2007.

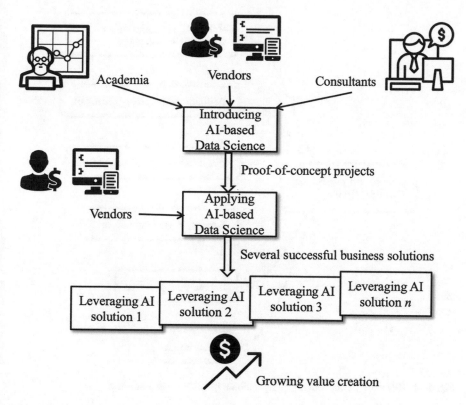

Fig. 16.3 Key phases of a roadmap for applying AI-based Data Science in a business

16.2.1   Steps for Introducing AI-Based Data Science into a Business

The key steps for introducing AI-based Data Science into a business are shown in Fig. 16.4 and discussed briefly below:

- *Market the competitive advantage of AI-based Data Science.* The obvious first step in introducing AI-based Data Science into an organization with no or minimal knowledge of the technology is marketing. The objective is to demonstrate the basics and key competitive advantages of AI-based Data Science to a broad audience of potential users. Some guidelines for organizing the marketing efforts were given in Chap. 13, and examples of presentation slides and elevator pitches for the key AI-based Data Science methods were given in Chap. 5. It is possible to include external resources from leading researchers, consultants, or vendors. The ideal case is a presentation that illustrates use cases and the benefits of applying AI-based Data Science in areas, similar to those in the targeted business.

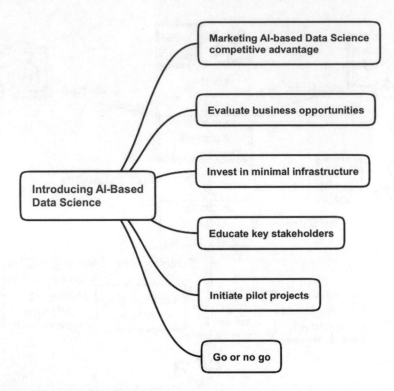

Fig. 16.4 Key steps for introducing computational intelligence in a business

- *Evaluate business opportunities*. The driving force for using AI-based Data Science is some business needs that are difficult to solve using existing methods. At this early stage, it is necessary to identify and prioritize the generic needs. Examples of such generic needs are improved process-monitoring systems, process control based on inferential sensors, and development of robust empirical models for new products. In many cases, the existing solutions for each of these needs, based either on first-principles models or on statistical models, have exhausted their potential. Most AI-based Data Science technologies offer capabilities, such as low-cost data-driven models or pattern recognition algorithms, which may deliver the desired solutions at low cost.
- *Invest in minimal infrastructure*. Selecting appropriate vendors and developing the needed database, hardware, and software infrastructure has to be done carefully, focusing on low-cost options. The key purpose of this minimal infrastructure is to allow a demonstration of the power of AI-based Data Science on the selected pilot projects. Another objective is to train and educate key stakeholders.
- *Educate key stakeholders*. It is good practice to introduce the technology in more detail to future developers and users. The resources given in this book are a good starting point. Many introductory presentations and webinars for each of the AI-based Data Science technologies are available on the web as well. Most of

the vendors and leading AI-based Data Science conferences also offer tutorials on the technologies. It is strongly recommended to attend these conferences (most of them are business-oriented) and to begin building a network with the growing business community of companies applying AI-based Data Science. This is a very dynamic research area with fast transfer to the business world, and needs to be continuously on the radar screen of the AI-based Data Science leadership.

- *Initiate pilot projects.* The most important step in the introduction stage is the problem of identifying the top-priority business needs, demonstrateting the competitive advantage over the existing solutions, and implementing this in a relatively short period of time, preferably less than 6 months. An example of a pilot project is the development and deployment of a specific inferential sensor in a manufacturing process, similar to the emissions estimation project, discussed in this book. It is normal practice to transfer knowledge from external resources during the development and deployment of the pilot project.
- *Go or no go.* The results from the pilot project are the litmus test for triggering the next phase of the application strategy—applying AI-based Data Science in different large-scale projects. The decision has to be made based on the performance metric of the pilot project and acceptance of the solution by the users.

16.2.2   Steps for Applying AI-Based Data Science in a Business

The key steps for applying AI-based Data Science in a business are shown in Fig. 16.5. It is assumed that the pilot project has raised interest in and demonstrated the unique technical capabilities of AI-based Data Science. The objective of this step is to demonstrate the value creation potential of the technology by effective applications for solving appropriate business problems:

- *Define application strategy.* It is strongly recommended to direct the AI-based Data Science application efforts in a systematic way. As a good starting point, a more detailed survey than in the introductory phase of the current business needs, and a value assessment, is recommended. Often some of the selected application areas are linked to the pilot application (for example, focusing on the development of inferential sensors). In this case, it is preferable to gradually increase the number of applications based on the value delivered. Identifying other application areas is always a plus, for example, predictive maintenance.
- *Identify business problems related to AI-based Data Science.* Selecting appropriate projects must follow the application strategy. However, it is not recommended to jump immediately from pilot to large-scale projects. For example, an ambitious project to develop 50 inferential sensors for a wide range of manufacturing processes creates a significant risk of failure due to limited experience and insufficient internal resources.

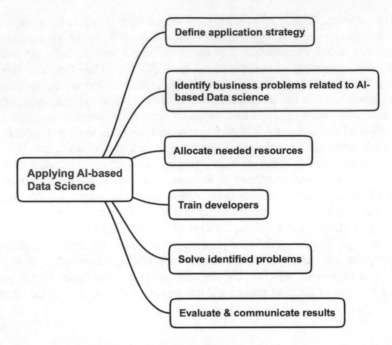

Fig. 16.5 Key steps for applying AI-based Data Science in a business

- *Allocate needed resources.* The biggest challenge in the application phase is the limited capacity available for development, deployment, and support of current and potential new projects. On the one hand, it is too early to estimate the long-term demand and to commit substantial internal resources. On the other hand, relying only on external resources does not lead to a strategic presence of AI-based Data Science in the business. The key to the decision-making process about this topic is an assessment of the need for internal development. If the results from the first few pilot applications demonstrate value creation, it is recommended to evaluate the option of gradually allocating resources for internal development of the projects. For example, in the case of growing demand for different inferential sensors, it is better to transfer the model development from the vendor to several data scientists supporting internal experts in the area of process monitoring.
- *Train developers.* An unavoidable step when resources are allocated for internal model development is comprehensive training in AI-based Data Science methods and the corresponding tools. Due to the rapid growth of the technology, the training mode is almost continuous. This training could be done by using publicly available webinars, by vendors and at key conferences related to AI-based Data Science.
- *Solve identified problems.* The most important step in this phase, however, is showing effective problem solutions. It is recommended to document all details in each steps of model development and deployment. Tracking the performance and

value creation for some period of time after deployment and comparison with accepted benchmarks is the suggested way to demonstrate the quality of the applied AI-based Data Science solutions.

- *Evaluate and communicate results.* Nothing speaks louder in support of AI-based Data Science than a series of successful applications with documented value creation. Of special importance is the demonstration of competitive advantage over established solutions. For example, the overall savings from applying several inferential sensors may exceed millions of dollars relative to established but expensive hardware analyzers.

In order to build credibility, it is recommended to communicate also the issues with and lessons learned from failed applications. For example, concerns about support for inferential sensors have to be addressed, and illustrated with comments from the final users. Cases of unsuccessful inferential-sensor development due to insufficient or low-quality data have to be given as examples of the limitations of the proposed technology.

16.2.3   Steps for Leveraging AI-Based Data Science in a Business

The best-case scenario of the roadmap is when the business finds opportunities for sustainable value creation by using AI-based Data Science in different areas. The key steps for leveraging the technology are shown in the mind map in Fig. 16.6 and discussed briefly below:

- *Mandate from management.* The critical step in moving ahead with AI-based Data Science across the business must be clearly supported by the top management. Ideally, a top manager takes the lead and forms a steering committee with key experts and executives. The mandate for leveraging the technology assumes strategic support for several years, and corresponding organizational decisions. The most important factor is the vision and commitment of the top manager in charge. Unfortunately, there is always a risk of management changes and the replacement of the visionary sponsor with a narrow-minded bureaucrat. In this case, there are no guarantees that the mandate will be fulfilled with the previous enthusiasm and support. The best protection is sustainable value creation from the existing applications.
- *Define statement of direction.* One of the important documents that navigates the leveraging efforts is the statement of direction. This explicitly defines the appropriate uses of AI-based Data Science in the business (for example, in process monitoring and control, preventive maintenance, and the supply chain). Another important section of the document is the scope of applicability (for example, a list of businesses, software environments, data warehouses, and control systems). The statement of direction also includes the appropriate actions that should be

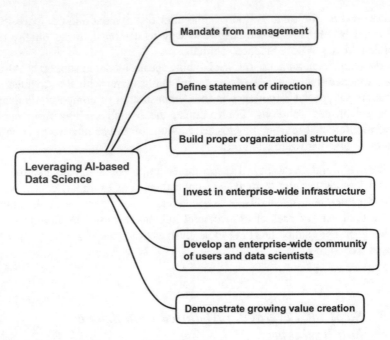

Fig. 16.6 Key steps for leveraging AI-based Data Science in a business

taken by the different stakeholders, the next steps to be taken to further leverage the technology in a specific period of time, the key milestones with specific dates, and the owners, stakeholders, and contacts.

- *Build a proper organizational structure*. A very important task in this final phase of integrating AI-based Data Science into the business is to select and develop an organizational structure that will drive effectively the future growth of the resulting capabilities. Some directions on how to accomplish this task are given in the next section.

- *Invest in enterprise-wide infrastructure*. Usually the small-scale infrastructure used for the introduction and initial application of AI-based Data Science is insufficient for scaling up to enterprise-wide level. Upgrading cloud services, big-data-type databases, and professional software packages is strongly recommended. Only in this way can the growing volume of projects and deployed solutions in production be effectively handled. Of special importance is resolving the issue of long-term maintenance and support for existing and future applications. In addition to support from vendors, internal resources are needed and must be allocated.

- *Develop an enterprise-wide community of users and data scientists*. In addition to the economic benefits, successful leveraging of AI-based Data Science creates a growing network of users and technology supporters who can initiate the next application wave in a natural way. Coordinating this community is a responsibility of data scientists. They can establish an appropriate form of regular communication between group members, exchanging experience from applied solutions and introducing new methods from this dynamic technical field.

- *Demonstrate growing value creation.* The definition of success for leveraging is self-sustainability, i.e., the value created from implemented projects fully covers the total cost of ownership of the AI-based Data Science organization. At that point the technology growth is based on project demand not on management support. In our example of inferential sensors, after several applications in different manufacturing processes, the demand and, correspondingly, the generated value and support began to grow and cover the development cost.

Unfortunately, not so many businesses have reached this important phase. A recent McKinsey survey found that only 8% of 1000 respondents with analytics initiatives engaged in effective scaling practices. More boards and shareholders are pressing for answers about the scant returns on many early and expensive analytics programs. Overall, McKinsey has observed that only a small fraction of the value that could be unlocked by advanced-analytics approaches (built on AI-based Data Science) has been unlocked—as little as 10%.[3]

16.3 Key Organizational Structures for AI-Based Data Science Capabilities

Defining an appropriate group size and gradually developing an effective organizational structure within a business are two of the most important tasks in implementing AI-based Data Science. The first critical decision is about the size of the group. Unfortunately, there is no magic, generally accepted number that can be suggested. The best approach is to evaluate the future demand first and then define the size of the group and its growth over the next 3–5 years. It is also recommended to clarify which capabilities could be shared with the IT groups. Effective coordination can eliminate functional duplication (especially in data management and collection) and avoid an inflated group size.

The second important decision is selecting the appropriate organizational structure. Various options are possible but they can be generalized into four key categories: (1) a centralized group, (2) decentralized groups, (3) a network of individual data scientists, and (4) using external options, such as AI-based Data Science as service. These options are discussed briefly below.

---

[3]Oliver Fleming, Tim Fountaine, Nicolaus Henke, and Tamim Saleh, *Ten Red Flags Signaling Your Analytics Program Will Fail*, McKinsey Analytics, May 2018.

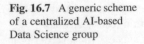

Fig. 16.7 A generic scheme of a centralized AI-based Data Science group

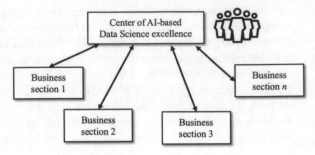

16.3.1   Centralized AI-Based Data Science Group

In this option, the business has one corporate organization that owns the AI-based Data Science capability, often called the center of excellence (COE). An example of such a structure is shown in Fig. 16.7.

The COE concentrates all data scientists, supporting staff, and available infrastructure in one central organization. The leader of this organization reports to a top executive. During the introduction and application phase, the COE is funded by a centralized business budget. It is expected, however, that during the leverage phase the center will gradually become self-funded. The COE helps all sections in the enterprise with different services, related to AI-based Data Science, such as: developing and deploying solutions, training, and taking care of the infrastructure. It also provides enterprise-wide functions, such as introducing new related technologies, continuously updating the statement of direction, and organizing a community of data scientists and users of AI-based Data Science.

There are clear benefits from this centralization approach. It reduces the need for hiring multiple data scientists and eliminates duplication of effort in different parts of the business. The efficiency of operation and the breadth of knowledge of the available data scientists is high because they can share their experience more effectively. Another advantage of a COE is that the high concentration of resources justifies purchasing of the most powerful infrastructure, including expensive software packages for enterprise-wide applications. The most important advantage, however, is that a single AI-based Data Science solution can be easily leveraged across the enterprise.

The key disadvantage of the centralized group structure is its high cost. Only large, profitable businesses can afford this option.

16.3.2   Decentralized AI-Based Data Science Groups

This is a functional structure where each sector in the enterprise has its own Data Science team. An example of such a structure is shown in Fig. 16.8.

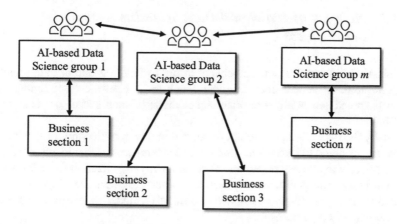

Fig. 16.8 A generic scheme of decentralized AI-based Data Science groups

These teams of three to ten data scientists are linked to one or several business departments. They are relatively independent, but their scope of activity is limited in comparison with the centralized group option. The specialized AI-based Data Science groups are mostly focused on developing, deploying, and leveraging projects related to their corresponding business sector. Usually it is not expected that they will perform training on AI-based Data science methods and introduce new technologies.

Decentralized AI-based Data Science groups are similar in many ways to a COE in their needs but tend to get less management visibility and financial support. As a result, they are often limited to using cheaper and less advanced tools and capabilities. Another side effect is that they are likely to get less support from IT, and they need to allocate group' resources for data management and collection.

The key advantage of a decentralized AI-based Data Science structure in a business is its lower cost relative to the COE option. For many companies, the degree of centralization may change over time. In the early phases of the long journey of implementing this great technology, it might make sense to work more centrally, since it is easier to build and run a COE and ensure the quality of the solutions derived. But over time, as the business becomes more proficient, it may be possible for the center to step back to more of a facilitation role. The businesses are given more autonomy, and project development is distributed across the key business sectors.

16.3.3 Network of Individual Data Scientists Across the Business

Medium-size and small businesses cannot afford to fund a group of data scientists. They are looking at low-cost solutions, such as a network of individual data scientists spread across the enterprise. An example of such a structure is shown in Fig. 16.9.

Usually the data scientists are not organized in any specific way. In general, this option is considered to be the least mature, and lacking in systematic guidance about how effectively to allocate resources. There are many organizational challenges, but from a technical standpoint, the emphasis is on a self-service mode of operation. This is based on using relatively automated and intuitive modeling platforms for model development and deployment.

The key issue with this option, however, is the high requirements for the skillsets of the individual data scientists. In contrast to the group options, where differences in knowledge, experience, and skillset are balanced among the group members, individual data scientists need to have broader than average professional capabilities. As a bare minimum, knowledge of key AI-based Data Science methods, and programming and database skills are expected. Other requirements are good communication capabilities, teamwork, and leadership. It is extremely difficult to find such "walk on water" data scientists, though. In reality, these high requirements are distributed with different levels of success among the available data scientists. For example, one data scientist could be very knowledgeable about machine learning but have bad communication skills and a lack of leadership. Another one could be a brilliant programmer but have minimal awareness of AI-based Data Science methods and no database knowledge. As a result, individual differences could be very big, and operating such a network of individual data scientists could be very challenging. It takes some time

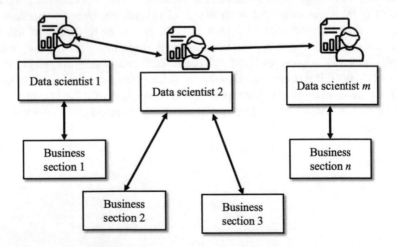

Fig. 16.9 Generic scheme of a network of individual data scientists

and patience to align these skillset differences and to achieve high-level performance consistently.

Another challenge with this structure is the performance assessment and career development of the individual data scientists. They are usually part of a business department and report to a manager with no knowledge of AI-based Data Science. As a result, the assessment of the data scientists' performance is very subjective and their career development is at a disadvantage. After realizing this issue, some data scientists leave.

In theory, a network of individual data scientists is a low-cost and affordable option for medium-sized and small businesses. In practice, some issues related to big differences in the data scientists' skillsets and a lack of an effective mechanism for their engagement may raise significantly the cost due to additional expenses for training.

16.3.4   Using AI-Based Data Science as a Service

Many organizations see the potential value of using the great opportunities of AI-based Data Science but are not willing to allocate internal resources to learn, develop, and maintain the technology. Fortunately, a new implementation option, AI-based Data Science as a service, has become available recently. An example of such a structure is shown in Fig. 16.10.

In this scenario, it is assumed that the business makes only a minimal investment in infrastructure and people related to the technology. The expectation is that most of the AI-based Data Science capabilities and activities will be delivered by a

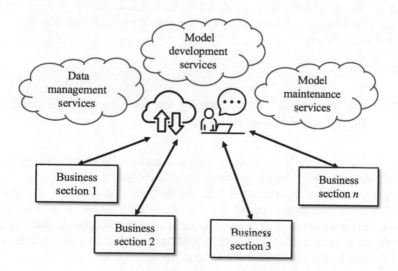

Fig. 16.10 Generic scheme of AI-based Data Science as a service

cloud-based service, and virtual interactions with data scientists at the service provider. It is still necessary, however, that the business hires at least one data scientist to coordinate these interactions and manage the developed project inside the organization. A suggested condition for selecting this option is the availability of a strong IT team that can handle the internal data.

The key advantage of using AI-based Data Science as a service is that, due to its low cost, it can open the door for mass-scale implementation of this technology to middle-size and small businesses.

16.4   Common Mistakes

16.4.1   Introducing AI-Based Data Science by Top Management Campaigns

One of the worst-case scenarios for implementing AI-based Data Science is when it is done by enterprise-wide campaigns pushed by ignorant "visionary" top executives. This usually includes big investments in infrastructure and people for a short period of time that are not justified by the business demand. It is expected, however, that these investments will be returned as soon as possible as a result of introducing the technology in a military style, following the top executives' orders. Unfortunately, due to its complex nature, AI-based Data Science is not appropriate for accelerated integration into a business's mode of operation. It takes time for training and understanding the technology, and evaluating the best opportunities for value creation. Attempts to deliver enterprise-wide "fast-track" solutions via taking shortcuts in the AI-based Data Science workflow lead to low-quality "burger" models with all related negative consequences. They pave the way for an inevitable total implementation fiasco.

16.4.2   Ignoring of the Readiness of the Business Assessment for AI-Based Data Science

Another, not so drastic form of the previous mistake is when top executives are pushing for technology implementation without analyzing the real needs in the business. Usually this is combined with avoiding an assessment of the level of readiness of the enterprise to absorb the cost and to use AI-based Data Science. As a result, the introduction of the technology could be very inefficient due to wrong investments in infrastructure and people, and inappropriate use. The guidelines in Sect. 16.1 can be used to avoid this common mistake.

16.4.3   No Gradual Application Strategy

Some businesses implement AI-based Data Science with no clear application strategy, relying on the project demand only. This "random walk" approach leads to an unpredictable trajectory of introduction of the technology. As a result, the planning process is difficult and the potential for future growth could be either mishandled or missed. Following the gradual application strategy described in Sect. 16.2 is the best way to avoid this mistake.

16.4.4   Inappropriate Organizational Structure

The most common mistake in implementing AI-based Data Science is the inefficient way it is structured. There is no generic solution, unfortunately, but businesses can test which of the structures discussed in Sect. 16.3 fit their needs the best.

16.5   Suggested Reading

- A. Burgess, *The Executive Guide to Artificial Intelligence*, Palgrave Macmillan, 2018.
- T. Davenport and J. Harris, *Competing on Analytics: The New Science of Winning*, Harvard Business School Press, 2007.
- T. Davenport, *The AI Advantage: How to Put the Artificial Intelligence Revolution to Work*, MIT Press, 2018.
- Ng. *Transformation Playbook: How to lead your company into the AI Area*, https://landing.ai/ai-transformation-playbook/, December 13, 2018.

16.6   Questions

Question 1
When is AI-based Data Science an appropriate technology for a business?

Question 2
When is a business ready for AI-based Data Science?

Question 3
When can a business move from applying AI-based Data Science to leveraging it?

Question 4
What are the key steps in applying AI-based Data Science?

Question 5
What are the advantages of a decentralized AI-based Data Science group?

Chapter 17
How to Become an Effective Data Scientist

> *Talented data scientists leverage data that everybody sees.*
> *Visionary data scientists leverage data that nobody sees.*
>
> *Vincent Granville*

This last chapter in the book is focused on the professionals who drive the broad field of Data Science in general and its AI-based augmentation in particular. Finding an acceptable and clear definition of a data scientist is a challenging task. The broad opinion varies from a magician doing voodoo in translating data into value to somebody who plays with numbers in Excel spreadsheets. The common perception, however, is that data scientists apply statistics, machine learning, and analytic approaches to data to solve critical business problems. Their primary objective is to help organizations to turn their volumes of data into valuable and actionable insights.

The exponential growth of Data Science (especially AI-based Data Science) as a technology has created a big problem of educating and training an enormous number of people in a short period of time. Unfortunately, this mass-scale effort of incubating fast-track data scientists has created a community with vast differences in knowledge and in capabilities to solve business problems.

The first objective of the chapter is to clarify the different types of data scientist—from the top experts, called unicorns, to the masses of relatively less knowledgeable citizen data scientists. The second objective is to describe the key recommended capabilities for a data scientist, while the third is to define the skillset expected from an effective data scientist. The fourth objective is to suggest a roadmap for how to become an effective data scientist.

The analysis and recommendations in this chapter are for generic data scientists. However, due to the more complex nature of AI-based Data Science, they are valid to an even greater extent for data scientists using these more advanced methods.

Acknowledging the power of good humor, we give some satirical quotations and versions of Murphy's law related to AI-based Data Science at the end of the chapter. A good laugh is the best way to communicate and resolve issues even in a complex area like AI-based Data Science. It is the finest way to finish this book, too.

© Springer Nature Switzerland AG 2020
A. K. Kordon, *Applying Data Science*,
https://doi.org/10.1007/978-3-030-36375-8_17

17.1   Types of Data Scientists

One of the biggest misconceptions about data scientists is that they are a homogeneous community of talented geeks with uniform high-level technical knowledge. Nothing is farther from reality, unfortunately. Due to the mass-scale educational efforts of different schools with various levels of depth, the Data Science community has several distinct types of data scientist with big differences in their experience in Data Science. One of the first task in analyzing data scientists is to define these types and clarify their differences.

17.1.1   "The Sexiest Job of the Twenty-first Century" Hype

We'll start with the most popular image of data scientists, spread by the media. In 2012 the job of data scientist was dubbed by a popular *Harvard Business Review* article as "the sexiest job of the twenty-first century".[1] Data scientist is still (as of December 2019) ranked number 1 in Glassdoor's list of "best jobs in America."[2] That is a lot of hype surrounding one job: it represents the future of technology and is an exciting career path that pays handsomely and fits neatly into a variety of businesses. On top of that, it has the quite unusual attraction for a technical job, of sex appeal! However, devoted data scientists have to beware of some unpleasant surprises, like the one, shown in Fig. 17.1.

Beyond the hype, the reality is quite different, though. Part of it is due to the tremendous growth in demand—while schools and programs are cranking out legions of new data scientists, these are still not enough to satisfy the business needs for top technical talent. In many organizations, data science departments are becoming the bottlenecks for future growth.

The fundamental problem about high-quality training of future data scientists is the enormous deficit of well-prepared teachers in this very broad and vaguely defined discipline. The paradox is that due to the novelty of the discipline, Data Science teachers have not been trained as data scientists themselves! They have migrated to this discipline from various kinds of degree, most of them from computer science degrees. That leads to significant variations in the content of the training materials and the quality of teaching. In addition, training in nontechnical skills, which are critical for dealing with the business world, is dangerously absent. As a result, the first wave of solidly trained data scientists with corresponding bachelor's or master's degrees in Data Science have wide differences in technical knowledge and minimal preparation for industrial reality.

---

[1]T. Davenport and D. J. Patil, Data scientist: The sexiest job of the twenty-first century. *Harvard Business Review*, **90**, (10), pp. 70–76, 2012.

[2]https://www.glassdoor.com/List/Best-Jobs-in-America-2019-LST_KQ0,25.htm.

Fig. 17.1   An unexpected effect of working hard in the sexiest job of the twenty-first century

The big issue is that the growing immediate business needs for top talent in Data Science cannot be satisfied by solidly trained-with-degree data scientists only. One solution to this talent demand crisis is to significantly reduce the professional standards. Another name for this process is the democratization of Data Science.

This is a practice that allows the proliferation of data engineers with limited Data Science skills, or even nontechnical individuals who have passed online training classes with fashionable names, who perform some of the basic functions that are supposed to be the territory of data scientists. These people are a significant part of the growing type of low-qualified data scientists called citizen data scientists. The expectation is that the routine steps in the Data Science workflow may be increasingly handled by citizen data scientists with the help of automated tools, rather than by actual data scientists with deep technical and social skills. The problem is that the value creation steps in the Data Science workflow require such skills, i.e., the final result will be inefficiently solved problems with low quality and return on investment.

The ultimate idea for resolving the crisis in the demand for data scientists is to replace them with magical software. Some vendors are raising such high expectations with shiny promises of "Data Science that is Automating Itself." As we discussed in Chap. 6, this is the road to a McDonald's-style Data Science with inefficient "burger-type" models.

Fig. 17.2 The Troika of
Data Scientists unicorns,
data scientists, and citizen
data scientists ranked by
their experience in Data
Science

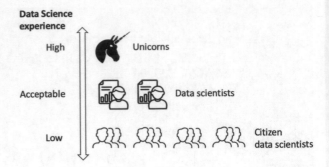

Not surprisingly, those trends are unpopular and controversial within the Data Science community at large. The biggest concern is about future career development issues. In most cases citizen data scientists have an inflated job title of "data scientist" that does not correspond to the expected skillset. When the next wave of solidly trained fresh data scientists comes from the universities, the corresponding positions in businesses will have already been occupied by such citizen data scientists pretending to be data scientists. Most probably they will compensate for their lack of deep knowledge with administrative power and become the new data scintists' bosses. . .

17.1.2   The Troika of Data Scientists

The drastic differences in data scientists' qualifications require special attention. In order to clarify the diverse capabilities of this growing professional community with the same job title, it can be divided into three key groups: (1) unicorns, (2) data scientists, and (3) citizen data scientists. These groups are defined based on their average experience in applying Data Science, which may be classified into three qualitative categories of high, acceptable, and low levels. This suggested professional grouping, called the Troika of Data Scientists, is visualized in Fig. 17.2, where the three different groups are shown with corresponding icons, and discussed briefly below.

Unicorns: The Leaders
This mythical group of data scientists have been called "unicorns" because finding the right person with the right skillset—including coding, statistics, machine learning, database management, visualization techniques, and industry-specific knowledge—is practically impossible. This unique combination allows unicorns to become the natural leaders of Data Science organizations. They are especially helpful in the introductory phase of AI-based Data Science in a business. Their integrated thinking is priceless in building the needed infrastructure, attracting and training talent, and selecting appropriate projects.

Unicorns are also the appropriate leaders in decentralized AI-based Data Science groups that require more universal knowledge about the technology due to their more limited resources. As typical generalists, unicorns have broad knowledge in almost all areas related to Data Science but are not masters of the details of all of the technologies. Their key strength is in understanding the forest, not the leaves.

Data Scientists: The Drivers
The second member of the Troika is the average data scientist. This is the group that drives most of the applications in businesses. Usually the average data scientist has a professional degree in the technology and a solid basis on math, statistics, and machine learning. It is generally expected that she/he is able to work with various elements of mathematics, databases, and computer science, although deep expertise in all of these subjects is not required. A data scientist is most likely to be an expert in only one or two of these disciplines and proficient in another two or three. As a result, the best practice in applying Data Science is to use a team, where the expertise and proficiency in the needed disciplines are balanced across the members of the team. It is recommended that the critical balancing act should be done by unicorns.

Citizen Data Scientists: The Builders
The third member of the Troika is the citizen data scientist. This is a large group that practices Data Science without specialized degrees or deep knowledge of related subjects. The core expertise of citizen data scientists varies from data engineering, and computer science, to different application areas of engineering, finance, economics, etc. The gaps with respect to the required Data Science skillset are gradually filled by training and project work. Some recommendations on how to fill these gaps effectively are given in Sect. 17.4.

This raises the key issue with citizen data scientists—where is the fine line between tolerating and not tolerating a lack of professional experience with respect to Data Science capabilities? The answer is not obvious and is business-, technology-, and individual-specific. A properly guided transition of a citizen data scientist to the level of an average data scientist will benefit both the organization and the individual. The alternative of too many compromises due to an obvious lack of technical skills can lead to opening the door to the dangerous species of fake data scientist...

Gartner forecasts that the need for citizen data scientists is going to grow five times faster than the need for highly skilled data science specialists.[3] They may become the dominant force for building future Data-Science-based systems. That is why the issue of their transition to effective data scientists is so important.

A special group of data scientists, outside these three types, are those who discover and advance the most cutting-edge methods and algorithms themselves. This elite group of top academics and industrial researchers includes a couple of thousands of people globally. With their creativity and vision, they are moving the

---

[3]*Predict 2017: Analytics Strategy and Technology*, Gartner Report, 11/30/2016.

technology ahead. They are the engine that will drive the future growth of Data Science and create opportunities for all types of data scientist.

17.2   Key Capabilities of Data Scientists

A typical feature of the data scientist's job is that it requires a mix of various skills, areas of expertise, and knowledge. On top of that, the wish list is continuously changing with the progress in technology. Five years ago, the requirements were more focused on the big data skillset. Recently, the focus has switched to AI-based Data Science capabilities, especially those related to machine/deep learning. It is expected this trend will continue, and that sends a clear message that data scientists should constantly update their technical capabilities.

A central question for any data scientist is about the key capabilities that can be realistically expected from her/him. There are different opinions and no full consensus on this issue, though. The most popular classification, although it is somewhat vague, is based on the Drew Conway's Venn diagram, suggesting three key capabilities: (1) math and statistics knowledge, substantive expertise, and hacking skills.[4] Another version of the same diagram recommends a mix of math, computer science, and domain expertise,[5] while a recent more detailed version by Stephan Kolassa includes a combination of communication, statistics, programming, and business capabilities.[6]

We suggest a different and more specific answer to this question. We recommend that a data scientist should focus on six important roles: (1) problem solver, (2) scientist, (3) communicator, (4) data analyst, (5) math modeler, and (6) programmer. The first three capabilities are related to social skills that are more complex and difficult to learn, while the last three are based on technical skills with a faster learning curve. A Swiss Army Knife has been selected as a suitable visual representation of the recommended classification, as shown in Fig. 17.3, where the technical roles are encircled.

The recommended capabilities of a data scientist are described below.

17.2.1   Problem Solver

The most important role of a data scientist is finding and solving business problems with appropriate Data Science methods. The problem-solving skills of a data

---

[4]http://drewconway.com/zia/2013/3/26/the-data-science-venn-diagram

[5]S. Palmer, *Data Science for the C-Suite*, Digital Living Press, New York, 2015.

[6]https://datascience.stackexchange.com/questions/2403/data-science-without-knowledge-of-a-spe cific-topic-is-it-worth-pursuing-as-a-career

Fig. 17.3 A Swiss Army
Knife representation of the
key capabilities of a data
scientist

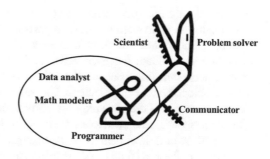

scientist requires an understanding of traditional and new data analysis methods to
build statistical and machine learning models or discover patterns in data. Of equal
importance, however, is understanding business operations and key issues. A data
scientist should gain knowledge about related business processes by training or from
SMEs. Some recommendations about dealing with Data-Science-based business
problems have been given in Chap. 2.

Another important factor for solving problems related to Data Science is to use
effective work processes. An example of such a process for the development and
deployment of AI-based Data Science systems has been given in Chap. 6, with
detailed descriptions of the key steps in Chaps. 7, 8, 9, 10 and 11. Following this
methodology can help a data scientist to deliver a solution in an effective way
according to the defined objectives.

The performance and career development of a data scientist will be mostly
decided by the success or failure of solved problems. It is strongly recommended
that preparing well for this critical capability should be the top priority of any data
scientist. The formula for success is a good balance between three factors: (1) understanding the business problem, (2) finding the best technical solution relevant to the
problem, and (3) solving the problem with an appropriate work process.

It has to be made very clear, especially to inexperienced data scientists, that it is
impossible to solve business problems with technical skills only.

17.2.2   Scientist

The most challenging role of a data scientist is to have the mindset of a scientist. A
data scientist needs to have an intrinsic high level of curiosity and a strong appetite
for intellectual challenges. It is expected that the data scientist will question everything related to the defined problem. Above all, it is her/his responsibility to translate
these questions into well-formulated hypotheses. Validating or rejecting these
hypotheses is the key step in any project related to Data Science. If there is no
hypothesis to be proved in solving the problem, there is no science but only the
application of some well-known process, i.e., this is not a Data Science project. Data

scientists should not spend time on such projects and should recommend that the known solution is diverted to different parts of the organization.

Data scientists need to be able to pick up the best methods and mathematical techniques to validate the hypotheses and solve the problem. They need to be creative and often think outside the box in order to find the optimal combination of Data-Science-based approaches. The more options they have, the higher the chances of success. Often this requires a broad understanding of a variety of methods with different scientific principles, such as machine learning, evolutionary computation, and swarm intelligence. It is strongly recommended that data scientists gradually add these methods to their technical capabilities.

A very important duty of a data scientist is to draw the final scientific conclusion about the validity or rejection of the defined hypotheses. It is expected that this purely scientifically driven result should be communicated to the project team as clear recommendations about the next steps to be taken. The arguments about validation or rejection of the hypotheses should be well defined in plain English.

The formula for success of a data scientist as a scientist is a good balance between three factors: (1) formulating proper hypotheses about a defined business problem, (2) applying related methods to validate the defined hypotheses, and (3) giving recommendations for solutions based on proof of these hypotheses.

17.2.3   Communicator

It is important for data scientists to communicate effectively with business users, utilizing business language and at the same time avoiding a shift toward a conversation that is too technical. The basis of good communication skills is that data scientists can clearly and fluently translate their technical findings to a nontechnical team. This requires some efforts to improve communication between the problem and solution spaces, related to what is defined in the book as the lost-in-translation trap. Several recommendations for how to increase communication skills related to AI-based Data Science and to avoid the lost-in-translation trap are given in Chap. 5.

Good communication skills for a data scientist can be expressed in various ways, including:

- *General communication.* Working as a data scientist almost always means working as a team with different stakeholders—business SMEs, engineers, operators, managers, etc. Good general communication can help facilitate trust and understanding, which is incredibly important for the final success of the project.
- *Communicating insights.* The biggest challenge for data scientists is to communicate discovered insights in a clear, concise, and acceptable way, so that others in the business can effectively translate those insights into actions and potential value. This is the critical component by which the communication skills will be assessed.

- *Data and results visualization.* Often an appropriate visualization is the best way to make or convey a point. Various techniques for preparing informative presentations are given in Chap. 13 and many books on the topic.

The best data scientists have the ability to communicate effectively with senior-level management and translate the everyday language of business into the quantitative language of Data Science modeling. This communication skill is a big benefit to the Data Science group and can be effectively used.

The formula for success of a data scientist as a communicator is a good balance between three factors: (1) communicating flawlessly with all problem stakeholders, (2) explaining insight clearly without technical jargon, and (3) understanding and using problem-related vocabulary.

17.2.4   Data Analyzer

It is assumed that a data scientist will be responsible for translating data into insight and value but not for collecting or managing the data. From that perspective, it is not required to have deep knowledge of database management, handling distributed storage of big data, and the related languages, such as SQL, MySQL, and MongoDB.

In a professional Data Science group, data scientists are not involved with data collection. However, they have to be responsible for defining the size of the data and the metadata structure, as well as to have full knowledge of how to access the collected problem-related data. It is their duty to decide if the available data collection is complete.

The data analyzer' role of a data scientist requires two key skills: (1) professional data preparation and (2) professional data analysis.

Professional data preparation includes deep knowledge of the following topics: manipulating data, dealing with missing data, handling outliers, data transformation, and balancing unbalanced data. It is expected that the data scientist will be familiar with several methods and tools for performing these tasks. These topics are discussed in detail in Chap. 8. Of special importance is personal experience in defining "clean" data sets, ready for analysis, which is a decisive factor for modeling success.

Professional data analysis requires high-quality skills in the following topics: multivariate analysis, dealing with multicollinearity, variable selection, feature extraction, and effective data visualization. The data scientist should have fundamental knowledge of the principles of multivariate statistics and variable reduction. It is also expected that several different methods will be used, especially for variable selection. These topics and key principles are discussed in details in Chap. 9.

The formula for success of a data scientist as a data analyzer is a good balance between three factors: (1) knowledge of how to access the collected problem-related data, (2) professional skills in data preparation, and (3) professional skills in data analysis.

17.2.5   Math Modeler

The most important technical role of a data scientist is to deliver math models. In order to accomplish this task, expertise in academic disciplines such as math, machine learning, advanced statistics, probability theory, artificial intelligence, and others is desirable. Covering all potentially useful disciplines on the list, however, would be a life time achievement for any data scientist and is not a requirement for high performance. Rather, a data scientist needs to have a healthy mix of selected modeling methods to succeed. The mix is business- and problem- specific and should be continuously updated and broadened.

It is recommended that the shortlist of methods for a prospective math modeler should include: statistics and statistical modeling, machine learning, deep learning, decision trees, and evolutionary computation. It is not expected that the data scientist will develop new methods or algorithms. As a result, there is no need to sink into the nitty-gritty of the math details of the algorithms. What matters is to understand the principles of the methods, their capabilities and applicability. Examples of the right level of detail for the most popular AI-based Data Science modeling approaches are given in Chaps. 3 and 4.

The most important capability of the data scientist as a good math modeler is the ability to fight successfully the overfitting problem by selecting models with appropriate complexity that will deliver robust performance on new, unseen data. This task can be accomplished when model development is done in a systematic way. An example of an effective sequence for the generation of mathematical models is given in Chap. 10.

Recently, the use of model ensembles as a solution to problems has been gaining popularity across the Data Science community. It is strongly recommended that data scientists should add the capabilities of bagging, boosting, and related model ensemble methods to their skillset.

The formula for success of a data scientist as a math modeler is a good balance between three factors: (1) knowledge of a wide range of modeling methods, (2) experience in developing and selecting robust math models, and (3) ability to understand and apply model ensembles.

17.2.6   Programmer

The last role, but not the least important one, of a data scientist is as a programmer. The importance of this capability is gradually decreasing with the growing abilities of integrated enterprise-wide software packages for automatic code generation. However, it is expected that programming proficiencies will need to be part of a data scientist's skillset portfolio for the foreseeable future.

As programming languages are changing fast, it is recommended that data scientists focus on the fundamental principles of programming, which are the same for almost any language. That will allow easy adaptation to the next fashionable programming language.

In order to have full freedom in analyzing the data and building models, beyond the limitations of integrated Data Science software packages, data scientists have to be proficient in at least one scripting language. Recently, the obvious choice has been Python. The popularity of this open source language has been widely observed. It is easy to learn, with many options for webinars, classes, and books. The vast majority of the new Data Science tools, especially those related to deep learning, are compatible with it. That makes Python the primary language for data scientists.

Often, in implementing data analysis and model development, data scientists need other Data-Science-specific languages. The most typical example is R, which is a popular programming language for solving statistical problems. Another example is the SAS language, which is the software glue of all SAS enterprise-wide systems, based on statistics, forecasting, and machine learning.

The formula for success of a data scientist as a programmer is a good balance between three factors: (1) understanding of programming fundamentals, (2) good programming skills in a generic script language such as Python, and (3) knowledge of other Data-Science-specific languages such as R, SAS, or Matlab.

17.3   Skillset of an Effective Data Scientist

The other critical issue of high practical importance to data scientists is defining the desired skillset. On the one hand, this is needed for data scientists, to guide their professional development. On the other hand, it is also needed for their managers, to evaluate the data scientists' performance more accurately.

We define an effective data scientist as a professional who solves business problems by efficiently applying Data Science to deliver sustainable value. It is our belief that the skillset shown in Fig. 17.4 and discussed below is the basis of an effective data scientist.

17.3.1   Understand Business Problems

This is a critical skillset for finding, selecting, and defining business problems appropriate to be solved by Data Science. Data scientists need to have a good understanding of business processes in order to develop solutions that can be deployed and ultimately used by the respective business users. In terms of Data Science, being able to discern which problems are important to solve for the business is critical, in addition to identifying new ways in which the business should be leveraging its data. Out-of-the-box thinking helps data scientists to consider business problems from new angles, utilizing Data Science methods that their business users do not know about. Last but not least, many successful applications of Data Science

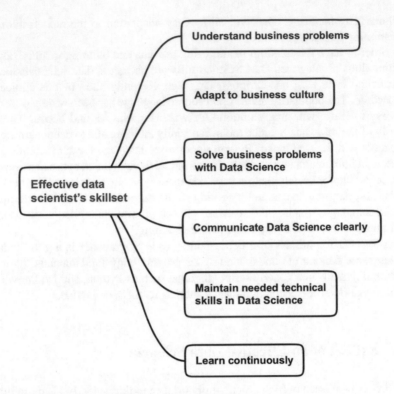

Fig. 17.4 Mind map of an effective data scientist's skillset

Table 17.1 Scale of data scientists' skill levels

Skill level	Scale
No skills	0
Elementary	1
Average	2
Acceptable	3
Professional	5

have been achieved through joining the dots by data scientists working on different projects across the enterprise.

In order to make Data Science actionable and to make it deliver value, data scientists must have sufficient expertise regarding how businesses operate. To accomplish this task, they can use the suggested techniques for knowledge acquisition discussed in Chap. 7. Another important requirement for achieving a good understanding of business processes is to gradually build and maintain a broad network of business experts.

An important issue in developing an effective data scientist's skillset is evaluating the different skill levels. A simple common-sense qualitative scale of five levels, shown in Table 17.1, is suggested for this purpose.

The meaning of each level is skill-specific, however, and will be explicitly defined below. Every level includes the requirements of the levels below it as well. For understanding business problems, the recommended skill levels are as follows:

- *Elementary.* This includes a generic awareness of basic business operations.
- *Average.* This assumes knowledge of the principles of basic business operations and the meaning of key business-related terminology.
- *Acceptable.* This involves detailed knowledge of defined business problem and identified related SMEs.
- *Professional.* This requires a deep understanding of business operations related to defined problems, effective knowledge acquisition from available internal and external references, and a broad network of associated SMEs.

17.3.2   Adapt to Business Culture

The next skillset allows data scientists to transform their technical solutions effectively into actions in business settings. It is very difficult to succeed in implementing Data Science projects and delivering value without adapting to business culture. By ignoring the significance of knowledge of business culture, data scientists risk becoming "paper tigers" who shine in delivering models on paper but fail to move them into production. Understanding the importance of adapting to business culture is a litmus test for distinguishing between an inexperienced, technically focused data scientist and a skilled, business-focused one.

This skillset is underestimated by the Data Science community at large. Usually it is not a top priority for organizations in the introductory phase of Data Science in the business where showing technical capabilities in small pilot projects is of critical importance. During the application and leveraging phases, however, when value creation at the enterprise-wide level becomes a top priority, organizations have started to understand the need for data scientists who know how to move technical solutions into the business environment.

The problem is that this skillset is difficult to build and that it takes a long time to learn and improve the needed skills. On top of that, it is outside the comfort zone of data scientists and there are almost no literature and training classes to help them.

We'll focus on an important task that data scientists need to do to adapt to business culture: to require a knowledge of office politics.[7]

Office politics is a complex topic, outside the scope of this book. In order to help data scientists, however, we'll focus on two practical components that drive office politics: (1) perceptions and (2) hidden agendas.

---

[7]Most of this subsection is based on D. Stevenson, *Office Politics: A Survival Guide for DataSscientists*, Data Science Central, 06/13/2017.

Companies typically aim to be fair and to use objective criteria for measuring success. As fallible humans, however, we are often forced to rely ultimately on culture, intuition, perceptions, and incomplete information. These perceptions, in turn, are dictated by our backgrounds and past experiences and are largely influenced by often unsubstantiated input from others around us.

The input from others, in turn, may come from their own set of perceptions, and it may also have its source in the second component of office politics: the hidden agenda. These hidden agendas may be fueled by any one of a number of factors, including ambition, revenge, and insecurity.

Why are data scientists especially vulnerable?

What makes office politics so difficult for data scientists, especially those just coming from school, is that they are used to working with clearly defined standards of success. They succeeded at school when they were able to solve clearly defined technical problems, and were evaluated in a fair and precise manner by their instructors. The decisive factor to their performance assessment was their technical skills, which they have continuously improved. Unfortunately, schooling didn't develop skills in perceiving the perceptions and agendas of the people around them, i.e., the future data scientists had insufficient awareness of the subtle interpersonal signals that could give them a clue about the presence of office politics.

In business settings, however, their technical skills and talents are no longer the only things that matter. If data scientists don't have the support of the people around them, they will fail. Getting the support of their colleagues requires awareness of the perceptions that they generate and about the hidden agendas operating around them.

Perceptions

A key piece of advice to data scientists with no experience in business settings is to be aware that everything they do makes an impression on their colleagues. Coming to work late, leaving early, taking long breaks, or spending time on social media can all damage their career by creating a poor perception. There is a thin line to follow, however; if they come across as too smart, too hard-working, or too energetic, people will start to perceive them as a threat, and they can run afoul of one of the most common types of hidden agenda.

Hidden Agendas

For many people, the most important item on their agenda is to preserve or increase their own power (possibly by hiding their incompetence). These people will not support you or your work if they perceive you as a threat, and some people feel threatened whenever someone else is successful. Being a focus of attention, data scientists are especially vulnerable to this perception and need to be extremely careful. Some generic suggestions start with are: Do yourself a favor and stay humble. Pass credit to colleagues whenever possible. Always publicly acknowledge the accomplishments of your colleagues.

Understanding office politics is the first step in adapting to business culture. Other steps, such as analyzing the company power structure,[8] and knowing the corporate incentives and project development expectations, are needed to gradually complete this long process.

The recommended skill levels for this skillset are as follows:

- *Elementary.* An understanding of office politics.
- *Average.* Knowledge of project development expectations.
- *Acceptable.* A basic understanding of the company power structure and of mechanisms for running Data Science projects efficiently.
- *Professional.* A detailed knowledge of business culture, to develop effective approaches to move Data Science projects into production.

17.3.3   Solve Business Problems with Data Science

This is the skillset with which data scientists act as problem solvers and scientists. In addition to the key requirements for these capabilities, discussed already in Sect. 17.2, the most important expectation placed on data scientists is to protect the technology from management pressure. In order to accomplish this difficult task, they need to learn how to make some compromises between the principles and methods of Data Science and pleasing managers. The most frequent compromises data scientists are pushed to make are the following:

- *Lack of a work process.* The most dangerous compromise is when data scientists are pressed to develop projects in an ad hoc manner. Data scientists should resist and give solid arguments, supported by references, for the need for a relevant Data Science work process. Recommendations for such arguments are given in Chap.6.
- *Accelerated project development.* The most frequent compromise is in the response to a management request for fast project delivery. Data scientists must estimate the time needed for each step of the work process (as described in Sect. 6.3) and explain these times to management. If the requested acceleration is more than half of the estimated time, it should be rejected with appropriate numbers and arguments. If the time reduction is between 20 and 50%, potential changes to the project objectives should be discussed to allow the time to be shortened. Acceleration of less than 20% could be handled with revised project management and prioritization.
- *Starting projects without problem definition.* There is a very simple rule that data scientists should follow: "no charter—no project"! We strongly do not

---

[8]A recommended reference on the nature of power is R. Greene, *The 48 Laws of Power*, Penguin Books, 1998.

recommend starting any technical activities without a clear problem definition and expected deliverables. Any compromise on this topic could lead to the dangerous path of modeling to nowhere …

- *When to stop analysis due to low-quality data.* Handling the well-known GIGO effect is one of the frequent challenges for data scientists. We strongly recommend stopping a project when the data is of low-quality, and suggest either seeking a different opportunity or changing the data collection process to use new sources of data.
- *When to stop modeling due to low correlation.* Another simple rule "no correlation—no equation," should be followed by data scientists without compromise, especially if the highest absolute correlation coefficient of all potential inputs with the target variable is less than 0.3.
- *Deploying low-quality models.* The push to put into production low-quality models should be firmly resisted by data scientists.

The recommended skill levels for this skillset are as follows:

- *Elementary.* Delivers only technical problem solutions under supervision.
- *Average.* Basic project management skills with low responsibilities.
- *Acceptable.* Good project management skills with demonstrated leadership and ability to handle project issues with correct compromises.
- *Professional.* Demonstrated leadership in solving diverse problems with minimum compromises.

17.3.4 Communicate Data Science Clearly

This skillset corresponds to the communicator role of data scientists and was discussed in Sect. 17.2.3.

The recommended skill levels for this skillset are as follows:

- *Elementary.* General communication skills.
- *Average.* Communication of Data Science without technical jargon.
- *Acceptable.* Communicating clearly insights discovered.
- *Professional.* Communicating flawlessly with all stakeholders and management.

17.3.5 Maintain Needed Technical Skills in Data Science

The requirements for this skillset of data scientists are defined by their roles as math modelers (Sect. 17.2.5) and programmers (Sect. 17.2.6).

The recommended skill levels for this skillset are as follows:

- *Elementary.* Knowledge of Python and statistics.
- *Average.* Programming in Python and knowledge of machine learning.

- *Acceptable*. Experience in developing math models by using multiple methods with several software packages.
- *Professional*. Experience in deploying robust math models with sustainable value creation.

17.3.6 Learn Continuously

The fast growth of AI-based Data Science requires data scientists to update constantly their technical skills. Their efficiency and career growth depend significantly on how well they accomplish this task. Fortunately, there are gazillions of training opportunities in machine learning, deep learning, and AI. Not all of them are of the needed professional quality, though. Some of them promise miracles with titles like "Machine learning in 10 days" even "Machine learning in 7 lines of code"! We strongly recommend data scientists not to fall into this trap. There is no way to learn and understand the fundamentals and principles of such a complex technology as machine learning either in 10 days or in seven lines of code.

In addition to appropriate training, the following steps may contribute to making the learning process continuous:

- *Be part of related Data Science communities*. Becoming a member of respected online communities in Data Science, machine learning, advanced analytics, etc. gives a unique opportunity to be continuously in touch with the latest developments in these technologies.
- *Attend key conferences*. Another efficient way to fill knowledge gaps, especially in relation to potential business applications, is attendance at business-related Data Science conferences.
- *Grow professional network*. Of special importance to maintaining the mode of continuous learning is to broaden your professional network. One of the best ways to do this is by attending conferences.

The recommended skill levels for this skillset are as follows:

- *Elementary*. Become a member of key online Data Science groups.
- *Average*. Analyze gaps in Data Science knowledge and attend corresponding training classes.
- *Acceptable*. Attend key conferences and build a professional network.
- *Professional*. Present at key conferences, become an active member of key online Data Science groups, and share learning with colleagues.

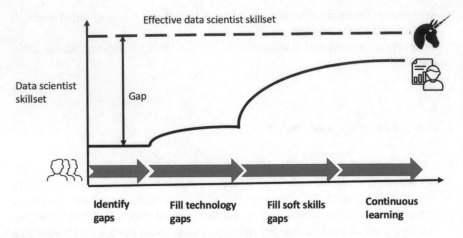

Fig. 17.5 A roadmap to the skillset of an effective data scientist

17.4   Roadmap to Becoming an Effective Data Scientist

Data scientists are highly educated −88% have at least a master's degree and 46% have Ph.D.s—and while there are notable exceptions, a very strong educational background is usually required to develop the depth of knowledge necessary to be a data scientist. The expectation is that to become a data scientist, one needs to earn a bachelor's degree in computer science, the social sciences, the physical sciences, or statistics. The most common fields of study are mathematics and statistics (32%), followed by computer science (19%) and engineering (16%). A degree in any of these fields gives the basic technical skills needed to prepare, analyze, and extract insight from data.[9]

The space of Data Science is so broad, however, that even this very high educational level is insufficient for becoming an effective data scientist. As we have discussed several times in the book, a continuous learning mode is recommended. It is extremely difficult to just learn everything needed in huge depth, though. A better strategy for professional growth is to learn the key skillsets to an acceptable level of acumen and then deepen your knowledge in the topics you need.

A generic roadmap to accomplishing this strategy, shown in Fig. 17.5, is recommended. It illustrates the common case of a skillset trajectory from a newly hired citizen data scientist position to data scientist level. The roadmap includes four steps: (1) identifying skillset gaps, (2) filling technical gaps, (3) filling nontechnical gaps, and (4) continuous learning. These steps of gradually decreasing the skillset gaps to become an efficient data scientist are discusses briefly below.

---

[9]https://www.kdnuggets.com/2018/05/simplilearn-9-must-have-skills-data-scientist.html

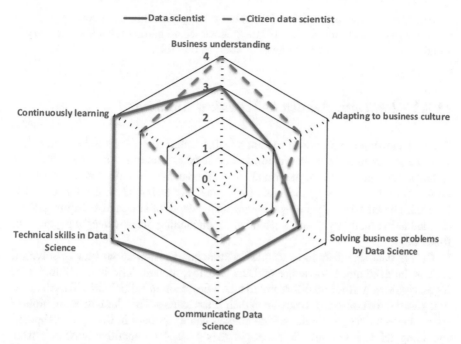

Fig. 17.6 Identifying skillset gaps by a radar chart

17.4.1   Identify Gaps in Personal Data Science Experience

The first step in this long journey is an assessment of one's current skillset level and comparison with the expected skills of an effective data scientist, as defined in Sect. 17.3. This could be done as a self-assessment by the data scientist or done by her/his manager, based on the suggested scales in Sect. 17.3. A convenient way to represent the evaluation of effective data scientist's skillset is by a radar map with axes corresponding to the skillsets and scaled according to the levels in Table 17.1. An example of such a radar chart is shown in Fig. 17.6. It includes two typical cases of skillsets, for a data scientist (represented by the thick continuous line) and for a citizen data scientist migrating from her/his business role to the new role of citizen data scientist (represented by the thick dashed line). This radar chart visualizes the clear advantage of the data scientist in terms of technical and continuous learning skills and the advantage of the citizen data scientist knowledge in terms of business operation and culture.

The biggest benefit of the radar map, however, is in identifying the gaps in each skillset. As shown in Fig. 17.6, the data scientist has a big gap in the skillset of adapting to business culture, where she/he has only an average level of proficiency. She/he also needs improvement in business understanding, solving business problems with Data Science, and communicating Data Science.

As expected, the citizen data scientist has more gaps to fill. Of special importance is the big gap in technical skills in Data Science where her/his level is elementary. It should be improved to average level as soon as possible.

17.4.2   Fill Gaps Related to Technical Knowledge

The idea in using the proposed roadmap is that technical gaps should be filled first. The reason for this recommendation is that it takes a shorter time to accomplish technical training, based on the broad range of courses available. There are a lot of books and related training materials for almost any technical skill related to Data Science. The duration of training depends on the depth of knowledge required and is decided individually by the data scientist, depending on how deep the gaps to be filled are.

Citizen data scientists have to make a strategic decision about how to solve the lack of fundamental knowledge of Data Science, though. The best solution is to invest significant time and effort and to take a degree in this discipline. This decision will give them a strategic advantage in their future career. The alternative solution, of fixing the technical gaps with several courses can keep them in their jobs temporarily. Long term, however, these people may be not competitive with the well-educated future data scientists with corresponding degrees.

17.4.3   Fill Gaps Related to Nontechnical Skills

Improving nontechnical skills is a much longer process and may take years until they reach professional level. One factor that may significantly shorten the process of filling the gaps is finding a good mentor. Such a person's help in adapting to business culture could be priceless.

Unfortunately, the options for training courses and materials for nontechnical skills are very limited. However, many companies offer internal training in the basics of their business operations, which could be used to fill in the gaps in this critical skillset. Courses on teamwork, communication skills, and project management at different levels are also available and strongly recommended. It is suggested to use the techniques for knowledge acquisition discussed in Chap. 7, too.

After identifying the gaps, the best strategy is to develop a skillset improvement plan. An example of such individual plan is shown in Fig. 17.7. In this case, the data scientist is focused on filling the gaps in her/his understanding of the business by attending the corporate trainings available, learning about and applying knowledge acquisition, and broadening her/his network of domain experts. The gaps in adapting to business culture will be filled by finding a mentor (with the help of the person's manager) and by understanding the power structure of the business. The data scientist has identified that she/he needs to make more efforts to explain the

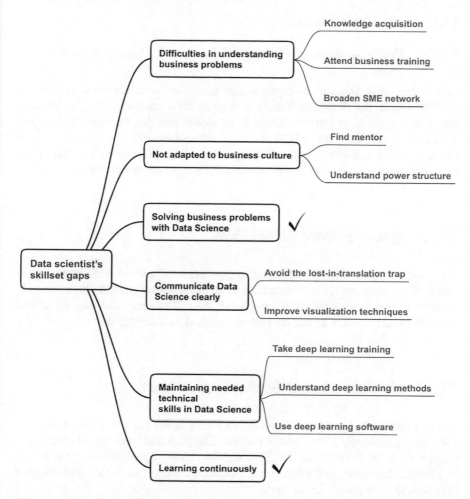

Fig. 17.7 Example of an individual data scientist's skillset improvement plan

AI-based Data Science technologies clearly and to gain knowledge about data visualization. The biggest technical skills gap that this data scientist is trying to fill is about deep learning. She/he plans to gain comprehensive knowledge of the fundamentals of the technology and learn the needed software tools in several appropriate training classes.

17.5   Common Mistakes

17.5.1   Programming in Python Makes you a Data Scientist

One of the most frequent perceptions among citizen data scientists is that the decisive factor for their fast transformation to data scientists will be learning programming skills in Python. There is no doubt that this proficiency is a key technical requirement, but only at the low, elementary level. In general, no one becomes a data scientist because of completing a Data Science course. This is an even more valid observation in the case of a Python programming class.

17.5.2   Ignoring Nontechnical Skills

The most common mistake is defining the job expectations and performance evaluation of a data scientist based on technical skills only. One possible explanation is the lack of precise criteria for nontechnical skills. The suggested data scientists' capabilities and skillsets in this chapter can be used as a starting point.

17.5.3   No Clear Plan for Skillset Improvement

Another mistake is ignoring the process of filling in the gaps in a data scientist's knowledge, especially in nontechnical skills. The proposed sequence of identifying the specific holes in the person's skillset and developing a plan to correct them, first in technical areas and, in the long term, in nontechnical areas, is the recommended approach to addressing this big issue.

17.5.4   Not in Continuous Data Science Learning Mode

Some data scientists with excessive self-esteem and arrogance think that they have reached the top of the world, working in the "sexiest job of the twenty-first century." The hype and high attention give them the confidence of semigods in Data Science who know the ultimate knowledge and do not need additional learning. As we have discussed several times in the book, this technology is continuously in development and data scientists need to be in touch with new progress.

17.6   Data Scientists and Humor

Humor is the universal language that opens all doors. If used appropriately, it can help data scientists to communicate their messages in an attractive and funny way. The paradox is that very often jokes, quotations, and forms of Murphy's law stay longer in the memory than any technical or business information. Some examples of useful quotations and versions of Murphy's law related to AI-based Data Science, that could be used in communicating the technology are given below. They are a good starting point for inspiring data scientists to use them and develop their own humor libraries.

17.6.1   Useful Quotations

"All models are wrong; some models are useful." (George Box).
  "Don't fall in love with a model." (George Box).
  "Complex theories do not work, simple algorithms do." [10] (Fig. 17.8; Vladimir Vapnik).
  "Nothing is more practical than a good theory." (Vladimir Vapnik).
  "A silly theory means a silly graphic." (Fig. 17.9; Edward Tufte).
  "The complexity of a model is not required to exceed the needs required of it." (D. Silver).
  "It is better to solve the right problem approximately than to solve the wrong problem exactly." (Fig. 17.10; John Tukey).

Fig. 17.8 Complex theories and simple algorithms

---

[10]The caricatures in this section are by the Bulgarian cartoonist Stelian Sarev.

Fig. 17.9 A silly graphic

Fig. 17.10 The wrong problem

17.6.2 Murphy's Law Related to Data Science

Murphy's Law
If anything can go wrong, it will.

Murphy's Axiom
When something goes wrong, it won't be the thing you expected (Fig. 17.11).

Fig. 17.11 Murphy's axiom

Box's Law
When Murphy speaks, listen.

Corollary
Murphy is the only guru who is never wrong.

Law of the Ubiquitous Role of Data mining
In God we trust, all others bring data (Fig. 17.12).

Fig. 17.12 In God we trust

Murphy's Law of Machine Learning
The number one cause of computer problems is computer solutions.

Kordon's Law of Artificial Intelligence
Computers amplify human intelligence.

Positive Corollary
Smart guys get smarter (Fig. 17.13).

Negative Corollary
Dumb guys get dumber.

Murphy's Law of Rule-Based Systems
Exceptions always outnumber rules (Fig. 17.14).

Fig. 17.13 Smart guys get smarter

Fig. 17.14 Exceptions outnumber rules

Murphy's Law of Sources of Fuzzy Logic
Clearly stated instructions will consistently produce multiple interpretations (Fig. 17.15).

Fig. 17.15 Clearly stated instructions

Fig. 17.16 Pessimistic
forecasts

Effects of Murphy's Law on Data
If you need it, it's not recommended.
If it's recommended, it's not collected.
If it's collected, it's missing.
If it's available, it's junk.

Knuth's Law of Evil Optimization
Premature optimization is the root of all evil.

Murphy's Law of Evolutionary Computation[11]
Simulated evolution maximizes the number of defects which survive a selection
process.

Murphy's Law of Business Forecasting
Pessimistic forecasts are more valid than optimistic forecasts (Fig. 17.16).

17.7   Suggested Reading

- R. Greene, *The 48 Laws of Power*, Penguin Books, 1998.
- D. Stephenson, *Big Data Demystified*, Pearson, 2017.
- J. Thompson and R. Shawn, *Analytics: How to Win with Intelligence*, Techniks
  Publications, 2017.

17.8   Questions

Question 1
What are the differences between a citizen data scientist and a data scientist?

Question 2
What are the key technical capabilities of a data scientist?

---

[11]R. Brady, R. Anderson, R. Ball, *Murphy's Law, the Fitness of Evolving Species, and the Limits of Software Reliability*, University of Cambridge, Computer Laboratory, Technical Report 471, 1999.

Question 3
What are the key nontechnical capabilities of a data scientist?

Question 4
Why data scientists need good communication skills?

Question 5
What are the gaps in your Data Science skillset?

Glossary

Accuracy When applied to models, accuracy refers to the degree of fit between the model and the data. This measures how error-free the model's predictions are.

Activation function A function used by a node in a neural network to transform input data from any domain of values into a finite range of values.

Agent-based modeling See **Intelligent agents**.

Agent-based system A system with a number of agents which have social interactions, such as cooperation, coordination, and negotiation.

Analytics A broad category of inquiry that can be used to help drive changes and improvements in business practices.

Ant colony optimization (ACO) A population-based method for finding approximate solutions to difficult optimization problems. In ACO, a set of software agents called *artificial ants* searches for good solutions to a given optimization problem.

Artificial general intelligence (AGI) A form of artificial intelligence that can accomplish any intellectual task that a human being can do.

Artificial intelligence (AI) The part of computer science concerned with designing computer systems that exhibit characteristics we associate with intelligence in human behavior—understanding language, learning reasoning, solving problems, and so on.

Artificial intelligence-based Data Science The use of key AI methods in addition to statistics to transform data into insight and value.

Artificial neural network (ANN) A system of processing elements, called neurons, connected in a network that can learn from examples through adjustment of their weights.

Assumption space The conditions under which valid results can be obtained from a specific modeling technique.

Autonomous vehicle A vehicle that uses a combination of sensors, cameras, radar, and artificial intelligence to travel between destinations without a human operator.

© Springer Nature Switzerland AG 2020
A. K. Kordon, *Applying Data Science*,
https://doi.org/10.1007/978-3-030-36375-8

Back-propagation A machine learning training method used to calculate the weights in a neural network from the data.

Bagging A technique where multiple models are created on a subset of the data, and the final predictions are determined by combining the predictions of all the models.

Big data This refers to quantities of data that are too big, complex, fast-moving or poorly structured to be evaluated using manual, traditional data-processing methods.

Black-box model A model that is opaque to its user.

Boosting A sequential process where each subsequent model attempts to correct the errors of the previous model.

Bot A computer program that operates autonomously, to carry out tasks for a user and/or to mimic the behavior of a person.

Bottom-up modeling Bottom-up modeling includes piecing together systems to give rise to larger systems, thus making the original systems subsystems of the system that emerges. In a bottom-up approach, the individual base elements of the system are first specified in great detail. These elements are then linked together to form larger subsystems, which then in turn are linked, sometimes in many levels, until a complete top-level system is formed.

Branches in a decision tree Arrows connecting nodes in the tree, showing the flow from question to answer.

Chatbot A chat robot (chatbot for short) that is designed to simulate a conversation with human users by communicating through text chats, voice commands, or both.

Chromosome A string of genes that represents an individual (entity).

Citizen data scientist A professional who is not considered an expert in Data Science but is capable of using its tools to obtain various insights from data that can be useful for a business.

Classification Separation of objects into different groups.

Cloud Computer services that are remotely hosted and accessed through the Internet.

Clustering This refers to algorithms that find similar groups of items. For example, clustering could be used by an insurance company to group customers according to income, age, types of policies purchased, and prior claims experience.

Complex system A system featuring a large number of interactive components whose aggregate activity is nonlinear and which typically exhibit self-organization.

Confidence limit (or confidence) The region containing the limits or band of a model parameter with an associated confidence level (for example 95%).

Confusion matrix A matrix that shows the counts of the actual versus predicted class values.

Convergence In evolutionary algorithms, the tendency of individuals in the population to become the same.

Conversational corpus A body of text, images, or sounds used to "train" a neural network (giving it labeled examples from which to learn).

Convolutional neural network (CNN) A type of neural network that uses convolutions to extract patterns from the input data in a hierarchical manner.

Cross Industry Standard Process for Data Mining (CRISP) A popular work process for developing projects in data mining and analytics.

Cross-validation A technique which involves reserving a particular sample of a data set that is not used to train a model. Later, the model is tested on this sample to evaluate the performance.

Crossover Exchange of genetic material in sexual reproduction.

Curse of dimensionality The problem caused by the exponential increase in volume associated with adding extra dimensions to a (mathematical) space.

Customer churn analysis This profiles the customers who are likely to stop using a company's services or products and identifies those whose churn is likely to bring the biggest loss.

Data Values collected through record keeping or by polling, observing, or measuring, typically organized for analysis or decision making or unorganized in social media. More simply, data is facts, transactions, images, videos, and voice records.

Data analysis The process of looking at and summarizing data with the intent to extract useful information and draw conclusions.

Data mining An information and knowledge extraction activity whose goal is to discover hidden facts contained in databases. Typical applications include market segmentation, customer profiling, fraud detection, evaluation of retail promotions, and credit risk analysis.

Data preparation A sequence of actions to prepare raw data for data analysis by increasing the informational content.

Data Science An interdisciplinary field that combines scientific methods, statistics, information science, and computer science to provide insight into phenomena via either structured or unstructured data.

Data scientist A professional who uses Data Science to extract insight from data.

Data stories A methodology for communicating insight obtained from data, tailored to a specific audience, with a compelling narrative.

Decision-making An outcome of mental processes leading to selection from among several alternatives.

Decision tree An algorithm to learn decision rules inferred from data.

Deep learning Deep learning systems use multiple layers of calculation. The first layers look at very simple features (lines in an image, for example), while the later layers abstract more complex features (such as faces).

Degrees of freedom The number of measurements that are independently available for estimating a model parameter.

Dependent variable The dependent variables (outputs or responses) of a model are the variables predicted by the equation or rules of the model using the *Independent variables* (inputs or predictors). In the typical representation of a model as $y = f(x)$, y is the dependent variable and x is the independent variable.

Deployment cost   This includes the following components: the hardware cost of running the application, the run-time license fees, and the labor cost of integrating the solution into the existing work processes (or creating a new work procedure), training the final user, and deploying the model.

Derivative-free optimization   An optimization method that is not based on explicit calculation of a derivative of the *objective function*.

Descriptive analytics   A type of analytics that provides insight about what has happened.

Design of experiments (DOE)   A systematic approach to data collection such that the information obtained from the data is maximized by determining the (cause-and-effect) relationships between the factors (controllable inputs) affecting a process and one or more outputs measured from that process.

Designed data   Data collected by a design of experiments.

Development cost   This includes the following components: the necessary hardware cost (especially if more powerful computational resources than PCs are needed), the development software licenses, and, above all, the labor cost of introducing, improving, maintaining, and applying a new technology.

Diagnostic analytics   A form of analytics which examines data or content to answer the question "Why did it happen?"

Digital twin   A digital model of a physical asset.

Disruptive technology   A technological innovation, product, or service that uses a "disruptive" strategy, rather than an "evolutionary" or "sustaining" strategy, to overturn the existing dominant technologies or status quo products in a market.

Distance to cluster   There are various ways to calculate a distance measure between a data point and a cluster. The most frequently used method is called the squared Euclidean distance and is calculated as the square root of the sum of the squared differences in value for each variable.

Domain expert   A person with deep knowledge and strong practical experience in a specific technical domain.

Elevator pitch   A condensed overview of an idea for a product, service, or project. The name reflects the fact that an elevator pitch can be delivered in the time span of an elevator ride, approximately 60 seconds.

Emergent phenomenon   Emergent phenomena can appear when a number of simple entities operate in an environment, forming more complex behaviors as a collective.

Empirical emulator   A data-driven representation of a first-principles model.

Empirical model   A model that is based only on data and is used to predict, not explain, a system. An empirical model consists of a function that captures the trend of the data.

Ensemble learning   A machine learning technique that trains multiple models with the same data with each model using a different learning algorithm.

Epoch   The presentation of the entire training set to a neural network.

Error   The difference between the actual value and the predicted value from a model.

Euclidean distance See **Distance to cluster**.

Evolutionary algorithm A population-based optimization algorithm that uses mechanisms inspired by biological evolution, such as reproduction, recombination, mutation, and selection.

Evolutionary computation This automatically generates solutions to a given problem, with a defined fitness, by simulating natural evolution in a computer.

Expert system A computer program capable of performing at the level of a human expert in a specific domain of knowledge.

Expressional complexity A complexity measure for mathematical expressions based on the number of nodes in their tree-structure representation.

Extrapolation Predictive modeling which goes beyond the known and recognized behavior of a system and the range of available data.

False negative A test result that is read as negative but actually is positive.

False positive A positive test that results from a subject that does not possess the attribute for which the test is being conducted.

Feature An individual measurable property of a phenomenon being observed.

Feature engineering This includes finding connections and relationships between variables and packaging them into a single variable, called a feature.

Feature space An abstract space of mathematically transformed original variables. The function that performs the transformation is called a *kernel*.

First-principles model A model developed by use of the laws of nature.

Fitness function A mathematical function used for calculating fitness. In complex optimization problems, it measures how good any particular solution is. The lower (or higher oin some cases) the value of this function, the better the solution.

Fitness landscape The evaluations of a fitness function for all candidate solutions.

Forecasting Estimation of some variable of interest at some specified period in time in the future.

Generalization ability The ability of a model to produce correct results from data on which it has not been trained.

Generation One iteration of an evolutionary algorithm.

Genetic algorithm (GA) An evolutionary algorithm that generates a population of possible solutions encoded as chromosomes, evaluates their fitness, and creates a new population of offspring by using genetic operators, such as crossover and mutation.

Genetic programming (GP) An evolutionary algorithm in which structures evolve.

Genotype The DNA of an organism.

GIGO 1.0 effect Garbage-In-Garbage-Out

GIGO 2.0 effect Garbage-In-Gold-Out

Global minimum The lowest value of a function over the entire range of its parameters.

Gradient descent A first-order iterative optimization algorithm for finding the minimum of a function. It is used in machine learning and deep learning algorithms to minimize a cost function.

Graphics processing unit (GPU) A single-chip processor with a highly parallel structure used to manage and boost the performance of video and graphics, as well as algorithms for large blocks of data, in order to lessen the burden on the CPU.

Hidden layer A layer in a neural network between the input layer and the output layer (the prediction).

Heuristics Rules of thumb.

Hyperparameter A tuning parameter used during successive runs of training a model.

Hyperplane A higher-dimensional generalization of the concepts of a line in a two-dimensional plane and a plane in three-dimensional space.

Hypothesis A possible view about a problem of interest or an assertion about it by an analyst. It may be true or not true.

Hypothesis test A statistical algorithm used to choose between alternatives (for or against a hypothesis) which minimizes certain risks.

Ill-defined problem A problem whose structure lacks definition in some respect. The problem has unknowns associated with the ends (the set of objectives) and means (the set of process actions and decision rules) of the solution, at the outset of the problem-solving process.

Independent variable The independent variables (inputs or predictors) of a model are the variables used in the equation or rules of the model to predict the output (dependent) variable. In the typical representation of a model as $y = f(x)$, y is the *dependent* variable and x is the independent variable. Statistically, two variables are independent when the probability of both events specified by the variables occurring is equal to the product of the probabilities of each event occurring.

Inductive learning Progressing from particular cases to general relationships represented by models.

Input See **Independent variable**.

Insight Knowledge in the form of understanding, vision, and deduction.

Intelligent agents Artificial entities that have several intelligent features, such as being autonomous, responding appropriately to changes in their environment, persistently pursuing goals, and being flexible, robust, and social by interacting with other agents.

Internet of Things (IoT) A wide-ranging network of electronic devices of various kinds exchanging data over the Internet and coordinating their activities with one another.

Interpolation Model predictions inside the known ranges of inputs.

K-means An unsupervised learning approach used to group (or cluster) different instances of data based upon their similarity to each other.

Kernel A mathematical function that transforms the original variables into features.

Kernel trick Mapping input data into a high-dimensional feature space.

Knowledge acquisition Extracting existing knowledge from domain experts and related references and arranging it for effective use in problem solution of problems.

Lack of fit (LOF) A statistical measure that indicates that a model does not fit the data properly.

Laws of nature A generalization that describes recurring facts or events in nature, such as the laws of physics.

Laws of numbers The rules or theorems that state that the average of a large number of independent measurements of a random quantity tends toward the theoretical average of that quantity.

Leaf nodes (in decision tree) Nodes that contain questions to be answered, or criteria.

Learning Training of models (estimating their parameters) based on existing data.

Least squares The most common method of training (estimating) the weights (parameters) of a model, by choosing the weights that minimize the sum of the squared deviations of the values predicted by the model from the observed values of the data.

Linear model This term refers to a model that is linear in the parameters, β_k, not the *input variables* (the x's), i.e., models in which the output is related to the inputs in a nonlinear fashion can still be treated as linear provided that the parameters enter the model in a linear fashion.

Local minimum The minimum value of a function over a limited range of its input parameters.

Long short-term memory (LSTM) A type or recurrent neural network composed of a cell, an input gate, an output gate and a forget gate. The cell is responsible for "remembering" values over arbitrary time intervals, hence the word "memory" in "long short-term memory."

Loss function See **Fitness function**.

Lost-in-translation trap A lack of understanding between data scientists and domain experts.

Machine learning A research field dialing with the design and development of algorithms and techniques that allow computers to "learn."

Maintenance cost of a model This includes assessment of the costs of long-term support efforts for model validation, readjustment, and redesign.

Margin In machine learning, this refers to the distance between the two hyperplanes that separate linearly-separable classes of data points.

Missing-data imputation Replacement of missing values with their median or mean or inferring missing data from existing values.

Model A model can be descriptive or predictive. A descriptive model helps in understanding an underlying processes or behavior. A predictive model is an equation or set of rules that makes it possible to predict an unseen or unmeasured value (the dependent variable, or output) from other, known values (independent variables or input). In the typical representation of a model as $y = f(x)$, y is the dependent variable and x is the independent variable.

Model-based decisions Business rules defined by models.

Model complexity This includes different measures related to the type of model and its structure. Examples are the number of parameters in a statistical model and the numbers of hidden layers and neurons in a neural network.

Model credibility Trustworthy model performance over a wide range of operating conditions. Usually the credibility of a model is based on its principles, performance, and transparency.

Model deployment The simplest meaning is the use of a developed model. Examples might include using a model for process control or making a model accessible on the Internet for others who will use it periodically. The deployment may be an offline, static use of the model, or be embedded online as part of a larger system.

Model exploration Investigating the behavior of a model for a broad set of operating conditions.

Model interpretability The possibility of explaining the structure and behavior of a model using problem knowledge.

Model maintenance This includes periodic model validation and corrections such as model readjustment or model redesign.

Model overfit Fitting a model that has too many parameters and, as a result, its performance deteriorates sharply on unknown data.

Model performance A measure of a model's ability to reliably predict or classify on known data according to a defined fitness function.

Model readjustment Refitting model parameters on new data.

Model redesign Redevelopment of a model from scratch to new data.

Model selection Choosing a model based on several criteria, mostly based on its performance on validation data and its complexity.

Model underfit Fitting a model that has too few parameters to represent properly the functional relationship in the given data.

Monte Carlo simulation Using random samples of parameters or inputs to explore the behavior of a complex process.

Multicollinearity This assumes a correlation structure among the linputs to a model. Multicollinearity among the inputs leads to biased estimates of model parameters.

Multilayer perceptron The most popular neural network architecture, in which neurons are connected together to form layers. A multilayer perceptron has a three-layer structure: an input layer, one hidden layer that captures the nonlinear relationship, and an output layer.

Multiobjective optimization The process of simultaneously optimizing two or more conflicting objectives subject to certain constraints. An example in modeling is deriving optimal models with the two objectives of model accuracy and model complexity.

Multivariate analysis A process of analyzing the relationships of multiple variables with each other.

Mutation A small random tweak in a chromosome to get a new solution.

Natural language processing (NLP) Applying computer algorithms to determine properties of natural human language in an effort to enable machines to comprehend spoken or written language.

Neural network See **Artificial neural network**

Nonlinear model A model in which the dependent variable y depends in a nonlinear fashion on at least one of the parameters in the model.

Objective function See **Fitness function**.

Objective intelligence Artificial intelligence based on solutions automatically extracted through machine learning, simulated evolution, and emergent phenomena.

Ontology A description of concepts and relationships among concepts in a domain.

Optimization An iterative process of improving a solution to a problem as effectively or functionally as possible with respect to a specific objective function.

Optimization criterion A positive function of the difference between predictions and data estimates that are chosen so as to optimize the function or criterion.

Optimum The point at which the condition, degree, or amount of something is the most favorable.

Outliers Technically, outliers are data items that did not come from the assumed population of data and fall outside the boundaries that enclose most of the other data items in the data set.

Output See **Dependent variable**.

Overfitting A tendency of some modeling techniques to assign importance to random variations in the data by declaring them important patterns.

Oversampling Increasing the size of minority class samples in unbalanced data by repetition, bootstrapping or other methods.

Pareto-front genetic programming A genetic programming algorithm based on multiobjective selection.

Parsimony The principle that the simplest explanation that explains a phenomenon is the one that should be selected. In empirical modeling, this generally refers to using as few variables as possible to develop a robust model.

Partial least squares A multivariate model that simultaneously summarizes the variation of the original inputs while being optimally correlated with the target variable.

Particle swarm optimization (PSO) A population-based stochastic optimization technique inspired by the social behavior of bird flocking and fish schooling.

Pattern A regularity that represents an abstraction. For example, a pattern can be a relationship between two variables.

Pattern recognition The automatic identification of figures, characters, shapes, forms, and patterns without active human participation in the decision process.

Phenotype The set of observable properties of an organism, mostly its body and behavior.

Pheromone A chemical produced and secreted by insects to transmit a message to other members of the same species.

Population   A group of individuals that breed together.

Population diversity   This includes the structural differences between genotypes and behavioral differences between phenotypes.

Predictive analytics   A form of analytics that uses both new and historical data to forecast activity, behavior, and trends.

Predictive model   A mathematical representation of an entity used for analysis and planning.

Premature convergence   This occurs when a population for an optimization problem converges too early.

Prescriptive analytics   The area of analytics dedicated to finding the best course of action for a given problem.

Principal component analysis (PCA)   A technique used to reduce multidimensional data sets to lower dimensions for analysis.

Probability   A quantitative description of the likelihood of a particular event.

Problem   A perceived gap between the existing state and a desired state, or a deviation from a norm, standard, or status quo.

Problem definition   A clear description of what has to be solved, fixed or corrected.

Problem space   A set of defined problems

Process monitoring system   A centralized system which monitors and controls entire sites, or complex systems spread out over a large area (anything between an industrial plant and a country).

Pruning   Eliminating lower-level splits or entire subtrees in a decision tree.

r-squared (r^2)   A number between 0 and 1 that measures how well a model fits its training data. A value of one is a perfect fit, and 0 implies the model has no predictive ability.

Random forest   An ensemble of decision trees (with each tree constructed by using a random subset of the training data) that will output a prediction or classification value.

Range   The range of a set of data is the difference between the maximum value and the minimum value. Alternatively, the range can include the minimum and maximum, as in "The value ranges from 2 to 8."

Record   A list of values of features that describe an event. Usually one row of data in a data set, database, or spreadsheet. Also called an instance, example, or case.

Recurrent neural network   A neural network architecture with a memory and feedback loops that is able to represent dynamic systems.

Regression   A technique used to build equation-based predictive models for continuous variables that minimize some measure of the error.

Regularization   A technique used to solve the overfitting problem by penalizing some coefficients such that the model generalizes better.

Reinforcement learning   A learning mechanism based on trial-and-error of actions and evaluating the rewards.

Residual   See **Error**.

Robust model   A model that is less influenced than others by a small number of aberrant observations or outliers.

Rule   A statement expressed in an IF (antecedent) THEN (consequent) form.

Rule base   A set of rules that uses a set of assertions, for which rules on how to act upon those assertions are created.

Search space   The set of all possible solutions to a given problem.

Self-organization   A process of automatically increased organization of a system without guidance or management from an outside source.

Solution   A means of solving a defined problem.

Solution space   A set of potential solutions to a defined problem.

Stacking   A combination of different models for the same problem.

Statistical learning theory   A statistical theory of learning and generalization concerning the problem of choosing desired functions on the basis of empirical data.

Statistics   A branch of mathematics dealing with the collection, analysis, interpretation, and presentation of masses of data.

Stigmergy   A method of indirect communication in which individuals communicate with each other by modifying their local environment.

Stochastic process   The opposite of a deterministic process.

Subject matter expert (SME)   See **Domain expert**.

Subjective intelligence   Artificial intelligence based on expert knowledge and rules of thumb.

Supervised learning   A type of machine learning that requires an external teacher, who presents a sequence of training samples to a neural network. As a result, a functional relationship that maps the inputs to the desired outputs is generated.

Support vector machines (SVMs)   A machine learning method derived from a mathematical analysis of statistical learning theory. The generated models are based on the most informative data points called support vectors.

Swarm intelligence   A computational intelligence method that explores the advantages of the collective behavior of an artificial flock of computer entities by mimicking the social interactions of animal and human societies.

Symbolic regression   A method for automatic discovery of both the functional form and the numerical parameters of a mathematical expression.

Target variable   See **Dependent variable**

Test data   A data set independent of the training data set, used to evaluate the performance of developed models.

Time series   A series of observations taken at consecutive points in time.

Total cost of ownership   The sum of the development, deployment, and maintenance cost of a model.

Training data   A data set used to estimate or train a model.

Tree induction   The process of generating a decision tree.

Tree pruning   see **Pruning.**

Underfitting   This occurs when a model that cannot capture the underlying trend of the data.

Undersampling   Balancing unbalanced data by reducing the size of the majority class by randomly selecting an equal number of samples from the majority class.

Unicorn A data scientist with a broad skillset in many areas of Data Science.

Universal approximation The ability to approximate any function to an arbitrary degree of accuracy. Neural nets are universal approximators.

Unsupervised learning A type of machine learning that does not require a teacher, and where the neural network is self-adjusted by the patterns discovered in the input data.

Validation set In supervised learning, a set of data used to fine-tune the estimates of model parameters and avoid overfitting a developed model that has been trained on a training set.

Value Estimate the monetary worth of (something)

Vapnik–Chervonenkis (VC) dimension A measure of the capacity (complexity, expressive power, richness, or flexibility) of the space of functions that can be learned by a statistical learning algorithm.

Variable selection The process of reducing the number of random variables under consideration by obtaining a set of principal (most important) variables.

Printed in the United States
by Baker & Taylor Publisher Services